PRACTICAL FORMULAS FOR HOBBY OR PROFIT

Henry Goldschmiedt, Ph.D.

 CHEMICAL PUBLISHING CO., INC., NEW YORK. N. Y.

CONTENTS

INTRODUCTION

This section is presented in such a manner so that the novice, chemist or manufacturer will be able to read through the formulas in this book, and produce commodities in his kitchen or makeshift laboratory without an excessive amount of difficulty.

The formulas presented contain chemicals of a wide range: pigments, gums, resins, greases, waxes, emulsifying agents, dyestuffs, etc. To combine these ingredients into a reliable end-product requires careful study of the given procedure. Short cuts, chemical substitutions or deletions can result in complete failure or poor quality product. Follow directions precisely. Do not add A to B if B is specified to be added to A. In laboratory procedures A plus B is not always equal to B plus A as it is in mathematics, for it is the nature of the addition in the lab which determines the sum. Any experienced cook knows that in preparation of mayonnaise that if one adds the egg to the oil, one will get a mess, but that if one slowly adds the oil to the egg, stirring constantly, the result will be mayonnaise. The same holds true for the formulas presented in this book. Always use the specified grade of the chemical. If a cheaper end product is desired, don't look for cheap substitutions: use a different formula.

There is a limit to which this rule can be reasonably extended. In some cases successful substitutions can be made. When white beeswax is required yellow wax may be substituted (color not withstanding) but paraffin won't do, though its light color would lead one to think otherwise. One must experiment to achieve successful substitutions. As mentioned before, grade or type of a chemical must be followed precisely. Lanolin is not the same as anhydrous lanolin, and thus they are not interchangeable. If one is in doubt about a formula or a part of it, seek the help of a consultant or disgard the formula. Grades are specified on many chemicals. Use only those specified, and where none is given, understand that the best grade is to be used. In most cases a specific type of alcohol is referred to. When the general term alcohol is used, the experimenter should understand it to mean ethyl alcohol.

There are many terms and processes which one might find unfamiliar. They compose the language of chemistry. A glossary and an appendix is provided in this book to answer the most salient questions that the amateur chemist has with regard to terminology, processes and machinery.

Apparatus

In many instances you will not need anything more than the pots, glassware and china which are the normal equipment in every household. However, enameled pans are better than aluminum. For mixing emulsions, an electric mixer, blender or just plain egg beater will be satisfactory. For weighing, it might be worth your while to invest in a small scale from a laboratory supply house. Graduates and measuring glasses are found at drugstores. It would be best that you get a chemical thermometer from a laboratory supply house. Find out about the rental of, or purchase of, more specialized equipment for formulas that require it.

Heating

Always use a double boiler for heating liquids and dry materials when temperature called for is below 212°F. Be sure the pan does not go dry. To get uniform high temperature, use oil or grease in place of the water in the bottom of the double boiler. Stop heating if thick fumes are given off. They are flammable. Chemicals which melt uniformly, and are nonexplosive, may be heated directly over the flame. Use your thermometer to be sure you are working at the correct temperature.

Temperature Measurement

Both Fahrenheit and Centigrade are used. The boiling point of water is 212°F. or 100°C. The melting point of ice is 32°F or 0°C. (See conversion tables in the appendix.)

Temperatures of liquids are measured by inserting a glass thermometer as deep as possible into the liquid, but not against the side or bottom of the pan. Warm a cold thermometer (hold over surface of hot liquid) before immersion. Conversely a hot thermometer should not be placed on a cold surface. Sudden temperature change can crack the thermometer.

Mixing and Dissolving

This is accomplished easily by stirring and warming. For hand stirring, it is best to use a non-reactive material such as a glass rod.

Filtration

When undesirable particulate bodies are present in a liquid, they may be removed by settling (a centrifuge is an excellent device to speed up this process) or filtering.

The simplest way to achieve settling is to allow the solution to stand while the heavier than liquid particles sink to the bottom. Centrifuging speeds up this process by causing the denser particles to accelerate towards the end of the container because of the centrifugal force created by the machines spinning. In either case, when the particles have all collected at the bottom, the remaining liquid may be poured or siphoned off with care, and in some cases the liquid is then clean enough for use. If the particles cannot be removed by settling, they must be filtered out. Coarse or large bodies may be filtered through sterile muslin or cloth; if the particles are small, filter paper must be used. Filter papers are made in a wide range of fineness. Coarse paper will trap large particles but will let fine particles through; therefore, one must determine the fineness required to remove all particles. Sometimes these bodies are so small that no filter paper made is able to ensnare them; it is then necessary to add to the liquid 1 to 3 percent infusorial earth or magnesium carbonate. They aid the filter paper by clogging up its pores thus reducing their size and enabling them to hold back undissolved material of extreme fineness. In all such filtration, it is common practice to take the first portion of the filtered liquid and pour it through again.

Pulverizing and Grinding

Large masses are broken up by wrapping in a clean cloth, placing between two boards, and pounding with a hammer. This is repeated until the proper size is obtained. Fine grinding is done in a mortar and pestle.

Housekeeping

The manner in which you keep your materials will be reflected in your end products. Therefore, take care of them. Keep all containers tightly closed. This will prevent evaporation as well as contamination. Liquids should be kept in glass containers. They should be as full as possible. Corks or stoppers should be covered with foil or dipped in paraffin. Glues, gums, oils and certain other products may ferment or become rancid. To prevent this, use suitable antiseptics or preservatives. Cleanliness is of utmost importance. Clean all containers thoroughly before use.

Where to Buy Chemical and Apparatus

Many chemicals and glassware can be bought from your druggist. A list of suppliers will be found in the Appendix. Make use of the yellow pages of your telephone directory. It will list local suppliers.

Caution

Some chemicals are poisonous and corrosive. Check the labels. Do not inhale them. If smelling is necessary, sniff a few inches from the stopper. Work in a well-ventilated room. If anything spills, wipe and wash immediately. (See the Appendix for safety rules for your laboratory.)

Units for Measurement

The formulas presented in this book contain proportions by weight, volume, or percentages by weight; some formulas use specific volumetric or mass units, such as cc's or grams etc. Different industries and various countries use a variety of systems of weights and measures. It is therefore, impossible for one set of units to satisfy everyone. Since the formulas in this book have diverse unitary measures, a set of conversion tables is provided in the appendix.

Several examples of types of formulas found in this book follows:

Example No. 1

Mastic Adhesive

	pts./wt.
Water	180
SBR 2002 Latex (48% T.S.)	352
Whiting	336
Titanium Dioxide (pigment)	5
"Carbopol" 934	5
Ammonium Hydroxide	to mucilage pH 9.5

Proc.: Disperse the "Carbopol" 934 in the water with moderate mixing and neutralize to pH9.5 with ammonium hydroxide. Slowly add the latex to this preneutralized "Carbopol" 934 mucilage with moderate mixing, such as with a blender. Add the pigment, very slowly, with moderate agitation, and stir until the product is smooth.

Upon choosing this formula to experiment with, one should begin by gathering the ingredients together. A list of suppliers for trademark products such as "Carbopol" 934 and general chemicals will be found in the Appendix. The amounts of the ingredients to use are given in this formula in a weight ratio. You can decide to use grams, ounces, pounds, etc. But once you choose one of these, all the ingredients must be in this unit of measure. Because the amounts of the ingredients are given in a ratio, it is also possible for the experimenter to reduce the quantities proportionately. Thus, if one divides all the amounts in this formula by 5 the following values are obtained:

	pts./wt.
Water	36.0
SBR 2002 Latex (48% T. S.)	70.4
Whiting	67.2
Titanium Dioxide (pigment)	1.0
"Carbopol" 934	1.0
Ammonium Hydroxide	pH9.5

Now that the ingredients are established, the "Carbopol" 934 is added to the water. Using a pH meter or indicator, the pH of the solution is established. It will be less than 9.5. The ammonium hydroxide, which is a base, is used to bring the pH up to 9.5. This solution is put in a blender, and while mixing at a low speed, the latex is slowly added. The pigment is finally added to the mucilage, and the mixing continues until the product is smooth.

Example No. 2
Hair Lacquer (Non-Aerosol)

	pts./wt.
Refined (wax free)	
Bleached Shellac	15.00
Borax	3.45
Water	81.00

Proc.: Heat water to 145°F and add the borax. Add shellac, dissolving it at 145°F using a high speed stirrer. When the shellac is dissolved, cool and filter. Adjust pH with ammonia to about 8.5. This basic formula is compounded or reduced to the desired solids, to give either an all water hair lacquer, or water alcoholic hair sets.

Hair Lacquer Water Based Non-Aerosol

	pts./wt.
Basic Formula (see above)	80.0
Citroflex A-Z	1.0
Perfume	0.2

Water   18.8

Proc.: Mix all ingredients together.

Hair Lacquer Alcohol Water Based Non-Aerosol

	pts./wt.
Basic Formula (see above)	80.0
Citroflex A-Z	1.0
Perfume Oil	0.2
Alcohol #40	18.8

Proc.: Mix all ingredients together.

Here we have a base formula that is applied to two other formulas. The base formula is the Hair Lacquer (Non-Aerosol). The quantities of the ingredients are given in pts./wt. as in Example No. 1. Thus, the unit of measure of the materials is left up to the discretion of the experimenter.

The water is heated and a thermometer is used to determine when the temperature reaches 145°F. If your thermometer is in °C, use the conversion tables in the appendix to determine how many °C is equal to 145°F. Add the borax. Pour this solution into a blender and add the shellac. Turn on the blender and stir until the indredients are dissolved. After cooling the solution, take a piece of filter paper, fold it, and place it in a funnel. Place the funnel in a container. Pour the solution through the funnel lined with filter paper. Check the pH with a pH indicator or a pH meter. It will be below pH 8.5. The pH can be brought up to pH 8.5 using the base ammonia.

Now that the base for the hair lacquer has been formulated, it can be used in either the hair lacquer-water base formula, or the hair lacquer-alcohol water base formula.

Example No. 3
Nasal Drops

Menthol	2.5g.
Camphor	2.5g.
Eucalyptus Oil	6.0c.c.
Mucilage of	
Methyl Cellulose	6.0c.c.
Chloretone	5.0g.
Dextrose	45.0g.
Dist. Water q.s.	to 1000.00c.c.

Proc.: Liquify the menthol, camphor, and eucalyptus oil by trituration in a glass mortar. Add the mucilage under constant stirring until the oily drops disappear. Dissolve the dextrose and chloretone in the boiling water. When cool, mix the liquids, add water to make 1 liter and shake. Label: Shake well.

In this formula the exact units of measurement are given for each ingredient. Since grams are a weight measurement and cc's are a volumetric measurement, any conversions of units of measure should be done with this in mind.

The menthol, camphor and eucalyptus oil are placed in a glass mortar. With a pestle these ingredients are rubbed and ground up into very fine particles. The mucilage is added, and the stirring and grinding with the pestle is continued. Part of the distilled water is placed in a pot or beaker and heated until it boils. The dextrose and chloretone are then stirred into the water. When they are dissolved and cooled, the mixture of menthol, camphor, eucalyptus oil and mucilage are added to the solution. The entire solution is then placed into a calibrated glass container and distilled water is added to make 1000.00 cc or I liter.

Asterisks noted in the text of the formulas refer the reader to the equipment section of the book. The asterisk denotes processes using pieces of equipment illustrated and described in this section.

ABBREVIATIONS

°Bé	Baume
B.P.	boiling point
°C	degrees Centigrade
cc	cubic centimeters
conc.	concentration
cp.	centipoises
ctsks.	centistokes
dil.	dilute
dr.	dram
°F	degrees Fahrenheit
fl. oz.	fluid ounce
g.	gram/s
gal.	gallon
gr.	grain
hr.	hour
kg.	kilogram
l.	liter
lb.	pound
liq.	liquid
min.	minute or minim
ml.	milliter
M.P.	melting point
oz.	ounce
%	per cent
pH	hydrogen ion concentration
ppm	parts per million
pt.	pint
pts.	parts
q.s.	quantity sufficient to make
qt.	quart
sec.	second
vol.	volume
wt.	weight

GLOSSARY

abrasive — a substance used for grinding, polishing, etc., as sandpaper or emery.

accelerator — a substance that speeds up a reaction.

agglomerate — to gather into a cluster, mass, or ball.

agitator — an apparatus for shaking or stirring, e.g. blender.

alkali — any of a large class of compounds which can react with an acid to form a salt and water. Alkalies are in the pH range of above 7 to 14. Common strong alkalies are sodium and potassium hydroxides, ammonium hydroxide, etc.

anionic — negatively charged (atoms).

anodyne — anything that relieves pain or soothes.

atomize — to reduce a liquid into a fine spray.

bacteriostatic — that property which arrests the growth and multiplication of bacteria.

balsam — any of various oily or gummy aromatic resins obtained from certain of a family (Balsaminaceae) of trees.

bleeding — running together as dyes in a wet cloth.

bung — a cork or other stopper for the hole in a barrel, cask, or keg.

cationic — positively charged (atoms).

centipoise — unit for measuring viscosity; is equal to 0.01 poise c.g.s. system; 1 poise = 1g/cm sec.

centistokes — a unit obtained by dividing the viscosity (in poises by the density (grams/cubic centimeters).

colloid — a substance made up of very small insoluble, nondiffusable particles that remain suspended in a surrounding medium of different matter.

colophony — rosin.

crutcher — mixing device used in making soap.

cure — time necessary for a plastic compound to remain in a mold for complete reaction so that it becomes solidified and chemically inactive.

deaeration — any process used to remove air from a substance.

decoction — an extract produced by boiling down.

deionized water — water that has gone through a purification process.

dermatitis — inflammation of the skin.

dilution — something that has been thinned down or weakened by mixing it with water or other liquid.

dispersion — a medium in which scattered particles are suspended.

distilled water — water which has been purified through the process of distillation.

drier — a substance used to accelerate the drying of paints, varnishes, printing inks, etc.

ebonize — to finish wood.

emulsion — a fluid such as milk, formed by the suspension of a very finely divided oily or resinous liquid in another liquid.

entrainment — the suspension (a liquid in the form of fine droplets) in a vapor, such that the vapor will carry the liquid away, as during distillation or evaporation.

ester — an organic compound formed by the reaction between an alcohol and an acid.

evaporation — the change of a substance from the solid or liquid phase to the gaseous or vapor phase.

exothermic — chemical reaction in which there is a liberation of heat, as in combustion.

filtrate — a filtered liquid.

filtration — mechanical separation of solids from a liquid by means of porous media.

fungistatic — a term used to describe certain fungicides that inhibit fungus growth but do not kill or destroy the fungus.

gel — a jellylike substance formed by the coagulation of a colloidal solution into a solid phase.

gill — a liquid measure equal to ¼ pint.

grout — a fine plaster used for finishing surfaces.

homogeneous — uniform in composition and structure.

homologue — one member of a series of compounds in which each member has a structure differing regularly by some increment (as a CH_2 group) from that of the adjacent members.

impregnated — filled or saturated; permeated with a substance.

lesion — an injury or other change in an organ or tissue of the body tending to result in impairment or loss of function.

levigate — to grind to a fine, smooth powder.

metol — a trademark for a white soluble powder, C_7H_9ON, used in its hydrosulfate as a photographic developer.

mesh — one of the small openings in a woven material, e.g. a 400 mesh sieve has 400 openings per linear inch.

micronize — to reduce to particles of only a few microns in diameter.

mill — piece of equipment used for grinding and crushing.

mucilage — any watery solution of gum, glue, etc.; used as an adhesive.

neutralization — chemical reaction between an acid and a base in such proportions that the characteristic properties of each disappear, e.g., NaOH + HCl → NaCl + H_2O; the end products being a salt and water.
(base) (acid) (salt) (water)

organic acid — compound containing the carboxyl group, $-C = O$,

e.g. acetic acid, CH_3COOH. $\overset{|}{OH}$

osmotic pressure — the force exerted by a solvent passing through a semipermeable membrane in osmosis, equal to the pressure that must be applied to the solution in order to prevent passage of the solvent into it.

pastille — a small tablet or lozenge containing medicine, flavoring, etc.

pectin — a water-soluble carbohydrate, obtained from certain ripe fruits, which yields a gel that is the basis of jellies and jams.

pH — scale for measuring the acidity or alkalinity of a solution. From zero to seven readings indicating respectively very acid to neutral and seven to fourteen indicating increasing alkalinity (basicity).

plasticizer — any of various substances added to a plastic material to keep it soft and viscous.

polymer — a naturally occurring or synthetic substance consisting of giant molecules formed from smaller molecules of the same substance and often having a definite arrangement of the components of the giant molecules.

polymerization — the process of joining two or more like molecules to form a more complex molecule whose molecular weight is a multiple of the original and whose physical properties are different.

potency — having effectiveness or power in action, as a drug.

preservative — a substance that is added to a food to keep it from spoiling.

quaternary — designating a compound containing four different elements.

reagents — Any substance used in a reaction for the purpose of measuring, examining, detecting or analyzing other substances.

reduction — a chemical reaction in which hydrogen combines with another substance or in which oxygen is removed from a substance more generally, a chemical change in which the valence state of an atom of an element is decreased, due to gain of one or more electrons.

refluxing — continuous return of condensed vapor to the boiling liquid by the use of suitable apparatus.

residue — the matter remaining at the end of a process, as after evaporation, combustion, filtration, etc.; residual product.

saponitication — a process by which an ester heated witn an alkali (such as NaOH) reacts to form an alcohol and acid salt; this process produces soap.

saturation — process by which the maximum amount of gas, liquid, or solid is dissolved in a solution at a given temperature and pressure.

sequestering agent — added material in a solution that produces a stable, soluble complex.

shear (shearing force or stress) — the action or force causing two contacting parts or layers (liquid or solid) to slide upon each other moving in opposite directions parallel to their plane of contact.

slurry — a thin watery suspension; also a stream of pulverized metal ore.

solubilize — to render a substance soluble.

solution — a uniformly dispersed mixture, at the molecular or ionic level, of one or more substances (solute) in one or more other substances (the solvent).

solvent — a substance capable of dissolving another substance (solute) to form a uniformly dispersed mixture (solution) at the molecular or ionic size level.

specific gravity — the ratio between the weight of a given volume of any substance with the weight of an equal volume of water.

stabilizer — any substance which tends to keep a compound, mixture, or solution from changing its form or nature. Stabilizers may retard a reaction rate, preserve a chemical equilibrium, act as antioxidants, keep pigments and other components in emulsion form, or prevent the particles in colloidal suspension from precipitating.

supernatant — a liquid or fluid forming a layer on the surface of a solid or another liquid.

surfactant — any substance, such as a detergent, wetting agent, etc., that lowers the surface tension of the solvent in which it is dissolved.

suspension — a liquid having small solid or semisolid particles more or less uniformly dispersed through it. If the particles are small enough to pass through ordinary filters and do not settle out on standing, the suspension is called a colloidal suspension or solution.

therapeutic — serving to preserve health.

thixotropic — a gel or emulsion that has the property of becoming fluid when agitated and then setting again when left at rest.

titer, titre — standard of strength of a solution, as determined by the titration.

tung oil — a fast drying oil derived from the seeds of the tung tree, used in place of linseed oil in paints, varnishes, etc., for a more water resistant finish.

vehicle — a liquid such as water or oil, with which pigments are mixed for use.

viscosity — the internal friction of a fluid caused by molecular attraction, which makes it resist the tendency to flow.

volatile matter — gaseous products, except moisture, given off by a material.

wetting out — wetting or covering a surface completely; penetrating thoroughly.

1. Adhesives and Cements

"Gluing" or "pasting" isn't messy anymore. Remember when the flour and water, or the kind that came in a tube or jar was weak, sometimes lumpy and far from neat? Bottled glue was better, but in order to do a really good job on a hard-used chair you had to get the glue pot and heat up a batch of animal glue.

Today's technology gives us the ingredients to bond all types of material. It does however require a certain amount of thought, care, and experimenting.

The purpose of the following chapter is to give you the best recipe for the job you want to do. If the surface is properly prepared, and the right adhesive used, you are assured of a good, neat bond.

There are four basic types discussed herein:

1. Liquid or paste which makes the bond when the solvent or liquid evaporates or is absorbed by the material being bonded. A variation of this type is the "contact" cement. This is a solvent-containing cement which is coated on the two surfaces to be joined. When the solvent evaporates the two surfaces are brought together and the bond is made immediately.

2. The two-part system requires the measurement and mixing of two substances. One or both may be liquid. When they are mixed they react chemically to produce a substance which will make a firm bond. This is known as "curing."

3. A newer adhesive is the silicone rubber type. Silicone rubber adhesives cure by absorbing moisture from the air and releasing acetic acid which smells like vinegar.

4. Lastly is the "hot-melt" type. This usually requires a special "glue gun." It is applied in a hot, melted condition, and the bond is made as the material cools.

ADHESIVES

The term adhesive refers to any substance that bonds two similar or dissimilar substrates. There is no practical universal adhesive. The nature of the materials to be bonded . . . the character, condition and type of surfaces to be bonded . . . and the bonding purpose influence the choice of adhesive.

The requirements of a good adhesive are:

a. Maximum bond strength.
b. Minimum setting or drying time.
c. Ease and rapidity of application.
d. Ability of the bond to resist the stresses and strains encountered in use . . . and to retain its strength for long periods under unfavorable conditions.
e. Economy in use.
f. Good shelf stability and uniformity from batch to batch.
g. Freedom from health hazards and from objectionable corrosive or discoloring of the materials to be bonded.

Hot Melt Adhesives

I

	pts./wt.
Butvar B-76	10
Santicizer B-16	8
Castorwax	35
Ethylcellulose N-10	3
Poly-pale Ester #1	25
Staybelite Ester #10	19

Proc.: Heat all ingredients and stir until uniform.

II

	pts./wt.
"901 Compound"	52
"Nuba" No. 1	38
"X-743" Plasticizer	40

Proc.: The "901 Compound" and "Nuba" No. 1 are melted and mixed until uniform. The "X-743" plasticizer is then added under agitation.

(Leather to cloth, paper. Metal to cloth, paper. Paper to paper. Paper to tin.)

Cold Water Paste

	pts./wt.
Wheat Flour	8
Alum	1
Water	8

Proc.: Mix until smooth, evaporate to dryness and grind,

Iceproof Glue

	pts./wt.
(For labels on containers kept in ice water.)	
Animal Glue	120
Ammonium Thiocyanate	50
Tapioca Starch	120
Water	259
Phenol	1

Proc.: The animal glue is mixed with the water and allowed to stand for one hour without heating. The mixture is then raised to a temperature of about 55° to 60°C. to complete the dispersion of the glue, and the liquifying agent, the thiocyanate, is added to the mixture at about this temperature. The starch which may have been previously moistened with a portion of the water, is then added to the mixture at the elevated temperature. The mixture is then agitated and heated to a temperature of 65 to 70 C. and the remaining ingredients of the composition are added. The resultant adhesive is then strained, if necessary, and cooled.

Liquid Glues

Glue	3 oz.
Gelatin	3 oz.
Acetic Acid	4 oz.
Water	2 oz.
Alum	30 gr.

Proc.: Heat together for 6 hours, skim, and add:

II

Alcohol	1 fl. oz.
Animal Glue	2 lbs.
Sodium Carbonate	11 oz.
Water	3½ pt.
Oil of Clove	160 minims

Proc.: Dissolve the soda in the water, pour the solution over the dry glue, let stand overnight or till thoroughly soaked and swelled, then heat carefully on a water bath until dissolved. When nearly cold stir in the oil of cloves.

Gum Arabic Glues

	pts./wt.
Water	250.0
Calcium hydroxide	0.2
Glycerin	8.0
Gum Arabic	100.0

Proc.: The ingredients are added in the order mentioned to the cold water, with stirring. When everything is dissolved, the solution is allowed to stand in order to settle and clear. Supernatant liquid is filtered.

Pressure-Sensitive Adhesive

I

	pts./wt.
"Gelva" Multipolymer Solution 263, Acrylic Copolymer	100
Gantrez M-555	25
"Nirez" 2019 Terpene-Phenolic Resin	20

Proc.: Heat and stir all ingredients until uniform.

II

Natural Rubber (Crepe #1)	100
STA-TAC - B	90
Anti-Oxidant (Cyanamid 2246)	1
Hexane to 29% solids	375

Proc: Heat and stir all ingredients until uniform.

Mastic Adhesive

	pts./wt.
Water	180
SBR 2002 Latex (48% T.S.)	352
Whiting	336
Titanium Dioxide (pigment)	5
"Carbopol" 934	5
Ammonium Hydroxide	to mucilage pH 9.5

Proc.: Disperse the "Carbopol" 934 in the water with moderate mixing and neutralize to pH 9.5 with ammonium hydroxide. Slowly add the latex to this preneutralized "Carbopol" 934 mucilage with moderate mixing, such as with a blender. Add the pigment, very slowly, with moderate agitation and stir until the product is smooth.

An important step for the successful preparation of this product is that the "Carbopol" 934 solution must be neutralized to a pH of 9-9.5 before the latex is added. A lower pH will cause coagulation of the rubber and/or flocculation of the pigments.

Joint Sealant

I

	pts./wt.
Acrylic Polymer (25% xylene)	100.00
"Minusil" 5 (50% of polymer base)	37.50
"CAB-O-SIL" M-5	
(5.0% of polymer weight)	3.75
"Surfonic" N-300	
(20% CAB-O-SIL weight)	0.45

Proc.: Heat and stir all ingredients together until uniform.
Acrylic sealants may be applied in the same manner as any gun-consistency sealant.

II

	pts./wt.
Blown Soy Bean Oil	30.0
Whiting #1	67.00
"CAB-O-SIL" M-5	1.10
Ethylene glycol	0.90
Fatty acids	1.10
6% Cobalt drier	0.33

Proc.: Heat and stir all ingredients together until uniform.

Pipe-Joint Compound
(U.S. Patent 2,490,949)

	pts./wt.
Aluminum Silicate	10–50
Sodium Metasilicate	5–15
Propylene Glycol	10–50
Sodium Stearate	5–50
Calcium Stearate	2–20
Lubricating Oil	10–50

Proc.: Heat and stir all ingredients together until uniform.

Heat Seal

	pts./wt.
Alcohol-Soluble Cellulose Propionate	16.0
SAIB	16.0
"Kodaflex" *DOP*	8.0
"Tecsol" *C* (95%)	60.0

Proc.: Heat and stir all ingredients together until uniform.

Polysulfide Sealant

	pts./wt.
Polysulfide Elastomer	100.0
Wet-Ground Calcium Carbonate	20.0

Titanium Dioxide	10.0
Phenolic Resin Tackifier	5.0
Sulfur (curing aid)	0.1
Stearic Acid (wetting agent)	1.0
"CAB-O-SIL" M-5	1.0–3.0
Additives	0.2–1.2
Accelerator (50% lead peroxide in dibutyl phthalate)	15.0

Proc.: Heat and grind* all ingredients until uniform.

Excellent shelf stability and non-sagging properties during pot life can be obtained with this formulation.

Automotive Sealant

	pts./wt.
Polymer Package	
"Dion" 1002 Polymercaptan	100.0
Carbon Black, low oil absorption	150.0
"Ircogel" 900	5.0
Catalyst	
Zinc Dioxide	7.5
DOP	5.5

Proc.: Heat and grind all ingredients together until uniform.

Automotive Seam Sealer

		pts./wt.
A	Base Grease	2–3
	"Amine O"	2–3
	"Attagel" 40	10–15
	High Aniline Point Mineral oil	82–88

Proc.: The base grease is gelled in a Cowles dissolver and cooled to room temperature. The cooled grease is placed in a Z-bar type blender* and the Gilsonite is added in the proportions shown below:

B	Base Grease A (above)	60–70
	Brilliant Black A Gilsonite	40–30

Other ingredients such as asbestos, ground limestone, etc., can be blended in with Step B to modify the final bead characteristics.

Gasket Cement

I

	pts./wt.
Batu Scraped	50
Toluol	7.5
Mineral Spirits	42.5

*Refer to chapter on Laboratory Equipment.

Proc.: Mix the batu scraped with the toluol and half the mineral spirits. Agitate until a homogeneous mixture is obtained. Add the remainder of the mineral spirits and agitate.

No. 2

	lb.
Manila Resin	50
Solox (Denatured Alcohol)	35

Proc.: Mix the ingredients by agitation until a uniform mixture is obtained.

Asphalt Roofing Cement
(A Troweling Asphalt Composition)

Asphalt, Cutback	66
Slate Flour	16
"Short Fiber" Asbestos	21
"Long Fiber" Asbestos	7

Proc.: The materials are added in the order given, using a hoe and mortar box or mechanical mixing. The first three ingredients should be well mixed before adding the "long fiber" asbestos. All materials should be at room temperature. The ratio of reagents may be varied in accordance with the grade of asphalt used.

Heat Sealing for Folding Cartons

"Piccotex" 120	22.80
"Elvax" 260	27.30
Paraffin 155° AMP	40.80
Micro Wax 180° AMP	9.10
"Tenox" B.H.T.	0.10
"Armoslip" E	0.25

Proc.: Heat and stir all ingredients together until uniform.

Heat Seal Coating, Kraft

	pts./wt.
"Piccotex" 120	15.0
"Elvax" 260	15.0
Paraffin 155°F AMP	45.0
Micro Wax 180°F AMP	25.0
"Tenox" B.H.T.	0.1
"Armoslip" E	0.1

Proc.: Heat and stir all ingredients together until uniform.

Caulking Compound (Flexible Polyurethane)

	pts./wt.
Multrathane F-84	17.25
Atomite	17.25
"Niax" Triol LHT-35	28.50

Tributyl Phosphate	0.66
Dibutyl Tin Di-laurate	0.04
Thixcin R	2.74
Atomite	29.65.
Ti-Pure FF	3.91

Proc.: Heat and stir all ingredients together until uniform.

Oil Based Gun Grade Caulk

	pts./wt.
Blown Soybean Oil (Z-4)	14.0
Polybutene (Z6 to Z7)	6.4
Soya Fatty Acids	1.2
Atomite	68.7
Thixcin R	1.7
6% Cobalt Naphthenate	0.3
Mineral Spirits	7.7

Proc.: Heat and stir all ingredients together until uniform.

Butyl Caulk

	pts./wt
Butyl Rubber Solution (50% in mineral spirits)	35.700
Hydrocarbon Resin Solution (60% "Piccopale" 100 in mineral spirits)	6.770

Proc.: Mix the ingredients until uniform

Household Tub Caulk

	pts./wt.
30% Titanium-calcium	150
China Clay	225
Calcium Carbonate	750
"Keltrol" 1001	400
6% Cobalt Naphthenate (0.04%)	1.6

Proc.: Grind all ingredients together until uniform.

Waterproof Cement

	oz.
Zinc Monoxide, Yellow, Powdered	16
Glycerin	21/3
Water	1

or

Zinc Lead Monoxide	16
Glycerin	4
Water	1

Proc.: Mix the glycerin with the water. Add the lead oxide, and mix thoroughly. The amount of water may be varied, depending on the rate of set desired. The resultant product yields a fast-setting, hard, waterproof cement which will adhere firmly to china, glass, iron, stone, wood and other materials.

Waterproof Cement

(For cinder blocks and cement walls)

	pts./wt.
Water-Soluble Melamine Resin (*Melmac*)	20.0
Plaster of Paris	79.5
Ammonium Chloride	4.5

This mixture is added to an equal weight of water to form a paste which is brushed on cinder blocks or cement walls and allowed to dry to form a water-impervious coating.

Sulphur Thiokol Cements

I

	pts./wt.
Sulphur	58.8
Thiokol	1.2
Sand	40.0

Proc.: Mix all ingredients until uniform.

II

	pts./wt.
Sulphur	58.8
Thiokol	1.2
Sand	38.0
Carbon Black	2.0

Proc.: Mix all ingredients until uniform.

Refractory Cement

(U.S. Patent 1,952,119)

	pts./wt.
Magnesium Oxide, Powdered	
(Deadburned)	50
(Fused)	15
Zircon Sand	
60-mesh	25
300-mesh	10
Sodium Silicate (d. 1.3) sufficient to make paste.	

Proc.: Mix all ingredients until uniform.

Nitric Acid Resistant Putty

	pts./wt.
White Asbestos Powder	20
Blue Asbestos Fiber	10
China Clay	10
Linseed Oil	20

Proc.: Mix all ingredients until uniform.

Nonhardening Putty

This putty is useful as a permanently soft and flexible seal where some motion is to be expected, such as thermal expansion. It is also resistant to both water and oil.

	pts./wt.
Castor Oil	97
Magnesium Oxide	3

Proc.: Heat at 400 to 450°F. for 4 to 5 hours or until a cooled sample shows a very high body. Then mix in:

Short Asbestos Fiber	200

The amount of fiber can be varied depending upon its oil absorption. The putty should be made to a very stiff consistency in a small dough-type mixer.

Concrete Block Filler

	pts./wt.
Water	430
"Asbestine" 325	382
"Duramite"	190
"Ti-Pure" R-902	29
Asbestos 7RF-1	19
Emulsion (51.5% "Genepoxy" M195 in H_2O)	110
"Versamid" 265-WR70	

Proc.: Stir the pigments and filler into the water and continue stirring with low shear equipment for a few minutes. The epoxy emulsion is added last to complete this package. Curing agent is added and mixed by hand to provide the complete product.

Release Agent for Hot Concrete

	pts./wt.
"FT"-100H or FT-200	30.5
Paraffin 125/127 AMP	3.5
Emulsifier AP-6	6.0
Triethanolamine	2.5
Water	57.5

Proc.: Heat and mix all ingredients until uniform.

Stable-Viscosity Casein Adhesive

(British Patent 694,174)

	pts./wt.
Water	50
Casein	20
Barium Hydroxide	1
Zinc Oxide	1
Urea	15
Ammonium Sulfocyanate	10
Phenol	1/5

Proc.: Mix and heat to 195°F.

Pyroxylin Cement

Celluloid Scrap	40 g.
Amyl Acetate	350 cc.
Wood Alcohol	100 cc.
Ethyl Alcohol, Denatured	50 cc.
Gum Elemi	15 g.

Proc.: Mix all ingredients until uniform.

Dipping Organosol

(Vinyl)

	pts./wt.
Pliovic 40	48.0
Dioctyl Phthalate	16.6
Wetting agent Victor 85	1.0
Carbonox 1000 distearate	1.2
Apcothinner (Hi Flash Naphtha)	
(Petroleum Solv.)	11.0
Aliphatic Solvent Stand 350	12.0
Surfex	9.5
Thixcin R	0.7

Proc.: Mix all ingredients together until uniform.

Polyurethane Dipping Cement

	pts./wt.
"Estane" 5740 x 1	15.0
Tetrahydrofuran	42.5
Dimethylformamide	42.5
"Carbopol" 934	1.7
Di (2-ethylhexyl) amine	1.7
Water	4.2

Proc.: Dissolve the "Estane" 5740x1 in the THF and DMF. Slowly add the "Carbopol" 934 with vigorous agitation. When the

"Carbopol" 934 is completely dispersed, carefully blend in the amine. At this point, the cement will still be relatively fluid. Lastly, add the water rapidly with as much agitation as possible.

Adhesive for Polystyrene

(U.S. Patent 2,510,908)

Polystyrene	16
Monophenyl Di-*o*-Xenyl Phosphate	8
Ethylene Dichloride	76

Proc.: Mix all ingredients together until uniform.

Acid Proof Cement

I

(Silicate Type)

A cement capable of setting in the cold to a hard infusible mortar, resistant to acids at all concentrations and temperature; not resistant to boiling water or alkali.

A Powder
    1. Filler: 93 to 97
       Silica or pulverized quartz, 50 to 325 mesh, preferably.
    2. Setting Agent: 3 to 7
       a) Aryl or alkal-aryl sulfo chlorides such as benzene sulfochloride, p-toluene sulfochloride, etc.
    or b) Silico fluorides, such as sodium silico fluoride, etc.
    or c) Mixtures of a) and b)

B Binder
    Sodium or potassium silicate water glass, 30° to 40° Baume.

Proc.: Mix thoroughly 2 parts powder to 1 part binder, with allowable variations to obtain a desirable troweling consistency. Setting time 20 minutes to 1 hour.

II

(Synthetic Resin Type)

A cement capable of setting in the cold to a hard, infusible mortar, resistant to acids except the highly oxidizing ones, oils, solvents, greases, water and mild alkalies of temperatures up to 300 to 380 degrees Fahrenheit.

A Powder
    1. Filler: 93 to 96

a) Silica preferably 50 to 325 mesh
or b) Carbon preferably 50 to 325 mesh
2. Setting Agent: 4 to 7

Aromatic sulfonic acids, such as benzene sulfonic acid, p-toluene sulfonic acid, etc.

B Binder
1. A phenol-formaldehyde, or homologues of these two products, reacted by means of an acid or alkaline catalyst, and arrested at a stage wherein the partially polymerized resin has a viscosity of 25 to 70 seconds at 25°C. in a Gardner-Holt tube: 75
2. Glycerin: 10
3. Water: 15

Proc.: Mix thoroughly 2 to 2.5 parts powder to 1 part binder to obtain desirable troweling consistency. Setting time at 70°C. is under 1 hour. Final cure within a day.

"Hycar" Rubber Cement

	pts./wt.
"Hycar" 1001 x 225	15
Methyl Ethyl Ketone	250
"Carbopol" 934	6
Di (2-Ethylhexyl) Amine	6
Water	35

Proc.: Dissolve the rubber in the MEK by stirring.
Add the "Carbopol" 934 with good stirring, followed by addition of the amine. No thickening occurs but the cement may appear somewhat lumpy. Lastly, add the water with vigorous mixing, and the cement will immediately thicken to a smooth, easily spread compound.

Strippable Rubber Cement

	pts./wt.
"Hycar" 1042	10.0
Methyl Ethyl Ketone	79.0
Water	11.0
"Carbopol" 934	0.5
Di (2-Ethylhexyl) Amine	0.5

Proc.: All the ingredients, except the water, are charged to a can and rolled on a roller mill until complete solution and dispersion is effected. This mix is then stirred vigorously while the water is *rapidly* added. The cement immediately thickens to a viscosity of 3000 cps (Brookfield, 20 rpm).

14

Cement for Rubber to Metal

	pts./wt.
Rubber Cement (12% rubber in Benzol)	100.00
Latex	5.00
Zinc Oxide	0.12
Sulfur	2.40
Accelerator	0.12

Proc.: Mix the ingredients together until uniform.

Metal Container Sealing Dope

(British Patent 517,037)

	pts./wt.
Barytes	10.0
Asbestine	5.0

Proc.: Mix the ingredients together until uniform.

Metal-Metal Nitrile Adhesive

	I pts./wt.	II pts./wt.	III pts./wt.
"Chemigum N-5"	100.0	100.0	100.0
SP-8010	70.0	—	
SP-8014	—	70.0	
SP-8214	—	—	100.0
Methylethyl Ketone	510.0	510.0	510.0
Sulfur	2.0	2.0	2.0
Zinc Oxide	5.0	5.0	5.0
"Altax"	1.0	1.0	1.0

Cure 20–40 min. at 320°F.

Proc.: Heat and mix all ingredients together until uniform.

Brake Lining Adhesive

"Chemigum" N-5	100.0
SP-8219	100.0
SP-6600	20.0
Zinc Oxide	5.0
Sulfur	2.0
"Altax"	1.0
Diphenyl Guanidine	0.3
Methylethyl Ketone	600.0

Cure 8 min. at 350°F.

Proc.: Heat and mix all ingredients together until uniform.

Flame Retardant Neoprene Adhesive

"Neoprene" AC	100.0
Zinc Oxide	5.0
"Maglite" D	8.0
SP-134	45.0
Antimony Trioxide	50.0
Tricresyl Phosphate	3.0
Ionol	2.0
Methylethyl Ketone	213.0
Hexane	213.0
Toluene	213.0
Cure above 280°F.	

Proc.: Heat and mix all ingredients together until uniform.

Glass to Metal Seals

I

	pts./wt.
Iron	37
Nickel	30
Cobalt	25
Chromium	8

Proc.: Mix together.
The above is suitable for use with lead-glass.

II

	pts./wt.
Iron	54
Nickel	28
Cobalt	18

Proc.: Mix together.
Suitable for use with Corning glasses.

Paper to Glass Label Paste

		pts./wt.
A	Water	85
	Zinc Chloride	2
	Glycerin	10
B	Cassava Dextrin	10
	Water	15

Proc.: Boil (A) and add to (B), stirring well. The quantity of dextrin may be varied to give a paste of any desired consistency. Keep stoppered.

Glass Cloth Laminate (Wet Lay-Up) Adhesive

	pts./wt.
"Ricon" 100	57.00
Vinyl Toluene	38.00
Trimethylol Propane Trimethacrylate	5.00
Peroxide Catalyst	3.00
Antioxidant	0.25
A-172 Vinyl Tris (2-Methoxyethoxy) Silane	0.60

Type of Glass Reinforcement181 type—heat cleaned

Proc.: The reinforcing material—glass cloth—is impregnated with the resin formulation, press-cured for 30 minutes at 320°F, and then post-cured for 3 hours at 350°F.

Fire Retardant Mastic for Steel

	pts./wt.
Charge into mixer under agitation:	
Water	280.0
Potassium tripolyphosphate	4.0
"Phos-Check" P/30	235.0
Melamine	90.0
Dipentaerythritol	85.0
"Chlorowax" 70	45.0
"Resyn" 5000	26.0
"Igepal" CTA-639	3.5
Alkyd 100% N.V.	15.0
6% cobalt drier	0.5
24% lead drier	1.0
Asbestos S 33/65	50.0

Proc.: Heat and stir all ingredients until uniform.

Fabric-to-Metal Adhesive

(U.S. Patent 2,537,190)

	pts./wt.
Asphalt	44.3
GR-S Rubber	15.0
Rosin Soap	1.4
Methyl Cellulose	0.1
Carbon Black	6.0
Water	33.2

Proc.: Heat and stir all ingredients until uniform.

Paste for Labeling Tin

	pts./wt.
White Potato Dextrin	1

Gum Arabic Solution, 20%	5
Boiling Water	16 1/2
Flake Calcium Chloride	1 3/4
Glycerite of Starch*	1

Proc.: Dissolve the dextrin in the boiling water. Add the calcium chloride and shake well. Finally, stir in the gum arabic solution and the glycerite of starch.

Mordant for Handles of Kitchen-Knives

A	Potassium Bichromate	15 g.
	Water	1000 cc.
B	Ammonia (25%)	150-200 g.

Dissolve the chromate A and add B.

Proc.: Treat wood with solution, dry, rub over with a hard brush (horse-hair), optionally a thin polish.

Joining Stainless Steel in Knife Handles

I

A waterproof cement is used, made by mixing finely powdered litharge and glycerin. The glycerin should be added in an amount equal in volume to half the volume of the powdered litharge and mixed thoroughly. The end of hollow handle is filled with cement and then insert the blade. Setting time about 45 minutes. Mix only enough cement as needed as it sets quickly becoming hard and insoluble.

II

The stainless steel blade is first thoroughly tinned and then soldered in place. It is necessary to have all parts clean and free from scale. Solders used are either 50% tin and 50% lead or 66% tin and 34% lead. Flux used is made up of zinc chloride, commercial grade, 37 g.; glacial acetic acid 99.9%, 23 g.; hydrochloric acid (commercial), 34.5%, 40 g.

Aqueous Adhesive

(Bookbinders, Telephone Books)

	%/wt.
Gelatin	35.0
Glycerin	14.0
Glucose	5.0
Phenol	0.4
Water	45.6

Proc.: Add glycerin and glucose to water. Mix gelatin into solution. Let stand in the cold for several hours. Heat to 140°F. and maintain at that temperature until all the gelatin is dissolved. When clear, add phenol. Pour into suitable forms and allow to cool. This glue is melted when needed. Water may be added to suit requirements.

Upholsterer's and Bookbinder's Paste

A	Potato-Starch	50 kg.
	Water, Cold	140 l.
B	Caustic Potash (50° Bé.)	6 kg.
	Sodium Silicate	15 kg.
	Water, Cold	50 l.
C	Acid to neutralize to weak alkalinity	
D	Rosin Soap, Warm Fluid	5 kg.

Proc.: Stir A till smooth, warm and stir with B to form a mucilage. Stir 3/4 to 1 hour more, add C then D, and stir slowly.

Bookbinder's Paste

A	Rye or Wheat Flour	100 kg.
	Water, 25°C.	200 l.
B	Caustic Soda (35° Bé.)	24 kg.
C	Nitric Acid	until neutral
D	Alum, Cold Saturated Solution	20 kg.

Proc.: Stir A to dispersion, treat mildly with B neutralize with C and add D.

Glue for Bookbinding

Skin Gelatin	4.0 lb.
Boiling Water	3.0 lb.
Glycerin	1.8 lb.
Betanaphthol	10.0 oz.
Alcosolve 2	10.0 oz.

Proc.: Dissolve the gelatin in the boiling water. Mix the betanaphthol with the glycerin and heat gently to effect solution. Mix the two solutions and shake well. Finally, stir in the Alcosolve 2.

Glue

	pts./wt.
Urea	1
Casein	2
Hydrate of Lime	¼

Proc.: Mix together until uniform.

White Glue

	pts./wt.
A solution consisting of:	
Animal Glue	100
Zinc Oxide	50
Water	100

Proc.: Heat and mix together until uniform.

Waterproof Starch Adhesive

(U.S. Patent 2,487,448)

	pts./wt.
Starch	15–25
Polyvinyl Alcohol	15–25
Clay	50–70

Proc.: Heat with water to swell the starch.

Mucilage

Gum Arabic, Amber Sorts	100 lb.
Water	150 lb.

Proc.: Heat and stir until dissolved.

Strain and add	
Oil of Cloves	5 oz.
Oil of Wintergreen	5 oz.
Salicylic Acid	5 oz.

Proc.: Heat and stir until uniform.

Library Mucilage

I (Fluid)

Gum Arabic	25 g.
Saturated Lime Water	15 cc.
Glycerin	1 g.
Water	59 cc.

Proc.: Heat and stir until uniform.

II (Less Fluid)

Gum Arabic	40 g.
Lime Water, Saturated	20 cc.
Glycerin	2 g.
Water	38 cc.

Proc.: Heat and stir until uniform.

III (Viscous)

Gum Arabic	20 g.
Aluminum Sulphate Crystals	2 g.

2% Tragacanth Solution	15 cc.
Water	63 cc.

Proc.: Heat and stir until uniform.

Paste for Offices

White Potato Dextrin	5.0 lb.
Boiling Water	7.5 lb.
Borax	6.8 oz.
Glyceryl Borate*	3.2 oz.
Formalin	0.8 oz.
Phenol	0.8 oz.

Proc.: Dissolve the dextrin in the boiling water. Stir in the borax and glyceryl borate. Finally, add the formalin and phenol, and shake well.

Waxed Paper Adhesive

	I	pts./wt.
Gum Arabic		40.0
Potassium Hydroxide		34.0
Water		75.0

Proc.: The potassium hydroxide is dissolved in water. The solution, spontaneously, becomes very hot and should be allowed to cool to room temperature before the gum arabic is added. This glue is very caustic.

Wax Paper Adhesive

	pts./wt.
"Butvar" B-76	15.0
"Santicizer" B-16	5.0
"Castorwax"	36.0
"Armid" HT	0.7
"Acrawax" C	2.0
"Polypale" Ester 1	24.6
"Staybelite" Ester 10	18.4

Proc.: Heat and mix all ingredients together until uniform.

Envelope Adhesive

	pts./wt.
Gum Arabic	1
Starch	1
Water (sufficient to give the desired consistency)	q.s.
Sugar	4

Proc.: The gum arabic is first dissolved in some water, the sugar added, then the starch, after which the mixture is boiled for a few minutes in order to dissolve the starch, after which it is thinned down to the desired consistency.

Mucilage for sealing letters and documents, or for use in stamps and revenue tags which shows the effects of tampering and steaming is prepared by adding beta-methylum-belliferone to ordinary mucilage or glue base. When the stamp or seal has been tampered with or an attempt to steam open has been made, tell-tale traces of the attempt will be seen by a bright blue coloration on the paper or about the stamp which can be seen only in ultra-violet light and not in white light.

Label Gum

I—Fluid

Gum Arabic	30 g.
Saturated Lime Water	15 cc.
Glycerin	1 g.
Water	54 cc.

Proc.: Heat and stir all ingredients until uniform.

II—Less Fluid

Gum Arabic	35 g.
Aluminum Sulphate Crystals	2 g.
Glycerin	2 g.
Water	61 cc.

Proc.: Heat and stir all ingredients until uniform.

III—Viscous

Gum Arabic	30 g.
Aluminum Sulphate Crystals	2 g.
2% Tragacanth Solution	20 cc.
Water	48 cc.

Proc.: Heat and stir all ingredients until uniform.

Elastic Label Glue

(Swiss Patent 257,408)

	pts./wt.
Bone Glue	30
Turkey-Red Oil	10
Oxalic Acid	10
Water	50

Proc.: Heat and stir all ingredients until uniform. Dilute as desired for use.

Mailing Tube Adhesive

	pts./wt.
Glue, Ground Animal	40
Water	54.7
Nitric Acid	5.0
Phenol	0.3

Proc.: Heat and stir all ingredients until uniform.

Sealing of "Transparit," "Helioglas," or "Cellophane" Packages

		pts./wt.
A	Methyl Acetate	80
	Ethyl Lactate	20

B Collodion-Wool or washed film-scrap, as much as necessary to give a viscous solution (like 30-31° glycerin)

Proc.: Heat and stir all ingredients until uniform.

Cardboard Glues

I

	pts./wt.
Casein	13
Trisodium Phosphate	1
Ammonia (0.91)	2
Water	85

Proc.: Heat and stir all ingredients until uniform.

II

Casein	10
Borax	2
Glucose	2
Waterglass (30° Bé.)	15
Water	71

Proc.: Heat and stir all ingredients until uniform.

"Cellophane Adhesive"

	pts./wt.
Arabic, Gum	16.5
Glycerin	20.5
Glyceryl Bori-borate	9.0
Formaldehyde	4.5

Proc.: Heat and stir all ingredients until uniform.

Cellophane-Paper Hot-Melt Adhesive

	pts./wt.
Ethyl Cellulose (N-50)	8
"Hercolyn"	40
"Staybelite" Resin	52
Octylphenol	1

Proc.: Heat and stir all ingredients until uniform.

Cellophane-to-Paper Laminating Adhesive

	pts./wt.
Ethyl Cellulose (N-50)	8
"Staybelite" Resin, or "Staybelite" Ester 10	52
"Hercolyn"	40
Octylphenol	1

Proc.: Heat and stir all ingredients until uniform.

Wax (Laminating Compound)

	pts./wt.
Diethanolamine Stearate	3
"Attagel" 20	15
Microcrystalline Wax (m.p. 160-170°C.)	15

Proc.: Heat and stir all ingredients until uniform.

Padding Glue

	lbs.	oz.
1. Glue (Nat. Assoc. 8-10 Grade)	10	
2. Glycerin	10	
3. Water	12	2
4. Zinc Oxide	1	3
5. Beta Naphthol		¼
6. Methyl Salicylate		1

Proc.: Mix 2 and 4, then add 5 and 3, and then 1. Let stand over night, warm and stir until uniform; add 6 and pack.

In hot humid weather this glue may set too slowly. This may be corrected by

a. Using a higher grade of glue, or
b. Using less glycerin (which will, of course, reduce flexibility), or
c. Dusting surface after partial drying with talc or precipitated chalk.

Wallpaper Adhesive

Starch	10 kg.
Water	15 l.

Proc.: Mix until uniformly suspended. Add this, while stirring, to a boiling solution of:

Urea	5½ kg.
Water	90 l.

This gives a neutral product that does not penetrate the paper.

Paperhanger's Paste

A	White or Fish Glue	4 oz.
	Cold Water	8 oz.
B	Venice Turpentine	2 fl. oz.
C	Rye Flour	1 lb.
	Cold Water	16 fl. oz.
D	Boiling Water	64 fl. oz.

Proc.: Use a cheap grade of rye or wheat flour, mix thoroughly with cold water to about the consistency of dough or a little thinner, being careful to remove all lumps. Stir in a tbsp. of powdered alum to 1 qt. of flour, then pour in boiling water, stirring rapidly until the flour is thoroughly cooked. Let this cool before using and thin with cold water.

Soak the 4 oz. of blue in the cold water for 4 hours. Dissolve on a water bath (glue-pot) and while hot stir in the Venice turpentine. Make up C into a batter free from lumps and pour into D. Stir briskly, and finally add the glue solution. This makes a very strong paste, and it will also adhere to a painted surface, owing to the Venice turpentine in its composition.

Pregummed Wallpaper Adhesive

(U.S. Patent 2,524,008)

	pts./wt.
Sodium Carboxy Cellulose	3
Polyvinyl Alcohol	3
Water	94

Proc.: Heat and stir all ingredients until uniform.

Flock and Laminating Adhesive

I

	pts./wt.
"Rhoplex" K-3	79.0
Oxalic Acid (10%)	1.0
"Methocel" 65HG-4000 STD (4%)	21.0

Proc.: Heat and stir all ingredients until uniform. Following drying, a minimum cure of 1-½ minutes at 280°F is required for crosslinking.

II

	pts./wt.
"Rhoplex'. K-3	92.5
"Acrysol" ASE-60	3.5
Diammonium hydrogen phosphate (25% solution)	4.0

Proc.: Heat and grind* all ingredients until uniform.

This adhesive should be allowed to dry thoroughly before curing at 300°F for five minutes. A shelf life of three months can be expected for this formulation. The resulting viscosity will be approximately 50,000 cps and the pH will be about six.

Wall Size

	pts./wt.
Aluminum Stearate	4
Turpentine	25
Mineral Spirits (150-190°C.)	71

Proc.: Heat the turpentine to 180°F. and add the stearate slowly while stirring continuously. Add mineral spirits and stir until clear.

Joint Cements and Spackling Compounds

	Prepared Joint Cement pts./wt.	Prepared Spackle pts./wt.
Triethanolamine	2	2
Oleic Acid	4	4
Water	226	135
"Methocel 65 HG" 4000 cps (3% in water)	33	125
Polyvinyl Acetate (55% NV)	136	136
"Duramite"	900	1100
"Snowbrite Clay"	150	—
1K Mica	50	—
"Troysan PMA-30"	0.3	0.3
Mineral Spirits	26	26

Proc.: Mix in suitable equipment the triethanolamine and oleic acid until they are completely reacted. This is indicated by the formation of a clear gel. Water, methocel solution, and polyvinyl acetate emulsion are then added to the soap and thoroughly mixed. The pigments should be added slowly, followed by the preservative and the mineral spirits. The compound should be quite smooth and homogeneous throughout at the end of the mixing period. The material should be packaged in cans that have been lined to prevent rust and latex coagulation.

Dental Cement

(British Patent 430,624)

	pts./wt.
Lithium Phosphate	½
Phosphoric Acid	5
Zinc Phosphate	½
Aluminum Phosphate	1/3

Proc.: The above is added to a ground porcelain of following composition:

Alumina	30–50
Feldspar	10–20
Sand	25–40
Zinc Oxide	1–10

Antiseptic Dental-Plate Adhesive

(U.S. Patent 2,574,476)

	pts./wt.
8-Hydroxy Quinoline	0.2
Petrolatum U.S.P.	36.0
Mineral Oil U.S.P.	12.5
Karaya Gum	48.8
Food Color	1.5
Flavor	To suit

Proc.: Mix all ingredients until uniform.

Acrylic denture Cement

(U.S. Patent 2,471,501)

	pts./wt.
Ether	60
Acetone	30
Alcohol	10

Proc.: The edges of pieces are softened with this solvent and then pressed together.

Temporary Dental Cement

(U.S. Patent 2,406,063)

Eugenol	99.5
Acetic Acid	0.5

Proc.: Mix this in proportion of 0.1 cc. to 1.5 g. zinc oxide powder. It sets in 6 to 8 minutes.

Cement (Household) or Celluloid Adhesive

A waterproof cement for making repairs on many different materials (it is especially good for leather) is easily and cheaply made by dissolving celluloid in "Cellosolve," amyl acetate, or ethyl acetate, or a similar moderately high-boiling solvent. Use about one part of celluloid to two parts of solvent. A few days are generally necessary for the solution to form and the material, during making and storing, must be kept in tightly closed vessels. Many different easily available objects will furnish celluloid, e.g., tooth brush handles, parasol handles, some of the "horn" spectacle frames, some novelties, as balls, cards, etc. Some produce clear and some opaque cements, and many different colors are available. This cement is similar to various "household" cements marketed in metal tubes.

Adhesive Cement (For Fine Furniture)

	pts./wt.
Casein (fine ground)	12
Lime (powdered, unslaked)	13
Mica (dry, ground)	15
Barium Sulphate (barytes)	60

Proc.: Mix all ingredients. Keep in dry container. To use, mix with water until pasty. Hardens in about 24 hours.

Floor Crack Filler

	pts./wt.
Plaster of Paris	16
Silica	100
Dextrine Yellow	16.5

Proc.: Make into a stiff dough with water before use.

Wood Cement

I

	pts./wt.
Liquid Shellac	1
White Resin	1
Phenol	1

Proc.: The first two ingredients are mixed and melted together first, then the third ingredient is blended in.

II

(German)

	pts./wt
Chalk	420
Sawdust	150
Sulfite Liquor	400
Urea	30

Proc.: Heat and mix all ingredients until uniform

Panel Adhesive

	pts./wt.
"Ameripol" 1013 crumb	100
"Cumal" MH	25
"Pentalyn" H	25
"Dixie Clay"	5
"Amberol" ST 137X	15
"Primol" 355	15

Proc.: Heat and mix all ingredients until uniform.

Wood Veneer Glue

	pts./wt.
Blood Albumen	40
Casein	12
Slaked Lime	6
Sodium Fluosilicate	2
Wood Meal	40

Proc.: Mix and heat all ingredients until uniform.

Apply the adhesive by putting it on both sides of the middle piece of wood. If the adhesive is just too viscous, homogenize the adhesive layer. The wood pieces are put together, then pass through drying chambers* at 90-95°C., under a pressure of 12–18 kg. per cc. until the albumen is coagulated.

Wood Veneer Adhesive

(U.S. Patent 1,964,960)

	pts./wt.
Casein	1
Ammonium Sulphocyanate	2
Paraformaldehyde	.02–0.4
Water sufficient to make fluid.	

Proc.: Mix and heat all ingredients until uniform. This will remain fluid for several hours at ordinary temperature. Coagulates on heating to give strong bond.

Exterior Plywood Adhesive

	pts./wt.
"Unifine" Wheat Flour (100 mesh)	16.3
Idaho Clay	1.4
Water	62.3
Compregnite Resin	20.0

Proc.: The flour and clay are dry-mixed, then the water is added gradually under slow stirring (vigorous stirring produces foam!). When all the flour is wetted and no lumps remain, the resin is added with stirring. The adhesive is ready for use as soon as it begins to thicken, usually in about one-half hour.

Hot Press Plywood Waterproof Glue

	pts./wt.
M-3 Glue	100
Shell Flour	20
Water	55
Catalyst CP-10	¼

Proc.: Weigh all ingredients. Place 1/2 of water in mixer. Add all of the M-3 and shell flour. Mix until smooth. Add catalyst CP-10 to the remainder of the water and add to mixer. Continue to mix until a smooth, even mix is had. Glue is ready for immediate use.

NOTE: Use a mixer with double action* with a speed of about 60 RPM. The working life of this mix is about 8 hours at 75°F.

Vinyl Foam or Tile to Plywood-Concrete Adhesive

		pts./wt.
A	"Picco" 6070-3	145.0
	Mineral Spirits	5.0
	"Vistac" P	10.0
	Ammonium Oleate	5.0
	"Antara" BL0344	24.0
	Water	50.0
	"Nopco" 2271	15.0
B	"Rez-O-Sperse"	22.6
	Colloid Kaolin	150.0
	Acid Casein	24.0
	"Agerite" Spar	1.0
	Sodium Silicate 34	10.6
	"Marmix" 22334	192.0
	Natural Latex 2XB (low ammonium)	25.0
	"Ultrawet" DS	0.6
	"Alcogum" 6625	q.s.

Proc.: The materials in A and B and mixed separately using high shear. They are blended together and with the "Marmix" 22334 and natural latex using moderate to low shear agitation. The "Ultrawet" DS is added to the natural latex before addition to the compound. The "Alcogum" is added as required, to obtain a paste viscosity.

"Styrofoam" to Plywood or Concrete Adhesive

	pts./wt.
"Picco" 6070-3	106.6
"Piccolyte" S-10	15.0
Mineral Spirits	10.0
Ammonium Oleate	16.0

Water	15.0
Acid Casein	6.7
"Agerite" Spar	1.0
McNamee Clay	100.0
"Marmix" 21333	173.5
Natural Latex X2B (low-ammonia)	15.4
"Ultrawet" DS	0.6
ASE 60	q.s.

Proc.: The emulsion is prepared using high shear agitation and the other components are blended in. After the McNamee clay has been added, the shear is reduced as the "Marmix" 21333 and natural latex are added. The "Ultrawet" DS is added to the natural latex before addition to the compound. Viscosity is controlled by the judicious addition of the ASE 60.

Asphalt Mastic

		pts./wt.
A	Gilsonite Selects (pulverized)	46
	Calcium Carbonate or Silica (100-mesh)	46
	Asbestos 7r	8
B	Asphalt—100 Penetration	38
	Gilsonite Selects (lump or pulverized)	46
	Mineral Spirits	23
	Xylol	22

Proc.: Mix A, heat and mix B. Mix A + B together.

Mix ratio: 100 parts component A
70 parts component B

This mixture will be of heavy consistency suitable for trowel application, such as for floor patching. It will have a workable pot life of about 30 min. and will cure hard in 24 hr. at ambient temperatures.

Keying Plaster to Concrete

First secure a fast setting plaster which corresponds to Plaster of Paris, moulding plaster or something similar. This plaster is mixed thin enough so it can be whipped onto the wall with a brush. After this dash coat of plaster has thoroughly set, the wall, which now has a rough surface, may be plastered over in the usual way with ordinary gypsum plaster.

Retarding Setting of Cement
(U.S. Patent 3,317,327)

	%/wt.
Proc.: Add:	
Magnesium Silicofluoride	0.01 -1.0
Calcium Lignosulfonate	0.001-0.5
to cement.	

Adhesives For Rugs

	pts./wt. Dry	pts./wt. Wet
"Enjay" Butyl Latex 80-21	100	182
"Darvan No. 1"	1	1
"Piccopale A-22"	20	40
No. 10 Whiting	50	50
"Hi-Sil 233"	20	20
"Hercolyn"	20	20
"Cellosize QP-15,000" (3% Solution)	1.1	36
(Total Solids = 60%)		

Proc.: Mix all ingredients together until smooth.

Carpet to Concrete or Wood Adhesive

	pts./wt.
"Dow A" Antifoam	0.33
Ammonium Oleate	72.00
Triethanolamine	2.39
Ammonium Hydroxide	5.30
"Picco" 6070-3	428.00
Water	93.17
McNamee Clay	150.00
"Agerite" Spar	1.00
"Marmix" 22332	173.50
Natural Latex 2XB (low ammonia)	15.40
"Ultrawet" DS	0.60
"Acrysol" HV-1	30.00

Proc.: Blend the first components, then add to "Picco" 6070-3 under high shear agitation*. Add the water and clay slowly to obtain a uniform dispersion. Add the "Agerite" spar, then reduce the agitation while adding the "Marmix" 22332 and natural latex. The "Ultrawet" DS is added to the natural latex before addition to the compound. The "Acrysol" is added to increase the viscosity. Alternately, "Alcogum" 6625 may be used.

For Cracks in Wood & Masonry

	pts./wt.
"Parlon" (20-cp)	100
"Aroclor" 1254	30
"Aroclor" 5460	49
Asbestos	160
Titanium Dioxide (Rutile)	25
Toluene	300

Proc.: Mix all ingredients together until uniform.

Marine Linoleum Cement

Decks to be covered with linoleum should be thoroughly cleaned, and the linoleum stuck to the deck with the following adhesive:

To make 10 gallons, first cut 4 oz. of crude (ham) rubber into small lumps and dissolve in 4½ gallons of gasoline. It will require about two days to get the rubber into colloidal solution. When in proper condition it should string about two inches thumb and forefinger. Cut 19 lb. of gum shellac in 34 lb. of denatured (or wood) alcohol. Add 62 lb. of whiting then add the rubber solution. For best results this mixture should be ground in an iron or pebble mill.*

Linoleum Glue

A	Rye or Barley Flour	50 kg.
	Water, tepid	250 l.
B	Caustic Soda (20° Bé.)	20 kg.
C	Turpentine, Venice, melted	20–25 kg.

Proc.: Part A dispersed by stirrer is mixed with B (dissolved). The mixture is then boiled, and after cooling emulsified by adding C (while stirring add).

Painters' Glue (Cold)

Water (25° C.)	350 l.
Potato-Starch, Powder	100 kg.
Rosin, Finely Ground	21 kg.
Caustic Soda (24° Bé.)	56 kg.

Proc.: Mix altogether with strong stirring for 2-3 hours, let stand 1 hour, and neutralize with dilute nitric acid until red color with phenolphthalein disappears (in a sample). Stir 1/2 hour more.

Acoustic-Tile Adhesive

(U.S. Patent 2,610,924)

	pts./wt.
Limed Rosin	26

"Paraflow"	8
Clay	50
Naphtha (B.P. 300-375°F.)	11½
Water	2

Proc.: Mix and heat all ingredients until uniform.
This has high wet strength and uniform viscosity at 60 to 115°F.

Adhesive for Window Glass

	pts./wt.
Gum Arabic	18
Water	52
Glycerin	30

Proc.: Dissolve the gum arabic in water and add the glycerin.

Paper to Glass Adhesive

Water	150
Zinc Chloride	2
Glycerin	10
Cassava Flour or Tapioca	12½

Proc.: Mix thoroughly and heat to boiling, stirring vigorously, all the time, during the heating. If tapioca is in lumps, the mixture should soak overnight before heating. Other starches and flour can be used in place of the cassava and tapioca, but the products are much poorer in quality.

Tile Adhesive

I

"Ameripol" 1013 Crumb	2.5
"Ameripol" 1009 Crumb	7.5
"Calcene" TM	4.0
"Paragon" clay	12.0
"Cumar" MH 2½	9:5
Antioxidant 2246	0.1
Heptane	17.5

Proc.: Heat and mix all ingredients until uniform.

II

	pts./wt.
Rosin	17.0
Furfural	2.0
Drying Oil	6.8
White Spirit	9.8
Acetone	0.9
Dolomite (powder)	63–70

Proc.: Dissolve the ground rosin and turfural in white spirit at 80°C. Add oil and stir to a homogeneous mass. Cool and add the acetone adding the powdered dolomite slowly with continuous stirring to a workable consistency.

Abrasives to Cloth Adhesive

	pts./wt.
Zein	22
"Nevillac" Hard	7
"Nevillac" 10°	4
Methyl "Cellosolve"	54
Isopropanol	15

Proc.: The zein is dissolved in the methyl "Cellosolve" while the "Neville" Resins are dissolved in the isopropanol. The two solutions are then combined.

Cement for Leather Driving Belts

	pts./wt.
Rosin (light)	30
Rubber (Dry, Waste)	20
Linseed Oil Varnish	20
Benzine (High Boiling)	30

Proc.: Heat the rosin, rubber and linseed oil together until completely dissolved. Add the benzine, taking precautions against fire.

Leather Sole Cement

Nitrocellulose	22.5
Alcohol	22.5
Benzol	31.1
Ethyl Acetate	9.5
Camphor	1.1
Acetone Oil	0.09
Castor Oil	0.09

Proc.: Heat and mix all ingredients until uniform.

Cement for Leather or Leather on Rubber

	pts./wt.
Gutta-Percha	21.6
Carbon Bisulphide	17.7
Benzene	2.9
Turpentine Oil	23.5
Asphalt	34.3

Proc.: Heat and mix all ingredients until uniform.

Leather Cement

Celluloid	11.9
Naphthalene	1.2
Acetone	67.1

Proc.: Heat and mix all ingredients until uniform.

Leather Belt Cement

A	Glue, Hide	50 g.
B	Water	200 g.

Proc.: Soak A in B, pour excess water off, and melt the soaked A with:

C	Glycerin	2%
	Potassium Bichromate	2%

When cooled, pour into oiled metallic forms; pack the gelatinous product at once into grease-proof paper.

Apply on roughed surface, while the sharpened ends are pressed together for 6 to 10 hours.

Universal Putty for Wood, Stone, Glass, Porcelain

(Dries after 24–30 hours)

		pts./wt.
A	Alabaster Gypsum	4
	Gum Arabic	1
B	Cold Borax Solution, Saturated.	

Proc.: Stir until pasty.

Porcelain and China

I

	pts./wt.
Calcium Fluoride, Finely Powdered	2
Glass, Powdered	1
Sodium Silicate (36–38° Bé.)	to form a dough

Proc.: Mix all ingredients together.

II

(Not Heat and Solvent Resistant)

	pts./wt.
Venice Turpentine	10
Mastic	35
Bleached Shellac	50
Zinc Oxide	5

Proc.: Melt the resins on a water bath. Add the pigment slowly with stirring.

Apply like a sealing wax, then press the hot fractured parts together, let cool.

III

When a cement for porcelain, metals or stoneware is desired, casein mixed with sodium silicate and lime makes an efficient mixture.

Dried casein is soaked in an equal weight of water for two hours. The casein swells. Then the sodium silicate and lime are stirred in.

IV

This china glue is a vegetable product which is produced in the following manner: In a wooden vat which is provided with a wooden stirring device 10 kg. high grade potato flour and 40 litres of water are thoroughly mixed until a milky emulsion is formed. Then 2.4 kg. sodium hydroxide solution (40° Bé.) is gradually added in a thin stream under steady stirring whereby the mass is finally converted into a clear gelatinous glue. For the purpose of proper homogenation the stirring device should be operated at a speed of 60 revolutions per minute for a period of 1 to 2 hours whereupon through gradual addition of a mixture of 2.2 kg. nitric acid (36° Bé.) and 5 litres water, the starch glue which has been hydrolyzed by the sodium hydroxide solution becomes neutralized. After a further stirring for half an hour there is added to the uniformly white glue 40 g. phosphoric acid (sp. gr. 1.3) and finally for the purpose of preservation 400 g. formalin (formaldehyde 40%) and 50 g. betanaphthol. The finished glue may be slightly alkaline and neutralization need not be complete. If no stirring device is obtainable on which all the metal parts can be covered with wood, a wooden paddle may be used; however, the mixture will be rarely uniform as the stirring of the mass by hand is rather difficult.

Aquarium Cement

To 10 lbs. of glazier's putty add 1 lb. dry litharge, 1 lb. dry red lead, and 1 gill of asphaltum. Mix to a stiff consistency with boiled linseed oil and add sufficient lampblack to give a slate color.

Another well-known formula consists of 10 parts by bulk of plaster of Paris, 10 of fine sand, 10 of litharge, 1 part of powdered rosin, and sufficient boiled linseed oil to make a stiff putty. A third formula is as follows: Red lead 3 parts, litharge 7, fine sand 10, powdered rosin 1 part, and spar varnish sufficient to make a stiff cement.

In each case add the linseed oil or varnish little by little and mix the ingredients very thoroughly. If the putty should become too soft, merely add more of the dry materials as the exact proportions are not especially important.

Adhesive for Wigs

	pts./wt.
Damar	20
Rosin	20
Beeswax	40
Venice Turpentine	20

Proc: Heat to 90°C. and stir until uniform; cast in sticks.

*Decalcomania Adhesive

	pts./wt.
Glue	13.5
Water	28
Butanol	7.3
Toluol	9.7
Alcohol	26.8
Turkey Red Oil	14.7

Proc.: Heat and mix all ingredients.

2. Chemical Specialties

In olden days, when life was simple, one used to clean one's windows, pots and pans, clothes, carriages, horses, dogs, shoes, face, hair, mirrors, wash basins, spectacles, with much the same thing . . . soap and water. At present, all one has to do is look in one's sink closet, basement, garage or bathroom and observe the cluttered cornucopia of products which specialize in dealing with just one of the aforementioned items, and one will instantly comprehend the significance of the chemical specialty industry.

There are detergents, cleaners, pre-cleaners, polishes, floor care items, textiles and chemicals that alter and protect them, leathers, artificial leathers, industrial, automotive, aerospace and household maintenance products by the hundreds of thousands. For every additional complexity which our technological culture creates, corresponding chemical specialties are born.

Since the young science of ecology has begun to exert some influence in the direction of our technology, many prior "harmless" substances are being systematically eliminated from the environment. The list of these substances is growing larger, and chemicals which today are acceptable may not be so tomorrow. Federal, State and local prohibitions on the use of newly discovered pollutants do not always agree. Therefore, before beginning the preparation of any formula, however benign its ingredients may appear, you should consult the Food and Drug Administration, or the Environmental Protection Agency in your community. They will give you the latest list of forbidden chemicals.

Flexible Abrasive Sheet

A textile or paper is coated with the following ingredients:

	pts./wt.
Aluminum Oxide Calcinated	50
Dry Glue	70
Glycerine	12
Water	13
Linseed Oil	5

Proc.: Mix together all ingredients. Heat the mixture in a water bath until smooth, embed the aluminum oxide calcinated and coat the textile or paper with this mass.

Abrasive Grain

	%/wt.
Silicon Oxide	33
Aluminum Oxide	6
Barium Oxide	6
Calcium Oxide	4
Magnesium Oxide	2
Sodium Oxide	11
Potassium Oxide	15
Boron Oxide	23

Proc.: Mix together all ingredients until uniform.

Abrasive Wheels

Grinding wheels and other abrasive articles are made with latex as the binder for the abrasive material.

	pts./wt.
Aluminum Oxide Calcinated	600
Latex	200
Sulphur	40
Accelerator 808	40

Proc.: Add to the latex mix a solution of zinc acetate in the amount of 1%. The mixture is stirred until it has a cheese-like consistency and is then molded to shape, dried and vulcanized for two hours at a temperature of 287°F. To the above Formula you can add 25% of long fibre asbestos. Emery cloth or paper can be made from the above Formula to which Glue or Casein is added to make the binder more adhesive; 4% Glue (animal) is suggested. It is then spread on fabric or paper, dried and vulcanized.

AUTO PRODUCTS

Detergent For Spray-Type Automatic Washers
(Powdered)

I

	pts./wt.
"Plurafac" D-25 Surfactant	18
Sodium Tripolyphosphate L/D	20
Tetrasodium Pyrophosphate L/D	47
Sodium Metasilicate Pentahydrate	5
Sodium Sulfate	10

Proc.: Mix all ingredients together until uniform.

II
(Concentrated Liquid)

	pts./wt.
"Plurafac" D-25	3.4
"Klearfac" AA-270	0.6
Tetrapotassium Pyrophosphate	6.7
Sodium Benzoate	0.4
Water	88.9

Proc.: Mix all ingredients together until uniform.

Cleaner-Polish

I

	pts./wt.
Mineral Oil (60-80 sec.)	6
Petroleum Naphtha (95-140°C)	6
Oleic Acid	2
Carnauba Wax	1
Amino-methyl-propanol	1
Diatomaceous Earth	11
Water	73

Proc.: Melt the first four ingredients together and stir in the diatomaceous earth to make a thick paste. Then add the Amino-methyl-propanol and water—mixed together and warmed to approximately 75°C—to the warm melt with very vigorous agitation. The use of a colloid mill* for this operation gives excellent results. The proportions of diatomaceous earth, mineral oil, and naphtha may be varied within a reasonable range depending on the grades of the ingredients used and upon the characteristics desired in the cleaner-polish.

II
Detergent-Resistant

		%/wt.
1	DC-530 Fluid	0.75
1	DC-531 Fluid	4.50
2	San 80	1.00
	Kerosine (d)	10.00
	Stoddard Solvent	15.75
3	Kaopolite SF-O	10.00
	Water	58.00

Proc.: Mix all ingredients together until uniform.

Radiator Cleaner
I

	%/wt.
"Metso"	45
30% "Nullapon"	45
"Antarox" A-400	10

Proc.: First empty the radiator. Then dissolve the ingredients in warm water and use 1/3 ounce to gallon of radiator capacity. Run the engine for a few days. Repeat the treatment if necessary.

II

A combination of 3-5 parts by weight of "Plurafac" A-24 with 95-97 parts by weight of oxalic acid will effectively clean automotive radiators. It removes rust, suspends scale and prevents the cooling system from clogging. Unlike anionic surfactants, "Plurafac" A-24 is a low foamer.

Cleaner for Auto Radiators and Cooling System
(U.S. Patent 2,036,848)

	pts./wt.
Kerosene	4
Ortho-dichlorbenzol	7
Oleic Acid	1

Proc.: Add the above to water (while circulating). This deposits on and softens dirt and grease. Then add 2 oz. caustic soda and circulate to saponify grease and oil. Wash out with water.

Cooling System Cleaners

Washing soda, trisodium phosphate, and silicates form the basis of most alkaline type cleaners. Formerly, nearly all radiator cleaners were of this class, and they still find very extensive use. Although these materials cut grease and remove loose rust and dirt, they are deficient in certain respects. A cooling system cleaner must be capable of removing the hard adherent scale by a dissolving action. Alkalies have no solvent action on the rust itself, and this type of cleaner acts very slowly. They have, however, been described as being satisfactory for most cleaning jobs.

I

	%/wt.
Soda Ash	35.0
Sodium Dichromate, Dihydrate	65.0

Proc.: Mix all ingredients together until uniform.

II

	%/wt.
Trisodium Phosphate	12.0
Sodium Chromate	0.3
Water	87.7

Proc.: Mix all ingredients together until uniform.

III
(Aerosol Emulsion-Abrasive Type)

	pts./wt.
Silicone Oil, 1000 cs.	5.0
Silicone Oil 30,000 cs.	1.0
Duroxon J-324	2.0
Durmont E	2.0
Mintrol Spirits	12.0
Kerosine	8.0
TEA Tetraomyl-ammonium	2.5
Water	16.0
Snowfloss	14.0
Methocel 4000-cps. at 1%	37.0

Proc.: Mix all ingredients until uniform.

Aerosolization

Fill: 90 parts above/10 parts Freon 12
Valve: 3 x 0.40 Foam valve with inverted body.
Container: Lacquered 1/2 lb. ETP.
Filling Technique: Pressure

Paste Wax/Cleaner

		pts./wt.
A	GE Viscasil 10,000	5.0
	No. 2 Carnauba Wax	4.0
	Wax LP	1.8
	Wax E	7.2
	Ozokerite 77Y	2.0

B Mineral Spirits	55.5
Bayol 35	12.5
C Kaopolite SFO	12.0

Proc.: Heat components of A to 90°C (194°F) and hold at this temperature until all waxes are melted. Pre-heat B to 60°C (140°F) to prevent precipitation of the waxes. Add B slowly to A with constant low speed agitation. Continue agitation after B addition is complete. 4. Add C slowly to A-B mixture. 5. Continue agitation until the temperature reaches 40-42°C (104-108°F). Pour into chilled containers. Circulating cool air across the container and paste surface will aid solidification.

White Sidewall Tire Cleaner

I

	%/wt.
TSP*	30
Anionic	5
Soda ash	balance

*Sodium phosphate
Proc.: Mix all ingredients together until uniform.

TKPP*	10
Nonionic	5
Water miscible grease solvent*	5
Water	balance

*e.g.—Butyl Cellosolve
*Potassium phosphate

II

	%/wt.
*CONCO SULFATE WE	3.3
Sodium Metasilicate	6.0
Cellosolve	0.96
Hampene 100* Cheleting agent	0.09
Water	89.65
Coloring	Trace

*Sodium lamoyl ether sulfate Continental Chem. Co.
Proc.: Mix all ingredients together; heating until uniform.

Aluminum Truck Body Cleaner

	%/wt.
A 75% H_3PO_4 Phosphoric acid	35–40
Ethanol Alcohol	25
Noionic	5
$NaHF_2$ Sod. Bifluoride	1– 2

Abrasive (e.g., Ca SO$_4$)	2
Water	balance
B 75% Phosphoric Acid	10
Nonionic	4
48% Hydrofluoric Acid	15
Water	balance

Proc.: Mix all ingredients together; heating until uniform.

Silicone Based Liquid Polish

	pts./wt.
A "GE SF-96" (350) Silicone Fluid	1.5
Oleic Acid	2.5
"Isopar" M (solvent)	16.0
B Water (I)	49.1
Carnauba Wax Emulsion 15%	1.0
Morpholine	2.0
"Carbopol" 934	0.2
Triethanolamine	0.2
C "Isopar" L	5.0
"GE SF-1700"	1.5
"GE SF-1702"	1.5
D Water (II)	8.0
NH$_4$OH	q.s.
"Kaopolite" 73	6.0
"Kaolin" 1104	5.5

Proc.: Disperse "Carbopol" 934 in water and, when completely dissolved, add triethanolamine; then add the wax emulsion and morpholine. Add A to B with high speed agitation. Add C; continue high speed agitation or mill* to a stable emulsion. Neutralize the abrasive in D to pH 7.0; add this slurry to ABC. Adjust to pH 9.0 if necessary with additional NH$_4$OH.

Application: Apply a thin even coat with a damp cloth using circular motion. Allow the film to dry thoroughly. A little extra time will assure good detergent resistance. Buff with a clean dry cloth turning the cloth frequently to insure maximum ease of buffing.

Liquid Polish Water-in-Oil (W/O), Detergent Resistant

	%/wt.
A Durmont 500 Montan Wax	2.00
B DC 530 Silicone Fluid	0.75
DC 531 Silicone Fluid	4.25
SPAN 80	1.50

Kerosene	10.00
Stoddard Solvent	16.50
C Kaopolite SFO	10.00
D Water	55.00

Proc.: Melt wax in A (85-90°C). Add B ingredients to the melted wax and stir to blend well. Return temperature to 85-90°C. Add C to A/B blend and mix until uniform with medium agitation. Keep temperature in the 85-90°C range. Heat D to 95°C and slowly add to the blend with high speed stirring until emulsion is obtained. Cool to 40-45°C with continuous stirring. Homogenize.*

Asbestos Fiber Undercoating
(U.S. Federal Specifications—Coating, Underbody)

	pts./wt.
Blown Asphalt Cutback (60% solids)	50.0
Tackifier Resin (Cumar or equiv.)	0.5
Chlorinated Paraffin	1.0
Mineral Spirits	13.2
High Flash Naptha	1.8
Mica (325 mesh)	7.0
7T Asbestos Fiber	16.5
Weight per gallon	8.0–9.5 lb.
Total solids by wt	60% min.

Proc.: Mix all ingredients together stirring until uniform.

Undercar Sealer

	pts./wt.
Asphalt Cutback (Asphaltic bitumen)	50
Naptha–300°B.P.	40
Asbestos Powder	5
Mineral Filler (Calcium Carbonate and Pumice)	5

Proc.: Mix all ingredients together until uniform.

Stop-Leak Preparations for Automobiles

I

	pts./wt.
Soluble Starch (Amylodextrin)	4.48
Asbestos	2.56
Wood Flour	4.48
Water	107.52
Synthetic Gum	1.28
Isopropyl Alcohol	7.68

Proc.: Mix the water with the starch, asbestos and wood flour. Dissolve the gum in the alcohol, and mix the two solutions thoroughly.

II

	pts/wt.
Asbestos	2.56
Sodium Silicate ($Na_2O \cdot 3.5\ SiO_2$)	42.24
Water	83.20

Proc.: Mix thoroughly.

These products can be used in water jackets and radiators.

Radiator Antirust and Antifreeze

I

	pts./wt.
Ethylene Glycol	95.00
Water	4.85
Monoethanolamine	0.15

Proc.: Mix the monoethanolamine with the water. Then combine with the ethylene glycol.

II

	pts./wt.
Oxalic Acid	400 gm.
Sodium Acid Sulfate	50 gm.
Sodium Lauryl Sulfate	20 gm.
Kerosene	2 qts.
Water	5 gal.

Proc.: Mix all ingredients together until uniform.

Puncture Proofing Tires
(German Patent 589,394)

	pts./wt.
Ammoniacal Latex	40
Sesame Oil	50
Olein	10

Proc.: Mix all ingredients together until uniform. The mixture is introduced through the air valve of the tire, distributes itself over the inner surface and automatically seals any punctures which may develop.

Windshield Bug Remover

This concentrate should be diluted with up to 10 parts of water before use.

	%/wt.
Water	85.4

"Versene" 100	0.5
Triethanolamine	9.0
Monoethanolamine	1.0
"Triton" X-100	1.5
"Miranol" C2FM	1.5
Aqua Ammonia (28%)	1.1

Proc.: Mix all ingredients together using heat to about $120°F$., stir until smooth.

Deicer for Windshields

I
(German Patent 1,045,019)

	%/wt.
Ethylene Glycol	20.0
Denatured Alcohol	80.0
Silicone Oil	0.2

II
(Aerosol)

	pts./wt.
Ethylene Glycol	30.00
Isopropanol	70.00
Perfume	q.s.
Propellent:	
Carbon Dioxide	2.80

Proc.: Mix all ingredients together until uniform. Fill in an aerosol can and add the propellent.

Windshield Wash and Antifreeze

		%/wt.
A	Isopropylalcohol	44
	Glycerine	5
	Triton X100	1
B	Propellent 50/50 11/12	50

Proc.: Mix A stirring until smooth. Fill the liquid in an aerosol container and charge with the propellants (B).

HEAVY-DUTY CLEANERS

Acid Cleaner
(U.S. 3,218,260)

	%/wt.
Gluconic Acid	9.0
Alkylarylsulfonate	5.1
Non-ionic Detergent	4.0
Water to make	100.0

General cleaner for milk cans, glass and plastic containers, etc.
Proc.: Mix all ingredients together until dissolved and uniform.

All-Purpose Liquid Detergent (Pine Oil Type)

	%/wt.
CONDENSATE PS (Coconut oil diethanolamide)	5.0
CONCO AAS-35 (Sod. dodecyl-benzene)	5.0
CONCO SXS (Sodium xylene sulfonate)	7.0
Tetrapotassiumpyrophosphate	5.0
Isopropyl Alcohol (99%)	8.5
Hercules Pine Oil 317	7.5
Water	62.0

Proc.: Mix all ingredients together until dissolved and uniform.

Liquid Dishwashing Detergent

	%/wt.
Water @ 140°F.	50.5
VARSULF 60 L SOFT	35
VARAMIDE MA1	5
Linear Alcohol Ethoxylate (9 mole)	4
Ethyl Alcohol	5
Perfume, Sequestering Agents, color	0.5

Proc.: Add water and VARSULF 60 L SOFT to mixing tank. Add VARAMIDE MA1, mixing well, followed by addition of non-ionic surfactant. Cool, while mixing, to 100-105°F., then add remaining ingredients.

Bar Glass Cleaner

	pts./wt.
CLINDROL 200 CG diethanolamide of coconut oil	5
TRITON X-102	1
Chelate	0.1
Water	94

Proc.: Mix all ingredients together until uniform.

Aerosol Glass Cleaner

		%/wt.
A	"Veegum"	1.5
	Water	67.5
B	"Span" 20	2.0
	"Tween" 20	2.0
	Ammonia	2.0
	Deodorized Kerosene	15.0

Abrasive	10.0
Preservative	q.s.

Proc.: First add "Veegum" to the water slowly, agitating continuously until smooth. Then add components in B in the order listed, with mixing after each addition. To load aerosol:

Concentrate	100 g.
Propellant 12	20–30 g.
Container	Continental Can
Valve	Precision spray

Bottle Washing Compound

I

	pts./wt.
NaOH Sodium Hydroxide	48.5
Gluconic Acid	1.5

II

	pts./wt.
NaOH Sodium hydroxide	77.5
Gluconic Acid	2.5

Proc.: Mix the ingredients together until uniform.

Window Cleaners

I

	%/wt.
PETRO® BAF Liquid	20
Conoco Alfonic 1012-60	1
Water	79

Proc.: Mix all ingredients together until uniform.

II

	%/wt.
PETRO® BAF Liquid	20
Conoco Alfonic 1012-60	1
Con. aq. NH_3	2
Water	77

Proc.: Mix all ingredients together and heat until uniform.

For use purposes, either concentrate is cut one part by volume with 50 parts water.

III

	%/wt.
Isopropyl Alcohol	20–30
Nonionic Surface Active Agent, f.i.	
Fatty Alcohol Polyglycol Ether	0.5– 1

Water Soluble Color	0.01
Distilled Water	to make 100

Proc.: Mix all ingredients and stir until uniform.

Aerosol Window Cleaner

I

	pts./wt.
Perfume Compound #45019 or	
Perfume Compound #48424	0.1
Triton X-100	0.5
Silicone 555	0.2
Isopropyl Alcohol	64.2
Distilled Water	30.0
Iso Butane	5.0

Proc: Dissolve the perfume in the Triton X-100 followed by the remaining ingredients. Package in aerosol container.

II

	pts./wt.
Isopropyl Alcohol	35.00
S. D. Alcohol 40	35.00
Sodium Lauryl Sulfate	0.20
Perfume Compound #42252 or #43146	0.10
Distilled Water	19.70
Propellent 12	10.00

Proc.: Dissolve the perfume compound in the Isopropyl Alcohol and Specially Denatured Alcohol 40 followed by the remaining ingredients. Should be pressure filled. Package in aerosol container.

Glass Window Cleaner

I

	%/wt.
"Genapol LRO" liquid	1.5
Isopropyl Alcohol	35.0
Distilled Water	64.5

Proc.: Mix all ingredients until smooth.

II
(Glass wax-type)

		%/wt.
A	Stoddard Solvent	14.7
	Triethanolamine	0.6
	Oleic Acid	1.0

B	Water	75.0
	Ammonium Hydroxide (26° Bé.)	1.2
	Diatomaceous Earth	6.0
	Bentonite	1.5

Proc.: Prepare A and B separately to smooth mixtures before pouring A slowly into B with vigorous agitation. Add perfume and color to suit.

Lens Cleaner

	%/wt.
Isopropyl Alcohol	85.0
"Triton" X-100	0.1
Water	14.9

Proc.: Mix all ingredients together until smooth.

Eyeglass Cleaner

	%/wt.
Ammonium Soap	6.40 oz.
Glycerin	6.60 oz.
Water, enough to make	1.00 gal.
Fluorescein, soluble	
(D & C Yellow #7)	sufficient

Proc.: Dissolve the soap in the water, add the glycerin and tint it with the fluorescein. Shake well. Place a few drops of this preparation on the eyeglass and wipe off clean with a soft cloth.

A Chlorinated Detergent Scouring Powder

	%/wt.
Maprofix	
Sod. Laurylsulfate	5.0
Calcium Hypochloride	1.5
Soda Ash	5.0
Trisodium Phosphate Anhydrous	3.0
Tetrasodium Pyrophosphate	5.0
Bentonite or Silica Abrasive Powder	80.5

Proc: Mix all the ingredients together until smooth. Use one tablespoon to about 1 gal of warm water.

"Bucket" Soap, Liquid

	%/wt.
"Alipal" EO526 (57%)	10
NaOH (50%)	3
Tall Oil Fatty Acid	10
NTANa$_3$ Crystals	10
Water	67

Proc.: Mix until all ingredients are dissolved and homogeneous.

Liquid Dishwashing Detergent

I

	pts./wt.
CONCO AAS-40 or AAS-50M	65.0
CONCO NI-100	8.0
CONDENSATE PE or PO	5.0
Water	22.0
Perfume & Coloring	

Proc.: Mix all ingredients together until dissolved and uniform.

II

	%/wt.
"Maprofix" SXS	3.0
"Maprofix" 60N	20.0
"Ultrawet" 60K Soft	20.0
"Super Amide" GR	7.0
Water	48.0
Lemon Aroma LE-42	2.0

Proc.: The "Maprofix" 60N is slowly added to the water previously heated to 60°C. When the mixture is clear and homogenous, add the "Ultrawet" and the "Maprofix" SXS, continue agitation and add the "Super Amide" GR. The aroma oil is added after the mixture has cooled at 45°C. Color also may be added if desired.

Household Liquid Dishwashing Detergent

	pts./wt.
Sodium Dodecyl Benzene Sulfonate	20
Water	65
Triton X-102	10
CLINDROL Superamide 100LM	5

Proc.: Mix all ingredients together until dissolved and uniform.

Pearlescent Light Duty Liquid Hand Dishwashing Detergents

	pts./wt.
Sodium Alkyl Aryl Sulfonate (low salt, 100% active basis)	22
Triton X-102	8
CLINDROL Superamide 100LM	5
Water	63
CLINDROL SEG	2

Proc.: Mix all ingredients together and heat until uniform.

Powder Dishwashing Detergent

	%/wt.
Sodium Dodecyl Benzene Sulfonate[2] (40% flake)	25
CLINDROL 200-L	3
Sodium Tripolyphosphate	32
Sodium Sesquicarbonate	40

Proc.: Mix all ingredients together until uniform.

Automatic Dishwash Powder

	%/wt.
PETRO DW	5
Sodium CMC	2
$NTA-Na_3 \cdot H_2O$	20
Sodium Metasilicate Pentahydrate	20
Sodium Sesquicarbonate	53

Proc: Mix all ingredients together until uniform.

Germicidal Hard Surface Cleaner

		pts./wt.
A	SANAMID 29B	12.5
	Santophen I	4
B	Tetrasodium Pyrophosphate	2
	Trisodium Phosphate	3
	Water	78.5

Proc.: Mix A, mix B heat to 120°F and stir until uniform. The order of addition of ingredients is important in order to obtain a stable solution which will not separate.

Hospital-Strength Disinfectant Cleaner

1:64 aqueous dilution or 2 oz. per gallon of water

Active Ingredients	%/wt.
BARDAC-22	8.50
Hamp-ene 100	4.00
Sodium Carbonate	2.00
Sodium Metasilicate, Anhydrous	0.50
Inert Ingredients	
Tergitol 15-S-12[3]	4.50
Water	80.50

Proc.: Add the Tergitol to the water at room temp., stir until surfactant has been completely dissolved. During continued stirring, add Hamp-ene 100, Sodium Carbonate and anhydrous Sodium Metasilicate. When salts are dissolved and solution is clear, add

BARDAC-22. Continue stirring for approx. 15 min. or until solution is again clear.

General-Purpose Neutral Cleaner

Nacconol NR	2–5 lb.
Water	100 gal.

This is an excellent cleaner for washing floors, walls, woodwork, rugs, upholstery, automobiles, etc. This neutral solution is non-damaging to any surface which will 'not be harmed by water alone.

General-Purpose Alkaline Cleaner

I

	pts./wt.
Nacconol NR	15–40
Sodium Sesquicarbonate	50–65
Polyphosphate	10–20
Trisodium Phosphate	2.8
Water	66.3
Use 1 oz. to 5 gal. water.	

Proc.: Stir all ingredients until uniform.

II

Amerse	0.7 lb.
Soda Ash	0.5 lb.
Trisodium Polyphosphate	0.5 lb.
Triton X-100	1.3 lb.
Water	100.0 gal.

Proc.: Stir all ingredients until uniform.

Germicidal Sweeping Oil

	pts./wt.
Light Paraffin Oil	95.0
Oleyl Dimethyl Ethyl Ammonium Bromide	0.5
Isopropyl Alcohol	5–10

Proc.: Mix all ingredients until uniform.

Perfume may be added to this formula to give any desired odor. The product is used by either spraying over a floor or by impregnating a dust mop.

Household Cleaner

	pts./wt.
Abrasive	89.5
Powdered Soap or Synthetic Detergent	3.3
Ammonium Sulfate	0.3
Soda Ash	2.0

Trisodium Phosphate	5.0

Proc.: Stir all ingredients until uniform.

Antiseptic (Sanitizing) Detergent
(For glasses and dishes)

I

	pts./wt.
Trisodium Phosphate	50
Sodium Bicarbonate	25
Tetrasodium Pyrophosphate	25
Quaternary Compound	6

Proc.: Stir all ingredients until uniform.

The quaternary compound is perferably *Rodicide A*. It consists of cetyldimethyl-ethyl ammonium bromide plus an alkylated aryl poly-ether alcohol, in an aqueous-alcoholic solution.

Aerosol Surface Detergent and Sanitizer

	%/wt.
Solvay Cleanser 600	3.00
Hyamine #1622	0.05
Isopropyl Alcohol	10.00
Water	86.95

Proc.: Mix all ingredients together until uniform and package in aerosol container.

Pine Oil Disinfectants

	I	II
	pts./wt.	pts./wt.
Yarmor Pine Oil	80.00	65.00
Tall Oil Fatty Acids	9.25	23.5
Caustic Soda (NaOH)	1.25	3.15
Water	9.50	8.35

Proc.: The above disinfectants may be prepared at room temperature. The fatty acids and pine oil are first blended. A solution of caustic soda in all the water is gradually added to this blend with efficient stirring, which is continued for about 30 minutes. The respective products should be clear, amber solutions.

Pine-Oil Deodorant

Pine Oil	90 fl. oz.
Rosin FF	40 oz.
Solution of Caustic Soda	
(Sp. Gr. 1.275 at 25°C.)	20 fl. oz.

Proc.: Dissolve the rosin in the pine oil, heated in a steam-jacketed kettle at 80°C. When all is dissolved, add the caustic soda solution, stirring briskly and maintaining the temperature at 80°C.

Floor Cleaners

	I pts./wt.	II pts./wt.	III pts./wt.
Trisodium Phosphate or			
Sodium Tripolyphosphate	3	1–2	5
Tetrasodium Pyrophosphate or			
Tetrapotassium Pyrophosphate	2	–	3
Sodium Metasilicate	–	–	1.5–3.0
Water	87	86–87	77-79.5
Sodium Alkyl Aryl Sulfonate			
(low salt grade)	–	–	2
Sodium Xylene Sulfonate	–	–	2
CLINDROL 200-CGN	8	4	8
Potash Vegetable Soap	–	8	–

Proc.: Mix all ingredients together. Heat to 120°F. and stir until uniform.

IV
(Heavy Duty Liquid)

	%/wt.
Tetrasodium Ethylenediamine	
Tetraacetate (EDTA)	1.0
Water	76.5
Sodium Tripolyphosphate	2.0
Trisodium Phosphate, Anhydrous	2.0
"Solar" TE	9.0
"Solar" NP	2.0
Ethylene Glycol Monobutyl Ether	4.5
"Solar" S-2552	3.0

Proc.: Put water in vessel equipped with agitator. Dissolve EDTA. Add sodium tripolyphosphate and then trisodium phosphate with agitation. After phosphates are completely dissolved, add "Solar" TE and NP, continuing agitation. Then add ethylene glycol monobutyl ether and "Solar" S-2552. Agitate another 15 to 20 minutes for homogeneity. Package in glass or other suitable container.

This cleaner can be used as a wax stripper on heavily soiled surfaces at a concentration of four to eight ounces per gallon. At two to four ounce levels, it serves as an "excellent" general purpose cleaner. Label should carry warning against getting the product into the eyes and recommending copious flushing with water as treatment in case of accident.

LAUNDRY PRODUCTS

Heavy Duty Laundry
(Low Sudsing)

	pts./wt.
STPP	40
Sodium Silicate as SiO_2	5
Nonionic	8- 10
CMC	0.5-1.0
Optical Brightener	0.1-0.2
Salt Cake	balance

Proc.: Mix all ingredients together until uniform.

Heavy Duty Cold or Hot Water Laundry (Liquid)

	%/wt.
Water	19.8
CMC	0.5
50% KOH	2.2
39.4% Pot. Silicate Solution	12.7
(Water)	6.5
(Stabilizer ASE-108 *preblend)	6.5
60% TKPP Solution	41.7
Triton X-100	10.0
Optical Brightener	0.10
Add in order listed.	

Proc.: Mix all ingredients in order listed together until uniform.

Fine Fabric Detergent

I

	%/wt.
IGEPON T-73	30.0
Sodium Hexametaphosphate	35.0
Sodium Bicarbonate	34.2
BLANCOPHOR FFG Brightener	0.3
PVP K-30, Polyvinylpyrrolidone	0.5

Proc.: Mix all ingredients together until uniform.

II

	%/wt.
"Igepon" T-73	30.0
Sodium Hexametaphosphate	35.0
Sodium Bicarbonate	34.2
"Blancophor" FFG Brightener	0.3
Polyvinylpyrrolidone PVP K-30	0.5

Proc.: Mix all ingredients together until uniform.

Diaper-Rinse Tablets

		pts./wt.
A	"Hyamine" 1622	18.0
	Starch	47.4
	Milk Sugar	26.6
	Sodium Bicarbonate	9.0
B	Alcohol	6.0
	Water	90.0

Proc.: Mix A; Mix B. Add A to B stirring until uniform.

Diaper Rinse

"Cetab"	16.8
Water	73.2
"Triton" X 100	10.0

Proc.: Dissolve the "Cetab" in the water, with a little heat, if necessary. Add the "Triton" and stir until the solution is homogeneous.

Use 1 teaspoonful (about 8 cc.) of the antiseptic per gallon of warm water.

Biogradable Detergents Hand Laundry Powder

	%/wt.
Sodium Tripolyphosphate	15
"Sulframin" 4010	15
Sodium Carboxymethyl Cellulose	1
Foam Stabilizer	3
Sodium Sesquicarbonate	20
Sodium Sulfate	46

Proc.: Mix all ingredients together stirring until uniform.

Cold Water Wool Detergents

I

	%/wt.
Sodaphos	30.0
STPP	10.0
Anionic (85%)	25.0
Salt Cake	34.5
CMC	0.5

Proc.: Mix all ingredients together until uniform.

II

	%/wt.
STPP	45.7
NaHSO$_4$	7.3
Anionic	25.0
CMC	0.5
Salt Cake	Balance

Proc.: (For I and II) Mix ingredients together until uniform.

Stain Decomposers
(Prewash)

	pts./wt.
DDBS—Na (Sodium Dodecyl Benzene Sulphonate)	20.00
Sodium Tripolyphosphate	20.00
C.M.C. Technical 60%	2.50
Sodium Silicate	8.50
Optical Bleaching Agent	0.15
Sodium Sulphate	40.00
Borax or Sodium Bicarbonate	8.85

Proc.: Mix all ingredients together until uniform.

Laundry Pre-Spotter

	%/wt.
AEROTHENE TT	20.0
DOWANOL EB	6.0
Isopropyl Alcohol	10.0
Polyoxyethylene Glyceride Ester	22.0
Deodorized Kerosene	22.0
Propellant 12	20.0

Suggested Value: Precision 0.013" stem/0.013" body
Suggested Actuator: Precision 0.016" MBRT
Proc.: Mix all ingredients together until uniform. Fill this mixture in aerosol bottles or cans and charge it with the propellant 12.

Home Laundry Bleach

	%/wt.
Oxone	35.0
Light Granular Soda Ash	64.9
Tinopal RBS 200	0.1

Proc.: Mix all ingredients together until uniform.

Aerosol Spray Starch

	pts./wt.
#5541 Starch	4.00
Borax 5 mole (60% anhydrous)	0.48
GE SM-2061 emulsion (35% silicone)	0.35
Dowicide A	0.10
Sodium Nitrite	0.10
Morpholine	0.10
Water	94.87

Proc.: Dissolve the borax in approximately half the total water and then add the starch with moderate agitation to completely disperse the solids. Heat the mixture to the boiling point and hold at this temperature for 10 minutes with continued good agitation. Cool the bath by adding the remainder of the water with continued agitation. Add SM-2061 emulsion, Dowicide A, sodium nitrite and morpholine with *mild* agitation. Once these ingredients are thoroughly mixed, only occasional intermittent stirring is required.

Aerosol Ironing Aid
Concentrate

		%/wt.
A	CMC (Hercules, med. visc.)	4.0
	Water	94.0
B	"Atlas G-1300"	2.0
	Bactericide	q.s.

Proc.: With sufficient heat and agitation dissolve the CMC in the water (A). Add Part B and dissolve, stirring to blend well.

LEATHER

Bleaching Tanning Extract
(U.S. Patent 2,653,967)

	pts./wt.
Quebracho Extract	3000
Sodium Formate	60
Caustic Soda (in 25 cc. Water)	16
Sodium Hydrosulfite	45

Ammonium Vanadate
(in 25 cc. Water)   1/4

Proc: Heat slowly to 95°C. and keep at 90 to 95°C. for 1 hour. Then cool.

Syntan
(Synthetic Tanning Agent)

(French Patent 900,655)

	pts./wt.
Phenol	94
30% Formaldehyde	70
Oxalic Acid	1

Reflux for 45 minutes while stirring. Heat in vacuum at 75°C. Then add slowly:

Sulfuric Acid H_2O 50

at 50 to 60°C. and heat in vacuum at 70 to 75°C. for 2 hours until completely water soluble. Neutralize with alkali and adjust with an organic acid to a pH of 3 to 4.

Bleaching Deer Skin

	pts./wt.
Make a bath with	
Hydrogen Peroxide (30%)	5–8
Seignette Salt	0.5

Proc.: Put the skins into it for 1/2 hour. Dry them thereafter at 30°C. If the skins are not pale enough, repeat in the same bath.

Finish for Split Leather

	pts./wt.
First Coat	
"Hycar" 1561	100
Red Iron Oxide†	15
Clay†	15
Wetting Agent‡	1
Second Coat	
"Hycar" 1561	20
"Geon" Latex 652	80
Red Iron Oxide†	25
Clay†	25
Borated Shellac‡	5
Ammoniated Casein‡	2
Wetting Agent‡	1

Third Coat

"Hycar" 1561	20
"Geon" Latex 652	60
Red Iron Oxide†	25
Clay†	25
Candelilla Wax‡	4
Borated Shellac‡	5
Ammoniated Casein‡	2
Wetting Agent‡	1

Proc.: For all coats grind all ingredients together until uniform.
*All parts by weight on a dry basis.
†Added to the latex as a water dispersion.
‡Added to the latex as a water solution.

Non-ionic Fat Liquoring Stock Emulsion

I

	%/wt.
Ethofat 0/15 or 0/20	5
Oil	45
Water	50

Proc.: Dissolve the Ethofat and oil while agitating. Add water to produce an emulsion. This process is very simple and can be done with minimum equipment.

II

	%/wt.
35° Neatsfoot Oil	70
Mineral Oil	20
Ethofat 0/20	10

Proc.: The ingredients are mixed to form a solution. This mixture can be added directly to the fat liquoring drum containing the wet leather to form the emulsion.

Degreaser for Hides

I

	pts./wt.
Water	90
Benzene	5
Polyglycol 200 Monooleate	5

Proc.: Mix all ingredients together until uniform.

II

	pts./wt.
Water	87
Benzene	5
Polyglycol 200 Monostearate	2

Olive Oil	4

Proc.: Mix all ingredients together until uniform.

Hide and Skin Preservative

	pts./wt.
p-Chloro m-Cresol	1
Sodium Hydroxide	.175
Soda Ash	1
Water	500

Proc.: Spray both sides with this.

Leather Fungicide Protection

I

	pts./wt.
Salicylanilide	2.2
Isopropyl Alcohol	25.0
Paraffin Wax	33.0
Stoddard Solvent	39.8

II

	pts./wt.
Perchloroethylene	67.0
Mineral Oil	10.0
Neatsfoot Oil	10.0
Cyclohexanone	10.0
Paranitrophenol	1.5
Pentachlorophenol	1.5

Proc.: (For I and II) Warm gently and mix until dissolved. Impregnate the leather, squeeze, drain and dry.

Leather Stain Remover

A solution for removing stains from the flesh side of leather is composed of the following:

Water	250 cc.
Oxalic Acid	3 gr.

Proc.: Mix the ingredients together.

Fur Cleaning & Glazing

	I	II	III
	pts./wt.	pts./wt.	pts./wt.
Polyethyleneglycol (400) Monolaurate	5	5	5
Polyethyleneglycol (400) Monooleate	3	–	–
Silicone (D.C. 200-350)	–	3	–

"Acetulan"	–	–	3
Perchlorethylene	120	120	120

Proc.: Dilute 1 part of above with 7 parts of water; mix to emulsify and then mix into sawdust in tumbling barrel.

Oil for Softening Leather Goods

Glue	3
Montan Wax	5
Synthetic Wax (Glyceryl Monostearate)	10
Sulfonated Neatsfoot Oil	40
Glycerin	0.5
Water	41.5

Proc.: Soften glue in water, heat waxes just slightly above melting point, keep at same temperature while stirring in part of water; add sulfonated oil and keep stirring, add rest of water containing glue and softener.

For softening leather goods, preferably hard leather, like belts, etc., moisten soft cloth with mixture and rub in well. Go over oiled goods with dry soft cloth.

Fireproofing Emulsion for Leather

(French Patent 947,309)

	pts./wt.
Sodium Lorol Sulfate	10
Water	70

Proc.: Heat to boiling until dissolved. Cool to 50°C. and add slowly, with good mixing, a melt of:

Chlorinated Paraffin Wax	50
1-Chloronaphthalene	100

Use 3 parts of this emulsion in 150 parts of water to treat 100 g. leather at 50°C.

Mildewproofer for Leather Goods

	pts./wt.
Paranitrophenol	2
Silicone Resin	10
Rectified Tar Oil	5
Neatsfoot Oil	15
Cyclohexanone	10
Trichloroethylene	58

Proc.: Mix all the ingredients together, shaking well, and apply to the leather goods, allowing to dry thoroughly.

Synthetic Leather

I

	pts./wt.
Solvents	
Butyl Acetate	10
Ethyl Acetate	25
Gasoline	47
Synthetic Rubber	18

II
(U.S. Patent 2,163,610)

Solvents	
Benzene	1–3 g.
Trichlorethylene	5½–2 g.
Plastic Chloroprene (Polymerized)	1 g.

Proc.: Dissolve the rubber in the solvents.

Leather Finish

	pts./wt.
Pigment	5.0
U-3204, "Ubatol" 40%	1.5
U-3215, "Ubatol" 40%	1.5
Dye solution	4.0
Water	4.0

Proc.: 1. Grind pigment with the Ubatols (Dye Solution). 2. The dye solution should be dissolved with the water.

Leather Garment Treatment

	pts./wt.
"Hoechst" Wax OM	1.5
"Hoechst" Wax RT	0.5
Slab paraffin 52/54°	3.0
Benzine (100 to 140°C)	65.0
Methylene chloride	30.0

Proc.: Dissolve the waxes in benzene and methylene chloride.

SOAP & HAND CLEANERS

Floating Castile Soap

	pts./wt.
Refined and Bleach.~d Coconut Oil	475
Caustic Soda (37.4% Na_2O)72	172
Water	103
Perfume	1½
Color	As desired

Proc.: The oil, alkali and water are added to the crutcher and heated to approximately 140°F. to start saponification. Once started the heat of the reaction will drive the temperature up to 190 to 210°F., so that cold water circulation in the jacket may be necessary. When the saponification is complete, add the perfume and color and crutch downward until the desired specific gravity is reached. This may be anywhere between 0.60 and 1.0. The product has a fatty acid content of 60% and a superfat between 2 and 5%.

Coconut and Soya Hand Soap

	%/wt.
Distilled Coconut Fatty Acid (Saponification No. 263)	9.65
Soya Fatty Acid (Saponification No. 195	3.25
100% Caustic Potash	3.16
Water	To make 100.00

Proc.: Heat water to 120°F.; add fatty acids; stir until dissolved. Add caustic potash stirring until homogenized.

Perfume for Toilet Soap

A very fine perfume for violet soap is given by the following:

	pts./wt.
Ionone	20
Methyl Heptin Carbonate	10
Orris Resinoid	10
Bergamot	10
Cananga	10
Hydroxy Citronellal	10
Cinnamic Alcohol	10
Amyl Cinnamic Aldehyde	10
Violet Leaves Conc.	10
Bourbon Vetrivert Oil	10
Sandalwood	5
Styrax Tincture (25%)	100
Siam Benzoin (25%)	100
Distilled Olibanum Oil	50
Musk Tonquin (4 Oz. Per Gal.)	25
Civet	35

Proc.: After these oils are mixed, add 10–20% of orris as an impalpable powder. The orris is poured into a mixer, while the oil is dissolved in 500 g. of alcohol. The latter is added to the orris powder. The pony mixer is switched on and while it is turning, 90 lb. of soap are added.

To make this perfume cheaper, some of the expensive materials may be used more sparingly or eliminated. The formula can be toned down to any desired price but at the expense of quality.

Cold Process Soap

(British Patent 432,227)

Cold-process fat-resin soaps are made by treating fatty matter with just sufficient alkali for saponification, treating a mixture of rosin and fat or oil with alkali sufficient to saponify only the rosin, mixing the two products, and adding alkali to saponify the surplus fat. For example, 100 lb. of palm-kernel oil is stirred rapidly with 4.5 gal. of 36° Bé. caustic soda for 10-15 minutes, 4 gal. of a melt of rosin in an equal weight of palm-kernel oil is treated at 110-135°F. with 0.5 gal. of 36° Bé. caustic soda, the products are mixed, and immediately 1 gal. of 36° Bé. caustic soda is added, and the mixture stirred for a few seconds and run quickly into the frames, where it sets and saponification is completed.

Perfume for Cold-Made Soap

Citrene	54
Rosemary Oil	33
Spike Lavender Oil	13

Proc.: Mix until uniform.

Soya and Cottonseed Scrub Soaps

I

	pts./wt.
Soya Fatty Acids	
(Saponification No. 195)	13.70
100% Caustic Potash	2.67
Water	To make 100.00

II

Cottonseed Fatty Acid	
(Saponification No. 200)	13.20
100% Caustic Potash	2.64
Water	To make 100.00

Proc.: Heat water to 120°F add fatty acids, stir until dissolved. Add Caustic Potash stirring until homogeneous.

Addition of several percent of pine oil will give a pleasing odor and cause the soap to congeal slightly. The pine oil should not be added until the soap is fairly cool, as high temperature causes loss by evaporation. Furthermore, addition of pine oil at or near room temperature assures better control of viscosity of the finished soap.

Household-Soap Perfume

	pts./wt.
Citronella Oil	20
Rosemary Oil	32
White Thyme Oil	48

Use 0.20% perfume on the total soap stock.
Proc.: Mix all ingredients together.

Liquid Soap

I

	pts./wt.
Red Oil	10
Potassium Hydroxide	2
Trisodium Pyrophosphate	3
Aqua Ammonia (26° Bé.)	2
Pine Oil	5
Water	78

II
(20 to 22% Anhydrous)

	pts./wt.
Soya, Corn, Linseed, or	
Cottonseed Oil Fatty Acids	80
Coconut Fatty Acids	20
Potassium Hydroxide (45%)	47
Water	455

Proc.: Heat the fatty acids in a kettle until they become liquid
and hot. Slowly add the potassium hydroxide while mixing. There
will be a strong reaction which will subside when all the caustic is in.
Mix with a stick until you have a clear paste. At this point, the alkali
should be adjusted since the formula is an average and more or less
caustic may be needed. Add all of the water to the paste; then start
the mixing machine. In approximately 1/2 hour all the soap will
have dissolved in the water.

Waterless Hand Cleaner
(Germicidal)

		%/wt.
A	MAYPON® 4C	4.6
	Stearic Acid T.P.	11.8
	Cetyl Alcohol	0.8
	Triethanolamine	5.6
	Propylene Glycol	5.4
B	"Deo Base" (Deodorized Kerosene)	42.8
	Fragrance	0.3

C	Distilled Water	25.5
	Formalin 40%	0.2
	POLYPEPTIDE #37-S	3.0
	Color, as desired	

Proc.: Heat A gently until all parts are dissolved. Mix B and add to A with agitation. Warm C and add the A/B mixture slowly to it with good agitation. Continue the agitation until the product "firms up."

II

	pts./wt.
Carbowax 1500	20
Ultrawet (Wetting Agent)	8
Sodium Pyrophosphate	4
Carboxymethyl Cellulose	6
Lanolin	5
Glycerin	5
Dioxane	20
Water	32
Perfume	To suit

Proc.: The materials are mixed in a high speed stirrer in the following order: carboxymethyl cellulose, water, Carbowax, Ultrawet, sodium pyrophosphate, lanolin, glycerin and dioxane. In use the material is rubbed on the hands and removed with a paper towel.

Dry-Wash for Hands

	pts./wt.
Stoddard Solvent*	64.00
Stearic Acid	12.16
Triethanolamine	5.12
Hexalin	3.07
Lanolin	2.68
Water	40.97

*High flash point petroleum distillate.

Proc.: Melt the stearic acid with the lanolin. Heat the triethanolamine and water to the same temperature, and pour slowly with constant stirring into the molten fats. When cool, stir in the Stoddard Solvent and Hexalin. Continue stirring until cold.

This preparation will remove oil, paint, grease, grime, tar and printer's ink from the hands.

Hand Cleaner and Ink Stain Remover

	%/wt.
Stearic Acid	3.5
Triethanolamine	2.0

600-Distearate	5.0
Lanolin	3.0
Propylene Glycol	7.5
CARBITOL Solvent	7.5
Sodium Meta-Bisulfite	3.0
Deionized Water	67.5
Perfume	1.0

Proc.: Melt the stearic acid and lanolin and add the CARBITOL solvent and propylene glycol. Heat the water, triethanolamine, and distearate, and add to the other mixture. Cool with stirring to 40°C., add the sodium metabisulfite and perfume, and cool to room temperature with stirring.

Hand Stain Remover

Water	8 oz.
Soap Powder	4 oz.
Silica (150 Mesh)	½ oz.
Pumice Flour	¼ oz.
Chlorox (Chlorine Water)	1 dr.
Glycerin	1 oz.
Rubbing Alcohol	1 oz.

Proc.: Heat the soap in the water until dissolved, and add the Silica and Pumice Flour. When cool add the chlorox (or the like) the glycerin and alcohol.

Bottle and use in the same way as any liquid soap on well-wetted hands.

Hand Cleaners
(Mechanic's)

I

	pts./wt
Soap, 90% (Titre: 30°)	15
Tetrasodium Pyrophosphate	10
Silica Sand, Fine, Round Edged	40
Ash, Finest Volcanic	35

II

Borax	20
Sodium Silicate, "G" Type, Fine Mesh	15
Soap, 90% (Titre: 36°)	20
Sodium Metasilicate, Fine Mesh	5
Silica Sand, Fine, Round Edged	40

METAL CLEANERS AND POLISHES

Metal Cleaner Foam, Abrasive

pts./wt.	
Cab-O-Sil	3.50
Triton X-100	4.00
Butyl Cellosolve	7.50
Mapleton 325 Mesh Supersil	10.00
Water	74.90
Perfume	.10

Proc.: Dissolve the Triton X-100, perfume and Butyl Cellosolve in water. Add the abrasive, followed by the Cab-O-Sil, with enough agitation to insure even dispersion. Package in aerosol container.

Silver Polishes

I
(Paste)

Stearic Acid, Double Pressed	5 lbs.
Kerosine	2½ gals.
Silica #94	27½ lbs.
Ammonium Hydroxide, 28%	0.1 gals.

Proc.: Heat all ingredients until dissolved and stir until uniform.

II
(Paste)

	%/wt.
Anti-Tarnishing Agent	5.0
Brij 30	5.0
Natrosol 250 L at 5%	20.0
Kaopolite S.F.	14.5
Bentonite	5.0
Water	50.4
Methyl p-hydroxybenzoate	0.1

Proc.: Heat all ingredients until dissolved and stir until uniform.

Copper and Brass Polish

	pts./wt.
"Carbowax" Polyethylene Glycol 1500	35
"Tergitol" Anionic 7	3
Citric Acid	5
Sodium Chloride	5
Bentonite	8
"Multicel" 000 Abrasive	19
Water	25

Proc.: Heat all ingredients and stir until uniform.

Aluminum Polishes

I
(Paste-Type)

	%/wt.
Silica	55.0
Ammonia Tallow Soap	8.0
Stoddard Solvent	34.0
Water	3.0

Proc.: Heat all ingredients and stir until uniform.

II
(Liquid-Type)

	%/wt.
Silica	27.0
Ammonia Oleate Soap	5.0
Ammonia Oxalate	4.0
Ammonia Hydroxide, 28%	1.0
Isopropanol	6.0
Water	57.0

Proc.: Heat all ingredients and stir until uniform.

Aluminum Treatment

Polishing Bath
(U.S. 2,673,143)

	%/wt.
Sodium Gluconate	2.5
Sodium Hydroxide	62.0
Sodium Nitrate	25.0
Sodium Chlorate	10.5

Proc.: Mix all ingredients together until uniform.
Use 14 to 32 oz. of this mixture per gallon at 205°F. for 1-1/2 minutes.

Aerosol Chrome Cleaner-Polish

This chrome cleaner efficiently removes rust, dirt, tar and road film and wipes off easily leaving an invisible protective film.

	%/wt.
Union Carbide L-45, 350 cstks	1.85
Oleic Acid	1.85
Pine Oil	0.46
Morpholine	1.39
Water	39.35
Pumice	13.90
Snow Floss Powder	9.25

Naphthol Mineral Spirits	24.50
Triethanolamine Lauryl Sulfate	2.30
Hyonic FA-40	0.50
Carbopol 934 (2% dispersion in water)	4.65

Packaging	%/wt.
Concentrate	81.5
Ucon 12	18.5

Proc.:
1. Dissolve silicone oil, oleic acid and pine oil in the naphthol mineral spirits.
2. Dissolve the morpholine in the water.
3. With rapid agitation add the pumice and Snow Floss to the water-morpholine solution.
4. Slowly add the mineral spirits solution to the morpholine-water solution and continue rapid agitation.
5. Add the triethanolamine lauryl sulfate and Hyonic FA-40 and blend thoroughly.
6. Add the 2% Carbopol 934 dispersion and mix to uniformity.

NOTE: This cleaner-polish should be allowed to dry completely and then wiped off with a clean dry cloth.

Stainless Steel Cleaner

	%/wt.
Sodium Hydroxide	79
Tetrasodium Pyrophosphate	12
Sodium Gluconate	5
Alkylamine Polyglycol Condensate	3
2-Mercaptobenzothiazole	1

Proc.: Mix all ingredients together until uniform.
Dissolve in water to caustic concentration of between 0.5 to 5%. (16 to 160 times dilution).

Copper Cleaner

	%/wt.
Sodium Hydroxide	73.55
Sodium Metasilicate Anhydrous	15.0
Sodium Gluconate	5.0
Sodium Tripolyphosphate	5.0
Alkylamine Polyglycol Condensate	0.75
2-mercaptobenzothiazole	0.7

Proc.: Mix all ingredients together until uniform.
Dissolve in water to caustic concentration of between 0.5 to 5%. (16 to 160 times dilution).

Rustproofing and Cleaner
(U.S. Patent 2,587,777)

	pts./wt.
Kerosene	80
Triethanolamine Oleate	14
6% Cobalt Naphthenate	6
Water	5–5000

Proc.: Mix all ingredients together until uniform.

Rust Remover and Rustproofer
(U.S. Patent 2,493,327)

	pts./wt.
Phosphoric Acid	73.70
Formic Acid	6.00
Zinc	0.25
Water	10.10
"Naccanol"	0.50
Butyl "Cellosolve"	9.45

Proc.: Mix all ingredients together until uniform.

A 10 to 25% water solution of these is used with an immersion time of 2 to 5 minutes.

Degreasers—For Machine Parts, Floors, Etc.
Detergent Type

	%/wt.
MONAMINE ADS-100	5
Sodium Xylene Sulfonate, Powder	5
Tetra Potassium Pyro Phosphate	5
Butyl Cellosolve (U.S.C.)	1
Water	84

Proc.: Mix all ingredients together until uniform.

Solvent Type

	%/wt.
Aromatic Solvent	85
MONAMINE ADS-100T	15

Proc.: Mix the ingredients together until uniform.

This solvent concentrate may be sprayed on and washed off with water or a water emulsion can first be prepared. While the dilution will depend on the work to be done, 1 part in 10 parts of water is recommended for greasy garage floors.

Metal Degreaser-Cleaner

	%/wt.
CLINDROL 200CGN	3
Oleic Acid	1
High Flash Naphtha	96

Proc.: The above can be readily rinsed with water when used for cleaning engine blocks, garage floors, etc. Mix all ingredients together until uniform.

Deruster

	%/wt.
Trisodium Salt of N-Hydroxyethyl-ethylene-Diaminetriacetic Acid (3Na 'EDTA' OH)	13.8
Sodium Gluconate	25.0
Sodium Hydroxide	54.0
Phosphorus and Silica, total $PO_4 + SiO_2$	0.2
Others Including Foamers	7.0

Proc.: Mix all ingredients together until uniform.

Pickling and Cleaning
(U.S. 1,995,766)

	pts./wt.
Water	100.0
"Heavy Acid"	8.0
Gluconic Acid	0.8
Wetting Agent	0.1

"Heavy Acid" may be sulfuric acid, hydrochloric, nitric, etc.
Proc.: Mix all ingredients together until uniform.

HOUSEHOLD CLEANERS

Oven Cleaner

	pts./wt.
Olein, Distilled	40
Stearin	10
Mix warm.	
Spindle Oil	40
Tetralin	9
Ammonia (sp. g. 0.91)	1
Emery or Pumice or Tripoli sufficient to make pasty	

Proc.: Dissolve Stearin in Olein by heating. Then mix the other items and stir both together until uniform.

Aerosol Oven Cleaner

		pts./wt.
A	VEEGUM T	2
	Water	32
B	Ammonium Hydroxide	8
	1,1,1,-Trichloroethane	24
	Ethanol	10
	Tergitol NPX	24

Proc.: Add the VEEGUM to the water slowly, agitating continually until smooth. Add B to A and mix until uniform. Package in aerosol dispenser.

Aerosol package: Concentrate 75%; Propellant 12/114 25%.

Directions for use: Oven should be cool and empty. Do not use near flame. Contains alcohol. With can held upright, spray oven surface from a distance of 9 to 12 inches. Allow foam to work for 10 to 20 minutes. Wipe clean with wet cloth or sponge.

Aerosol Protective Oven Film

		pts./wt.
A	VEEGUM T	2.9
	Water	86.4
B	Pluronic F-127	4.3
C	Dow Corning 200 Fluid (60,000 cs)	6.4

Proc.: Add the VEEGUM T to the water slowly, agitating continually until smooth. Add B to A with agitation. Add C, mixing until uniform.

Aerosol package: Concentrate 70%; propellant 12 30%.

Directions for use: Spray evenly on clean unheated oven walls from a distance of about 10 inches. After cooking allow oven to cool, then wipe off film and grease splatters with damp sponge.

Refrigerator Cleaner

Many refrigerator manufacturers recommend cleaning refrigerators with a dilute sodium bicarbonate solution. The following solution, incorporates a quaternary ammonium compound, making it a germicidal, antistatic cleaner.

	%/wt.
"Gloquat" C	1
Sodium Bicarbonate	3
Water	96

Proc.: Mix all ingredients together until uniform.

Wall Cleaner

	%/wt.
Tergitol NPX	3.2
NACCONOL 90F	1.5
Tetrasodium Pyrophosphate	1.0
Diethylene Glycol Ethyl Ether	4.0
Water, Deionized	90.3

Proc.: Dissolve the ingredients in water.

Aerosol Formulation

	%/wt.
Concentrate	95.0
Genetron 12/114 (30:70)	5.0

Valve: Standard valve and mechanical breakup actuator.

Container: Side-seam, lined.

Directions for use: Shake well before using.

Hard Surface Cleaner

		%/wt.
A	VEEGUM T	0.45
	Kelzan	0.15
	Water	72.40
B	Monamid 150-ADD	0.50
	Tetrapotassium Pyrophosphate	2.00
	Water	21.00
	Plurafac C-17	2.50
C	Ammonium Hydroxide (28%)	1.00

Proc.: Dry blend the VEEGUM T and Kelzan and add to the water slowly, agitating continuously until smooth. Combine B, stirring slowly to dissolve the tetrapotassium pyrophosphate (avoid incorporation of air). Add B to A with mixing. Add C with stirring.

NOTE: This cleaner is ready to use and should not be diluted. It may be packaged in a pump dispenser bottle.

All-Purpose Hard Surface Household Cleaner

	%/wt.
PETRO BAF Powder	4.0
Tergitol NPX (or equivalent nonionic)	5.0
Metso 20 (or equivalent)	6.0
EDTA	0.5
Butyl Cellosolve	6.0
Water	78.5

Proc.: Dissolve by heating all ingredients and stir until uniform.

Paint and Wall Cleaners

Paint, enamel and varnish may be washed safely by using a good suds of neutral soap; the soap being dissolved in warm water and applied to the wall by means of a sponge. Strong soaps are best avoided, especially for glossy surfaces, since they may destroy the luster. When soaps are used, thorough rinsing is necessary to remove any residual film, since this, in itself, may act as an adhesive for dust and dirt. Although many people use hard white soaps, the general consensus appears to favor the use of potash vegetable soap. Using this idea, a simple soap paste for cleaning painted walls has been described containing:

I

	%/wt.
Potash Linseed Oil Soap	30
Potash Coconut Oil Soap	5
Water	65

Proc.: Heat all ingredients and stir until uniform.

II

A semi-solid, white permanent product, said to be very effective as a cleaner for painted surfaces may be made along the following lines:

	pts./wt.
Soap Chips	20.0
Mineral Spirits	10.0
Water	69.3
Oil of Sassafras	0.7

Proc.: Allow the soap to soak in water and then bring into solution with heat. With vigorous agitation, incorporate the mineral spirits and add the oil of sassafras.

Solvent Paint Remover

A solvent paint remover for stripping paint or lacquer from wood, plaster or wall board may be composed of the following:

		%/wt.
A	"Veegum" PRO	1
	"Klucel" M	1
B	Water	25
	N-methyl-2-Pyrrolidone	73

Proc.: Dry blend the "Veegum" PRO and "Klucel" M and add to B slowly, agitating continually until smooth. "Veegum" and "Klucel" are employed as thickeners to help the paint remover adhere to vertical surfaces without dripping and to reduce evaporation of the solvent during use. To use, apply evenly with a brush to painted surfaces. Allow to stand for 10 to 30 minutes. Remove old finish with a metal scraper or steel wool. Apply additional coats if necessary. Wipe wood surface with turpentine or denatured alcohol.

Paint-Brush Cleaner

I

	%/wt.
V.M. & P. Naphtha	93
Diglycol Laurate S	2
Dipentene	5

This not only cleans the bristles but leaves a protective softening film that is wetted by oil and water paints.

II

	%/wt.
Soda Ash	13
Sodium Metasilicate	40
Tetrasodium Pyrophosphate	40
Wetting Agent	7

Use 2 to 4 oz. per gal. warm water. Allow to soak long enough to loosen the dried paint, then flush with water, combing out loose particles if necessary.

Steam Cleaner Heavy Duty

I

	%/wt.
Gluconic Acid	3
Metasilicate Penta	10
KOH	10
Na EDTA	4
K Salt of Fatty Amino Polyethoxy Sulfate Derivative	10
Water	q.s.

Proc.: Mix all components together, and stir until uniform.

II

	%/wt.
Water	76.50
Tetrasodium Ethylenediamine Tetraacetate	1.50
Sodium Orthosilicate	4.50
"Solar" F-183	6.75
"Solar" F-382	6.75
Ethylene Glycol Monobutyl Ether	4.00

Proc.: Place water in agitator equipped vessel; add EDTA with agitation; continue mixing and add the orthosilicate slowly to surface. Avoid spattering. Break up all lumps of silicate. After silicate has dissolved, add "Solar" F-183 and F-382. Next, add the ether. Continue mixing for 15 to 20 minutes to assure homogeneity. During manufacture, avoid inhaling the alkaline silicate dust. Also avoid any skin or eye contact with the liquid.

Package as prepared in appropriate container; *do not use glass.* Label with caution against skin or eye contact. Eye contact should be followed by copious flushing with water. Usage level is two to four ounces per gallon of water in steam cleaning machine.

Liquid Tile Cleaner

I

		pts./wt.
A	VEEGUM T	1.50
	Water	71.25
B	DARVAN No. 7	2.00
	Triton X-102	5.00
	Santomerse 85	5.00
	Pine Oil	5.00
C	Mild Abrasive	10.00
D	VANCIDE TBS	0.25

Proc.: Add the VEEGUM T to the water slowly, agitating continually until smooth. Add the components in B to A in the order listed with mixing. Add C to A and B slowly with stirring. Mix until uniform. Add D to a small amount of the mixture with stirring until the VANCIDE TBS is dispersed. Add this concentrate to the remainder and mix until uniform.

Aerosol Package: Concentrate 91%; Propellant 12.9%.

II

	%/wt.
CONDENSATE PN	10.0
CONCO AAS-60	6.0
Sodiumtripolyphosphate	6.0
TSPP (Tetrasodiumpyrophosphate)	6.0
Water	61.9
CONCO SXS	10.0

Proc.: Mix all ingredients together until uniform. Use medium heat.

Toilet Bowl Cleaner

I

	%/wt.
A "Vancide" BN	10
Potassium Hydroxide (20% solution)	10
Sodium Gluconate	2
B "Vangard" BT	2
"Vanseal" CS	10
"Igepal" DM-970	66
Color	q.s.

Proc.: Dissolve the "Vancide BN and sodium gluconate in the potassium hydroxide solution. Heat the components of B to 75°C, stirring to dissolve the "Vangard" BT. Add A to B with agitation, mixing until uniform. Pour into dispenser and allow to cool until hard.

II

	%/wt.
20 Be HCl	73.0
Zinc Chloride	2.0
CONCO NI-100	10.0
CONCO Product 150	1.0
Water	13.0
Hyamine 1622	1.0

Proc.: Mix all ingredients heating until uniform.

Household Spray

	pts./wt.
Triethylene Glycol	2.0
Dipropylene Glycol	2.0
Isopropyl Myristate ("Deltyl" Extra by Givaudan)	.5
Perfume Oil	.5
Alcohol, Anhydrous #40	10.0
Propellants 11/12 (50:50)	85.0

A petroleum base spray might contain the following ingredients:

	%/wt.
Perfume Oil	0.5
Petroleum Distillate	12.5
Isopropyl Myristate ("Deltyl" Extra)	2.0
Propellants 11 and 12 (50:50)	85.0

Proc.: Heat all items on a waterbath until dissolved, add the perfume oil. Fill the mixture in an aerosol can. Charge the can with the aerosol mixture.

Garbage Can Deodorant

	pts./wt.
Ortho-Dichlorobenzene	40.0
Pine Oil, NF	25.0
Methylene Chloride	35.0

Proc.: Combine ingredients and stir. Package in aerosol container.

Wax Stripper

I

	pts./wt.
Oleic Acid	7.0
Monoethanolamine	2.0
Sodium Salt of Dodecylated Oxydibenzene Disulfonate	2.0
Potassium Hydroxide	1.0
Ammonium Hydroxide	1.0
Sodium Salt, Chelating Agent	1.5
Water	85.5

Proc.: Mix all ingredients together, heat to 120°F and stir until uniform.

		I	II	III
		%/wt.	%/wt.	%/wt.
A	"Petro" 11 powder	2	4-5	3
	LAS slurry (60%)			6
	"Gafac" RE-610	5		
	"Deriphat" 160		7	
	Lauric diethylamide (liquid)			3
	EDTA Na		1	
	Metasilicate Pentahydrate	10	20	
	Sodium tripolyphosphate	5		4
	Tetra potassium pyrophosphate			12
B	Water	to make 100%		

Proc.: Mix all ingredients of A. Heat to 120°F. Heat B to 120°F. Mix A and B until smooth.

WAXES AND POLISHES

Paste-Type Floor Wax

Yellow Beeswax	1 oz.
Ceresin	2½ oz.
Carnauba Wax	4½ oz.
Montan Wax	1¼ oz.
*Naphtha or Mineral Spirits	1 pt.
*Turpentine	2 oz.
Pine Oil	½ oz.

Proc.: Melt the waxes together in a double boiler. Turn off the heat and run in the last three ingredients in a thin stream, with stirring. Pour into cans, cover, and allow to stand undisturbed overnight.

*Inflammable.

Detergent-Resistant Non-Buff Floor Polish

I

	pts./wt.
RICHAMER R-747	75.00
Pentalyn 261	10.00
*Wax Emulsion "V"	15.00
Carbitol	3.60
KP-140	
(Tributoxyethyl Phosphate)	1.20
FC-128	0.30

Proc.: Add the above ingredients in the order listed. The Carbitol and KP-140 are premixed and added slowly, with agitation, to the preceeding components. Adjust pH to 8.3–8.7. The addition of 500 ppm of formaldehyde to serve as a preservative is recommended after the pH adjustment. Stirring for 30 minutes is recommended after the last addition. The pH of the floor polish should be readjusted to 8.3–8.7 after 30 minutes of stirring.

II

	pts./wt.
"Shanco" 6000 Resin (15%)	20.0
DW-855 (15%)	70.0
AC-680 (15%)	10.0
Tributoxyethyl Phosphate	1.0
Diethylene Glycol Monoethyl Ether	2.0
Ethylene Glycol	1.0
FC-128 (1%)	0.4

Proc.: Mix all ingredients together until uniform.

Floor Wax—(Liquid)

	%/wt.
MONAMINE ARA-100	2
Carnauba Wax	10
Rosin	2
Water	86

Proc.: Melt together MONAMINE, wax and rosin and add the required amount of boiling water.

This wax requires buffing.

Gym Floor Finish (Urethane)

	pts./wt.
13-307 Polyurethan	49.0
Mineral Spirits	18.0
"Amsco" No. 460	6.0
6% Cobalt Naphthenate	0.5
"Exkin" No. 2	0.5
Set-To-Touch—2 hr	Dry Hard—5 hr.

Proc.: Mix all ingredients together until uniform.

Furniture Polish

I

	pts./wt.
CLINDROL 200-S	3
Stearic Acid	1.2
Mineral Seal Oil	20
Heavy Mineral Oil	20
Water	55.8

Proc.: CLINDROL 200-S and stearic acid are dissolved in the oil by heating to 80°C. This solution is slowly added with rapid stirring to water previously heated to the same temperature (70-80°C.). The final product exhibits a consistency well suited to a furniture polish in that it pours readily, but at the same time permits control in confining the fluid to the desired area of application.

II

		pts./wt.
A	SF-96 (1000)	2.0
	"Duroxon" J-324 wax	2.0
	Esso Wax 5010	0.5
	"Isopar" G	15.0
B	"Isopar" E	80.5

Proc.: Heat components of A to 90°C (194°F) until all wax is melted. Add B (room temperature) with mild agitation and stir until cool. After mild shaking of the polish, apply a small amount to a clean dry cloth for application to the furniture. Allow the polish to dry to a white haze and then buff to a high gloss with another clean dry cloth.

Scratched-wood Polish

		%/wt.
A	VEEGUM T	1.50
	Water	25.50
B	Stearic Acid	4.50

	Carnauba Wax	16.50
	Beeswax, Yellow	16.50
C	Triethanolamine	2.75
D	Mineral Spirits	27.00
E	Color	0.50
	Water	5.25

Proc.: Add the VEEGUM T to the water slowly, agitating continually until smooth. Heat B to 90°C, add C and maintain temperature. Add D, maintaining 90° temp., avoiding open flames. Add A to B, C, and D, mixing until uniform. Disperse E in hot water and add to the mixture, continuing to mix until uniform by processing through a roller mill.*

Directions for use: Apply to damaged areas with a soft cloth and rub into area. Buff with a clean dry cloth.

W/O Aerosol Furniture Polish

		%/wt.
A	Concord CO-Wax in Water (10%)	17.0
	Dow Corning 922 Silicone Emulsion	8.0
	Water	54.0
B	VM&P Naphtha	20.0
	SPAN 80	1.0

Proc.: Add A ingredients together with simple stirring. Add B ingredients together with the simple stirring. Add A to B with high speed stirring until emulsion is obtained.

Package in aerosol container.

Cleaning and Polishing Cloths

Polishing cloths may be made of untreated fabrics of various kinds. Pieces of fine, soft wool make the most satisfactory dusting cloths and there are grades of wool made especially for the purpose. While they are fairly expensive, they may be washed in soap and water and reused. Next in usefulness are soft cotton, especially knitted materials, or cheesecloth, and linens. Oiled cloths may be made by saturating suitable fabrics with gasoline or other solvent solutions of paraffin, paraffin oil, linseed oil, or rapeseed oil, or a mixture of these oils. The cloth is then freed of excess fluid by wringing or passing through rollers, and is allowed to dry at room temperature. Sometimes, essential oils, like cedar oil, lemon oil or citronella oil, are added to the impregnating mixture. Occasionally, certain resins are included. With some methods, no solvent is required to thin the oils. This produces a more heavily impregnated dusting cloth, but it also eliminates the hazards of using volatile solvents.

I

	pts./wt.
Paraffin	30.0
Rapeseed Oil, double-refined	10.0
Benzoin	1.0

Proc.: Mix the paraffin and the oil, heat moderately and add the previously melted resin. Saturate the cloth with the liquid, and wring well. Dry in the shade.

II

Light Mineral Oil	3 gal.
Corn Oil	1 gal.
Odorizing Compound	3 oz.
Oil-Soluble Yellow Color	q.s.

Proc.: Same as for I.

Dust Control Oil

	pts./wt.
Mineral oil	96
"Variquat" K300	2
"Varonic" L202	2

Proc.: Mix all ingredients until uniform.

The composition exhibits a degree of germicidal and fungicidel activity because of the biocidal properties of the "Variquat K300."

LUBRICANTS

Aerosol Lubricant

I

	%/wt.
Molybdenum Disulphide Lubricant	50.0
Naphtha	50.0
Methylene Chloride	40.0
Aerosol Formulation:	
Concentrate	50.0
Genetron 12/11 (30:70)	50.0

Proc.: The ingredients in the concentrate are mixed at room temperature. Shake well before using. Package in an aerosol container.

	II	III	IV	V
	%/wt.	%/wt.	%/wt.	%/wt.
Silicone Fluid 200				
(50-300 cs)	5			

Molybdenum Disulfide	5			
Lubricating Oil		30		
Penetrating Oil			30	
Naphtha		20		
AEROTHENE® TT	70	40	35	35
Propellant 12	25	35	35	35

Proc.: Mix stirring all ingredients together until uniform. Fill in aerosol bottle or can and load with the propellant #12.

Glycerin-Graphite Lubricant

I

	pts./wt.
Colloidal Graphite	0.2
Glycerin, U.S.P.	49.8
Distilled Water	50.0

II

	pts./wt.
Concentrated Dispersion of Colloidal Graphite	2.56
Glycerin, U.S.P.	61.44
Distilled Water	64.00

Proc.: Stir the graphite into the glycerin, then add the water and stir well. This lubricant penetrates tight-fitting parts and leaves a durable film of graphite on rubber surfaces.

SHOE PREPARATIONS

Emulsion Shoe Polish, Neutral

	pts./wt.
"DUROXON" H-111	8
"EFTON" D or UC	3
Fully Refined Paraffin Wax 143/145	10
Stearic Acid, double pressed	2
"Lorol" 28	2
Mineral Spirits	25
Water	50

Proc.: Heat all ingredients and stir until uniform.

Black Shoe Polish

Carnauba Wax	5½ oz.
Crude Montan Wax	5½ oz.

Proc.: Melt together in a double boiler. (The water in the outer container should be boiling.) Then stir in the following melted and dissolved mixture:

Stearic Acid	2 oz.
Nigrosine Base	1 oz.

Then stir in

Ceresin	15 oz.

Remove all flames and run in slowly, while stirring,

Turpentine	90 fl. oz.

Proc.: Allow the mixture to cool to 105°F. and pour into airtight tins which should stand undisturbed overnight.

White-Shoe Dressing

Lithopone	19 oz.
Titanium Dioxide	1 oz.
Bleached Shellac	3 oz.
Ammonium Hydroxide	¼ fl. oz.
Water	25 fl. oz.
Alcohol	25 fl. oz.
Glycerin	1 oz.

Proc.: Dissolve the last four ingredients by mixing in a procelain vessel. When dissolved, stir in the first two pigments. Keep in stoppered bottles and shake before using.

White Shoe Polish

	%/wt.
A VEEGUM	1.0
Sodium Carboxymethylcellulose	
(med. visc.)	0.5
Water	77.7
B DARVAN No. 1	0.3
C Dibutyl Phthalate	3.0
Titanium Dioxide	15.0
Neatsfoot Oil	2.0
Triton X-102	0.5
Preservative	q.s.

Proc.: Dry blend the VEEGUM and CMC and add to 30 parts of the water slowly, agitating continually until smooth. Mix B in the remaining water and add to A. Add the components in C in the order listed to A and B and mix until uniform.

Naphtha Emulsion Shoe Cleaner

High Viscosity Methyl Cellulose	1 g.
Water	100 ml.
2-Amino-2-Methyl-1,3-Propanediol	2.3 g.
Stearic Acid	6.2 g.
Low-boiling Naphtha	200 ml.

Proc.: Disperse the methyl cellulose in the water and add the Amino-methyl-propanediol. Dissolve the stearic acid in the naphtha. Pour this solution into the methyl cellulose dispersion and agitate. Better mixing can, of course, be obtained with a homogenizer,* particularly in larger batches.

Aerosol Dry Bright Shoe Polish

	pts./wt.
A GE SR-82 Resin (60% Solids)	5.2
99% Isopropanol	42.0
Butyl Acetate	2.4
B Acryloid B-72 Resin	8.8
Toluene	41.6

Proc.: Dissolve Acryloid B-72 resin in toluene (B). Add remaining material in A to B solution.

Package in an aerosol container.

Waterproofing for Shoes

	pts./wt.
Wool Grease	8
Dark Petrolatum	4
Paraffin Wax	4

Proc.: Melt together in any container.

SPOT & STAIN REMOVERS

Powder Spot-Remover

	%/wt.
1. AEROTHENE TT	45.00
2. DOW-PER	15.00
3. Isopropyl Alcohol	9.95
4. Perfume	0.05
5. Microcel C	5.00
6. Propellant 12	25.00

Suggested Valve: Risdon #5832 0.020" stem/0.138" body
Suggested Actuator: Risdon #5832 Actuator (EH-16) 0.030"
Suggested Perfumes: Fleuroma Bouquet #686, Roure du Pont Bouquet #A5147.

Proc.: First mix the solvents and perfume. Then slowly mix in the powder. Mix thoroughly to get an even dispersion. Mix #1 and #4 then add the other ingredients. Stir until uniform. Fill in aerosol can or bottle and charge with the propellant #12.

TO APPLY: Place an absorbent cloth on the underside of the spot. Apply a liberal coating to the topside of the cloth. Allow the powder to dry thoroughly before brushing off.

Stain Remover

Tergitol N.P.X.	0.1 lb.
Carbitol	25.6 c.c.
Isopropylalcohol	19.2 c.c.
Cleaner's Naphtha	102.4 c.c.

Proc.: Mix Carbitol, Isopropylalcohol and naphtha. Add Tergitol and stir until a clear solution is obtained.

Warning: Avoid open flames, when mixing or using this formula. This product can remove lipstick stain from fabrics.

Dry Cleaning Compound

	%/wt.
"Triton" X-100	12.5
"Triton" X-45	5.0
"Dytol" B-35	2.5
Water	2.5
"Stoddard solvent"	77.5

Proc.: Dissolve the "Dytol" B-35 in the water. Add "Triton" X-100 and "Triton" X-45; then add "Stoddard solvent".

Use one part to 100 parts cleaning solvent.

Dry Cleaning Fluid
(Non-inflammable and quick acting)

	pts./wt.
Butyl Cellosolve	1
Diglycol Oleate	1
Water	1
Isopropyl Alcohol	10
Carbon Tetrachloride	14

Proc.: Mix all ingredients together until uniform.

Jewelry Cleaners
I

Warm soap and water is applied with a toothbrush. The article is held flat on a table with one hand while the other hand taps the article with the toothbrush so that the bristles get into and around the design of the jewelry. This method is suitable for removing ordinary dust and dirt, although regular cleaning may be necessary to remove all the accumulated film.

II

Kerosene	1 oz.
Carbon Tetrachloride	3 oz.
Citronella Oil	2 dr.

Proc.: Mix and apply to the parts to be cleaned with a soft cloth. Then polish.

RUG AND UPHOLSTERY CLEANERS

Rug and Upholstery Shampoos

I

	%/wt.
"Avirol" 103	15.00
Coco Diethanolamide (1:2)	2.00
Water, color, perfume, preservative	83.00

II

"Avirol" 103	25.00
Coco Diethanolamide (1:2)	3.00
Water, color, perfume, preservative	72.00

III

"Avirol" 103	35.00
Coco Diethanolamide (1:2)	4.25
Water, color, perfume, preservative	60.75

Proc.: In each of the three formulas, it is necessary to melt ingredients together to about 45°C. Some formulators may wish to add 0.5% of an optical brightener such as DBA–115 and 0.25% of "Santophen" 1 to give a superior type of product.

Rug Shampoo Concentrate

I

	pts./wt.
Water	91
EDTA type Chelate	1
Sodium Lauryl Sulfate (solid basis)	6
CLINDROL 101-LI	2
Preservative	q.s.

Proc.: Heat all items until dissolved.

The above concentrate is used at a dilution varying from 1:8 to 1:80. The residue remaining after shampooing can be picked up by vacuuming, thus reducing tendencies towards resoiling.

II

	%/wt.
VARONOL SLS	40
VARION CADG	4
Lauric Isopropanolamide	4
Water	52

Proc.: Dissolve the Lauric Isopropanolamide in hot water. Add VARONOL SLS and VARION CADG. Continue agitation until uniform.

Removing Shine from Garments
(U.S. Patent 2,459,236)

Wetting agents such as *Tergitol* or *Sulfatate B* in dilute solution, say 0.1%, are applied to the garment in the same manner in which a garment is customarily sponged with water before being pressed.

After the treating solution has been applied to the garment and the garment has been permitted to stand for a few minutes, the fibers or threads soften and the nap can then be readily "raised." This is preferably accomplished by brushing up the nap with a stiff bristle brush or a stiff brush made of some of the new synthetic plastics. In raising the nap, any abrading action should be avoided in order to prevent injury to the fibers. The garment is then pressed.

Nicotine Finger-Stain Remover

	pts./wt.
Hydrochloric Acid, Pure	0.4
Glycerin	10.0
Rose Water, Triple	90.0

Proc.: Combine ingredients with agitation until homogeneous.

To use the lotion, saturate a piece of cotton with it and rub gently over the stained area. Several applications may be necessary before desired results are obtained.

Soot Removers

	pts./wt.
Common Salt	85.0
Copper Sulfate	8.0
Zinc Dust	7.0

II

	pts./wt.
Coarse Rock Salt	125.0
Zinc Dust	2.0
Copper Sulfate	2.0

III

	%/wt.
Zinc Dust	50.0
Sodium Chloride	45.0
Sawdust	5.0

Proc.: Mix the ingredients of each formula until smooth
Throw approximately 5 oz. of soot remover onto fire.

Ink Remover

		pts./wt.
A	Ammonium Sulfide	1
	Water	19
B	Oxalic Acid	1
	Water	19

Proc.: Mix Ammonium Sulfide with the water until dissolved. Mix Oxalic Acid with the water until dissolved. Add A to B. Stir until homogeneous.

II

	pts./wt.
Lactic Acid	2
Isopropyl Alcohol	2
Hydrofluoric Acid	2
Water	10

Proc.: Mix Lactic Acid, Isopropyl Alcohol, Hydrofluoric Acid and water until homogeneous.

3. Cosmetics

Cosmetics are an essential aspect of our culture and their use date back as early as 50 B.C. when Cleopatra used fragrances and oils for her skin and colored "kohls" to beautify her eyes. The Bible refers in many passages to the practice of anointing the head or body with oils. Cosmetics were liberally used by Assyrians, Babylonians, Samarians, and Syrians. The Romans were ignorant of cosmetics until they came into contact with the Greeks. By the time Nero ruled Rome, pigments and powder were abundantly in use. English women were introduced to cosmetics by the returning explorers from the Near East.

Cosmetics have received censure and disapproval as well as praise. In Puritan New England cosmetics were banned on religious grounds; whereas in the French colonies they were enjoyed and used freely. It wasn't until as recently as World War I that cosmetics were generally approved by all classes of American society.

Since then cosmetics has become an intrinsic part of Western European and American culture. It was found that during World War II when cosmetics became a luxury and highly restricted in use, many women suffered from a lowered morale and decreased productivity. By providing women with make-up, as an experiment, a positive psychological reversal came about. Spirits were boosted and productivity of women defense workers reached a high.

The cosmetic industry enjoys retail sales of over two billion dollars annually and is rapidly expanding. There is always room for new products and great opportunity for profit in such a dynamic industry.

Presented in this chapter, is a compilation of a variety of formulas for every type of cosmetic product. Here are products for men, women, teenagers, children and infants. They run the gamut from simplicity to complexity—from kitchen type to commercial type, but they all share one thing in common—they can be made by you. Begin with formulas that are simple, and as you gain familiarity with working with the formulas advance to more complex ones. You can set your own pace as to your rate of progress.

Cosmetics are not only used for beautification but also for protection. Lotions and creams can moisturize dry and roughened skin. Astringents can be used to dry oily skin. Cosmetics can therefore help people stall the effects of age on their physical appearance.

Before beginning your production of cosmetics there are a few important facts about safety, laws, and the use of materials: Generally, do not use technical grades of materials. These are not pure enough for drug or cosmetic use. Do not make substitutions unless the formula under consideration lists an alternate ingredient to the one in question.

It is considered proper to use food coloring in face and hand creams, shaving preparations, clear lotions and bath products. However, don't use food coloring in cosmetics that require pigments, such as, lipsticks, face powder, make-up, and eye preparations. Pigments are United States government regulated and are certified for use in drugs and cosmetics by the Food and Drug Administration. Those colorants labeled D and C certified are used in lipsticks, rouges, face powders, nail preparations, and liquid make-up. These organic colorants are not to be used in foods or products applied to the area of the eye.

Inorganic colorants are approved and listed by the Food and Drug Administration. These colorants can be used in all the products mentioned above and also in eye preparations. They are labeled "Cosmetic," followed by the color of the pigment, e.g. Cosmetic Black.

CREAMS and LOTIONS

Creams and lotions are prepared for all areas of the body. They have a variety of purposes: keeping the skin clean and healthy, lubricating and softening it and leaving it smooth and supple but not excessively greasy. Skin care is really all-over body care. There are products for general body use and those made for specific parts of the body, e.g. hand and face creams and lotions. Some types that we have presented formulas for are:

Cold Creams — These are also known as all-purpose creams and are used for cleansing, softening and lubricating. There are two types: greasy (water-in-oil) and non-greasy (oil-in-water).

Cleansing Creams — These creams are made to penetrate the pores of the skin in order to rid them of build-ups of dirt and make-up. The lotion form is made with lighter oils which are suspended in water. The benefit of the lotion is that it does not leave a greasy feeling.

Vanishing Creams — These creams disappear into the skin without leaving a trace of oil or grease. They leave a protective film on the face.

Lubricating Creams — This type of cream is used by people who suffer from dry cracking skin. It has a rich amount of lanolin with a cold cream base.

Astringent Lotions — These lotions are used for closing the pores of the skin prior to the application of make-up. They leave the skin with a tingly refreshing feeling. They are delightful to use especially when a pleasing perfume is combined with it.

Cream Base

A soft, emollient cream, provides a versatile vehicle for a wide variety of cosmetic applications. It liquifies readily on the skin leaving an invisible film free from drag and stickiness. It is suggested for cleansing applications, as a lubricating and night cream, as a base for hormone preparations.

Oil Phase

	%/wt.
Beeswax, U.S.P.	2.20
Cetyl alcohol, N.F.	1.70
Glyceryl monostearate	9.00
Isopropyl myristate, "Deltyl Extra"	3.50
Mineral oil, "Carnation"	4.30

Water Phase

Fungitex R	0.05
Glycerin, U.S.P.	5.30
Lactic acid, U.S.P., 85%	0.08
Methocel 90 HG, 15,000 cps	0.25
Sapamine COB-ST	0.30
Water, distilled	73.12
Perfume	0.20

Proc.: Heat oil phase to 160°F. Heat water phase to 160°F. Add water phase to oil phase stirring until uniform. Add perfume.

Four-Purpose Cream

This cream has a cooling effect on the skin and is nongreasy. It serves equally well as a cleansing and as a massage cream. No oil or grease is left behind after the usual hot-towel application.

A	White petrolatum	.45 lb.
	Paraffin wax	1.8 lb.

	Mineral oil (white)	1.8 lb.
	"Glycostearin"	1.8 lb.
B	Water	7.1 lb.
	Triethanolamine	0.2 lb.
	Mold inhibitor	0.3 oz.
C	Perfume	q.s.

Proc.: Heat Part A to 166°F and stir until complete solution is obtained. Heat B to 166°F and add A to B, stirring continuously while cooling. Add Part C at 135°F and pour at 120-125°F.

A preservative, parahydroxybenzoic acid derivative, is generally used in the proportion of 18 oz. to 100 gal. of the finished product. It should be dissolved, by heating, in the water called for in the formula.

Avocado Oil Cleansing Cream

		%/wt.
A	Hydrogenated Oil (Cosmetic Grade)	11.0
	Beeswax U.S.P. White	5.0
	Stearic Acid Triple Pressed	0.5
	Sesame Seed Oil	60.0
	Avocado Oil	7.0
	Antioxident & Preservative	0.1
B	Distilled Water	15.3
	Borax U.S.P.	0.5
C	Perfume	0.6

Proc.: Melt A at 80°C. Heat B to 75° C.

When A has cooled to 75°C. add B with constant stirring. Add C at 60°C. Pour between 55°–50°C.

This cream is of the quick liquefying type. It does not melt or run in the jar even under extremely hot weather conditions. The quick liquefying property is mechanical not thermal.

Turtle Oil Cream

	%/wt.
Turtle Oil (Pale Deodorized)	10
Diglycol Stearate	10
Liquid Paraffin	30
Lanolin Absorption Base	9
Perfume	1
Distilled Water	40

Proc: Melt the fats and wax and stir in the hot water at a temperature of 170°F. The perfume is added with slow stirring when the batch has cooled to about 120°F., after which the stirring device may be switched off, and the batch allowed to cool.

Cold Cream (Inexpensive)

This is a good starter—both inexpensive to make and an excellent product.

	pts./wt.
Spermaceti	12.5
White Wax	12.0
Liquid Petrolatum	56.0
Borax	0.5
Distilled Water	19.0
Oil of Rose, Synthetic	q.s.

Proc.: Melt the wax and spermaceti on the water bath and add the liquid petrolatum. Heat the distilled water and in it dissolve the borax. Add this warm solution to the melted mixture while both are warm and at about the same temperature. Beat rapidly; as soon as it begins to congeal add the oil of rose and beat until congealed. Dispense preferably in pure tin tubes.

Theatrical Cold Cream

Basic Formula

	%/wt.
Beeswax	15.0
White Paraffin Oil	60.0
Rose Water	24.0
Borax	1.0

Proc.: Melt the wax, add the paraffin oil and continue to heat with constant stirring until they are well mixed. Use a water bath to avoid overheating. Dissolve the borax in the rose water with the aid of heat and while still warm gradually add to the melted wax and oil, stirring constantly until cold. If desired distilled water may be used in place of the rose water and any desired perfume added.

Facial Wash Cream

The creamy new way to wash the face without dryness, soap or grease. Smooth on wash cream, work up a froth skin with wet fingertips, rinse with water, no greasy residue.

	pts./wt.
Klearol Min. Oil	17
CRODA LIQUID BASE	5
POLAWAX A-22	5
STEARYL ALCOHOL 95%	2
Tegamine S-13	0.5
Tegamine P-13	0.5
VOLPO 20 Oleyl Ether	8

Propylene Glycol	12
Phosphoric Acid to pH 5-5.5	q.s.
Water, demin.	50

Proc.: Add the oils to the water phase at abt. 75-80°C, begin cooling to R.T. and adjust pH with Phosphoric Acid.

Vanishing Cream

I

They are named for the produce characteristic of leaving a matte and non-greasy residue on the skin.

	%/wt.
Arlacel 165 acid-stable	18.0
Lanolin	2.0
Cetyl Alcohol	1.0
Mineral Oil 70 Vis.	1.0
Sorbo 70% Sorbitol Solution U.S.P.	2.0
Water	76.0

Proc.: Heat the ingredients to 90°C., stir throughly to form an emulsion.

II

		%/wt.
A	Decaglycerol Decaoleate	2.0
	Decaglycerol Monolaurate	2.0
	Neobee M5	2.0
	Cetyl Alcohol	4.0
	Wecoline T.P.V.	16.0
	Spermacetti	2.0
	Parahydroxybenzoate	0.2
B	Glycerin	4.0
	Water	68.0
	Methylhydroxybenzoate	0.2
C	Perfume Oil	q.s.

Proc.: Heat A and B separately to 170°F. Add A to B with agitation. Cool. Add the perfume oil.

Dry Skin Cream

		pts./wt.
A	G-E SF-1091 Silicone Fluid	8.0
	Mineral Oil (65/75 SSU Viscosity)	8.0
	Lanolin	2.0
	Polawax	10.0
	Paraffin Wax	2.0
B	Water (Sterile)	69.8
	Cinaryl 200	0.2

C Perfume   q.s.

Proc.: Heat A and B components to 80°C (176°F). Add B slowly to A with rapid high shear agitation. Cool to 40°C (104°F) and blend in C.

Moisturizing Cream

		pts./wt.
A	"Veegum"	1.00
	Water	73.00
B	Glycerin	4.00
	Triethanolamine	1.00
	Polypeptide SF	5.00
C	"Lantrol"	10.00
	Stearic Acid	2.00
	Isopropyl Myristate	2.00
	Glyceryl Monostearate SE	—
	Cetyl Alcohol	2.00
	Perfume	q.s.
	Preservative	q.s.

Proc.: Add the "Veegum" to the water slowly, agitating continually until smooth. Add the components in B and heat to 70°C. Heat C to 75°C. and add to A and B. Mix until cool.

Cleansing Milk

A starting formulation for a cleansing milk may consist of the following:

	pts./wt.
Protagin (Absorption Base)	1.5
Glycerine	5
"Empicol" LQ	2.5
Water	91

Proc.: Mix all ingredients together until uniform.

Vitamin Ointment

	pts./wt.
Emersol 132	13.00
Lantrol	5.00
Glyceryl Monostearate, self-emulsifying	8.00
Tween 85	2.00
Span 85	1.00
Peach Kernel Oil	5.00
Vitamin A Palmitate	0.10
Vitamine E	0.02
Triethanolamine	1.00

Deionized Water	52.18
Methyl Paraben	0.20

Proc.: In a suitable mixer, melt 1-6 (80°C). Mix thoroughly. Add 7 and 8. In a separate mixer, mix 9-11, heat to 85°C. Add water phase to oil phase with rapid agitation. Cool slowly to 35°C. with stirring. (Use lightning mixer.)*

Hormone Cream

	pts./wt.
Natural Estrogenic Hormones	
(Estrone)	35,000 I.U.
Anhydrous Lanolin	40.0
Petrolatum	40.0
Water	19.8
Perfume	0.2

Proc.: Heat lanolin and petrolatum to 120°F. Heat water to 120°F. Stir until homogeneous. Add Hormone and perfume oil.

Enriched Skin Cream

		%/wt.
A	Propyl Paraben	0.1
	Tegacid	18.0
	Neobee M-5	3.0
	Cetyl Alcohol	1.0
	Amerchol CAB	5.0
	Solulan 98	1.0
	Robane	2.0
B	Propylene Glycol	5.0
	Distilled Water	64.2
	Allantoin	0.2
	Methyl Paraben	0.2
C	Perfume oil	0.3

Prod.: Heat A and B to 75°C. and with rapid agitation add B to A. Cool to 40° and add C.

Simple Hand Cream (Oily)

This formula from out of the past uses only three, readily available ingredients in its composition. You may wish to add color as well as perfume, for the finished product is not a snowy white but an off white.

Since the cream has over 70% oil plus two waxes, you know the finished product has to be oily in texture. For rough, dry hands it is soothing and softening.

Beeswax	1 qt.
Synthetic Spermaceti Wax	1 qt.
Almond Oil (sweet)	9 qt.

Proc.: Heat beeswax and spermaceti in top of double boiler to 158°F. (70°C). Warm almond oil to same temperature and add slowly to waxes while stirring. Remove heat and stir until temperature drops to 104°F (40°C), when you may add perfume. Pour into container at 86°F (30°C).

Deluxe Hand Cream

A lovely softening cream that disappears into the skin without leaving a greasy feeling. Texture is light and smooth.

Stearic Acid	2¼ qt.
Mineral Oil	1½ cups
Arlacel 60	1 cup
Tween 60	1 pt.
Sorbo	2¼ qt.
Distilled Water	5½ qt.
Preservative*	
Perfume, as desired	

Proc.: Heat stearic acid, mineral oil, Arlacel 60, and Tween 60 in top of double boiler over water until mixture reaches a temperature of 162°F. (72°C.). Heat Sorbo, water, and preservative in separate double boiler until it reaches the same temperature. Add the Sorbo solution to the stearic acid mixture. Stir rapidly while pouring. Stir until cream sets; then perfume, and pour into jar.

* This cream may tend to spoil before some others. Keep it refrigerated.

Protective Hand Cream

	%/wt.
Stearic Acid T.P.	10
Spermaceti	5
Polyoxyethylene Sorbitan Monooleate	2
Cetyl Alcohol	3
Methylparaben	0.1
Silicone Oil	10
Distilled Water	70

Proc.: Melt first five ingredients together and heat mixture to 80°C. Heat water separately to the same temperature and then add with steady stirring. Add silicone oil and continue stirring.

Waterless Hand Cleaner
Containing Pumice

This is a heavy-duty hand cleaner. In addition to acting as an emollient in the final formulation, hexadecyl alcohol affords a means of readily incorporating the pumice during manufacture.

		%/wt.
A	Hexadecyl Alcohol, C.G.	10.0
	Pumice	3.0
B	Isopar, C.G.	37.9
	Stearic Acid (triple pressed)	7.0
	Tridecyl Alcohol,	
	12 mole Ethoxylate	10.0
	Propyl p-hydroxybenzoate	0.1
C	Distilled Water	30.8
	Sodium Hydroxide, 50% (W/W)	1.0
	Methyl p-hydroxybenzoate	0.2

Proc.: Suspend pumice in hexadecyl alcohol with agitation. Combine the ingredients of B and add A to B. These combined phases are then added to C with vigorous agitation.

Pigmented Foundation Cream

This foundation cream works as do other foundations: it creates a protective and uniform base for the application of makeup or powder. The added advantage of a pigmented foundation cream is that it has light covering powder, concealing minor blemishes and imperfections in the skin. Used below the eyes, it is effective in hiding dark circles.

Mineral Oil	1½ pt.
Beeswax	½ cup
Ceresine Wax	¼ cup
Arlacel 186	¾ cup
Sorbo	3 pts.
Titanium dioxide	1 cup
Distilled Water	1¾ cups

Proc.: Heat mineral oil, beeswax, ceresine wax, Arlacel 186, and Sorbo in the top of a double boiler to a temperature of 158°F. (70°C.). Blend in the titanium dioxide until it is uniformly dispersed. Heat water to a temperature of 162°F. (72°C.), then add to other ingredients, blending until cool. You may "homogenize" by using an egg beater to increase the smoothness.

Astringent Lotion

	%/wt.
Menthol	0.05
Perfume Oil	0.50
Aluminum Chlorhydroxy Allantoinate	0.20
Propylene Glycol	3.00
S.D. Alcohol No. 40	50.00
Chlorhydrol 50%	5.00
Water Distilled	41.25

Proc.: Dissolve menthol and Perfume Oil in S.D. Alcohol No. 40. Dissolve rest of ingredients in water and add to the alcohol solution.

Milk Face Lotion

	pts./wt.
LANTROL®	2.0
Emersol 132	2.5
Mineral Oil, 70 vis.	33.0
Peach Kernel Oil	3.0
Sesame Oil	10.0
Tween 60	2.8
Span 60	2.2
Vitamin E (Alpha Tocopherol)	0.2
Deionized Water	q.s.
Propylene Glycol	4.0
Triethanolamine	1.0
Carbopol 941 (2½% aqueous sol.)	4.0
Non Fat Dry Milk	0.5
Methyl Paraben	0.1
Perfume	q.s.

Proc.: Heat oil phase to 80°C, water phase to 82°C. Mix oil into water. Cool to 45°C. Perfume and cool to 30°C. (Mixer can be used without beating in too much air.)

Protein Lotion

		%/wt.
A	Arlacel 165	9.0
	Neobee M5	3.0
	Cetyl Alcohol	1.0
	Amerchol L-101	5.0
	Solulan 98	1.0
B	Methyl P Hydroxy Benzoate	0.2
	Allantoin	0.2
	Propylene Glycol	5.0
	Distilled Water	73.0
	Germall 115	0.3
	Cosmetic Polypeptides B-315	2.0
C	Perfume oil	0.3

Proc.: Heat A and B to 75°C. and add B to A with rapid agitation. Add C at 40°C. and cool to room temperature.

Pearly Body Lotion

		%/wt.
A	Alcohol SDA 40	14.7
	Carbopol 941	0.3

B	Alcohol SDA 40	15.00
	Uvinul D-50	0.07
	Ceraphyl 50	2.50
	Ceraphyl 230	2.50
	Perfume	0.5–1.5
C	Water	20.00
D	Allantoin	0.2
	Diethanolamine	0.30
	Water	42.63
	Mearlin AC	0.30
	Color	q.s.

Proc.: Prepare A in advance. Dissolve B ingredients into the alcohol, then combine with A. Finally, add C, slowly, with high speed stirring. Phase 1 is now complete.

Combine D ingredients. D must be agitated continuously or pearl will drop out. Add D slowly to Phase 1 with continuous high speed mixer.

Face and Body Lotion

The germicidal properties of ethyl alcohol and quaternary bacteriacide are combined with the emolliency of hexadecyl alcohol to give an all-purpose lotion.

	%/wt.
70% Ethyl Alcohol No. 40	94.25
Hexadecyl Alcohol XX	5.00
Quarternary bacteriacide XXX	0.50
Perfume	0.25
Color	q.s.

Proc.: Mix together until uniform.

Aloe Body Spray

	%/wt.
Isopropyl Myristate	8.0
"Acetulan" or equivalent	1.5
Propylene Glycol	5.0
Perfume	0.5
Anhydrous Denatured Ethanol	30.0
Aloe (50% in Anhydrous	
Denatured Ethanol	10 0
Freon-12/Freon-114 (57/43)	
Propellant	45.0

Proc : Mix all the ingredients together and charge with the propel-lant *

Skin Balm

	%/wt.
Solan	0.5
Volpo 10	0.5
Isopropylmyristate	1.0
Perfume Oil	1.0
Alcohol No. 40	15.0
Water	67.0
Carbopol 941–1% Solution	8.0
Glycerin or Aloe	5.5
Diisopropanoamine	1.5
Preservative and Color	q.s.

Proc.: Combine Carbopol solution and water, add to the oil phase at 50°C. Dissolve perfume oil in alcohol and add to the above emulsion. Neutralize at 40°C.*

Hand Lotions

I

	pts./wt.
1. Mineral Oil	7.0
2. Olive Oil	0.8
3. Trihydroxyethylamine Stearate	1.4
4. Water	7.0
5. Perfume	0.2

Proc.: Heat Nos. 1, 2, and 3 together to 140°F. and stir until homogeneous. Add No. 4 slowly while stirring and then stir in the perfume. Continue stirring until cool. By varying the amount of water a thicker or thinner preparation will be formed. The thicker preparations are put up in tubes and are now carried by men and women, especially motorists, who, when water is not available, merely put a little of this cleaner on their hands, rub it in and then wipe off with it the grease, oil, paint or dirt present. Not only is this an excellent detergent but it leaves the skin smooth, and produces a cooling sensation and prevents chapping during cold weather.

II

		pts./wt.
A	"Veegum"	0.5
	Water	88.7
B	Stearic Acid	4.0
	Ceresin	1.0
	Paraffin	0.3
	Amerchol L-101	5.0
	Triethanolamine	0.5
	Preservative	q.s.

Proc.: Add the "Veegum" to the water slowly, agitating continually until smooth. Heat to 70°C. Heat B to 70°C. Add B to A, mixing until cool.

Note: This lotion can be varied with a number of additives such as a sunscreen, an insect repellent, or a bacteriostat.

Hand Balm Lemon Lotion

	%/wt.
CMC 7MF	2.0
Glycerine	15.0
Citric Acid	4.0
SDA 40 Alcohol 95%	8.0
Lemon Perfume	0.7
Deionized Water	69.3
LANTROL®	1.0
ETHOXYOL 16R	3.5

Proc.: Wet CMC with alcohol. Mix Glycerine and water together, add citric acid and heat to 90°. Pour heated LANTROL® and ETHOXYOL into water phase. Add water phase to CMC and alcohol and agitate by hand until homogeneous. Cool to 25°C. and perfume slowly.

Egg Oil Hand Lotion

		pts./wt.
A	Glyceryl Monostearate	18
	Stearic Acid	13
	Propylene Glycol	25
	Alcohol	50
	Triethanolamine	2.5
	Egg Oil	10
B	Water	376.7
C	Perfume	5

Proc.: Heat A. Heat B. Mix until uniform. Add C.

FACE PACKS AND MASKS

The use of packs and masks are the most ancient types of feminine beauty treatments. Their popularity is now on the upsurge for several reasons: With increasing pollution in the cities, people are finding the cleansing effects of face masks quite valuable. They are also used because of the stimulating, warming, tightening, and toning effect they have on the skin. When the mask is removed, the skin is left with a glowing, fresh look.

Face Mask Base

I

	%/wt.
Kaolin	15.0
Water	75.0
Ethanol	5.0
Glycerine	5.0
Preservative	q.s.
Perfume	1.0

Proc.: Disperse Kaolin in water. Agitate until smooth dispersion is produced. Add remaining ingredients and continue agitation until a uniform product is produced.

II

	%/wt.
Kaolin	10.0
Water	77.4
Solulan 98	5.0
Ethanol	5.0
Glycerine	2.5
Menthol	0.1
Perfume	1.0
Preservative	q.s.

Proc.: Disperse Kaolin in water. Agitate thoroughly until smooth dispersion is produced. Dissolve menthol in alcohol and combine with glycerine and Solulan 98. Blend with Kaolin dispersion until uniform product is produced. This formula produces a finished face mask that is a temporary wrinkle remover.

Sulphur Masks

	pts./wt.
Bolus alba	50
Borax	2
Sodium thiosulfate	8
Terra silicea	20
Talcum	20

Proc.: The sodium thiosulfate has to be milled fine and milled with other powders. This mask is therefore effective, because the sulfur is dropped on the skin only, when it comes in contact with water.

Strawberry Moisturing Mask

3 tbsp. pureed fresh strawberries
talcum powder

Proc.: Add enough talc to strawberries to thicken to a spreadable paste. Smooth on gently. Rinse off after 20-30 minutes with lukewarm water. Follow with a splash of skin tonic. Finish by applying a light moisturizer.

Peach Tightening Mask

1 small fresh peach, peeled
1 egg white

Proc.: Puree the peach in a blender. Add it to the egg white which has been beaten to a froth. Apply evenly with a shaving brush. Remove with cool water after 20-30 minutes.

Gelatin Mask

This thick, liquid mask is pleasant to use and rinses off with warm water. It serves to cleanse and tone the skin. Be certain to keep container tightly capped when not in use. Also, this mask needs to be shaken well before use so that it will pour.

Gum Tragacanth	1 cup
Glycerin	½ cup
Gelatin	1 cup
Distilled Water	5 qts.
Zinc Oxide	¼ cup

Proc.: Moisten the gum tragacanth with half the glycerin. (Add a few drops more glycerin if needed to make a runny paste.) Add gelatin to the water and warm in top of double boiler. Stir until gelatin is dissolved, then add to gum tragacanth and stir. Moisten the zinc oxide with the rest of the glycerin and add to the mixture. If lumps form, use an egg beater to blend. The final mixture will be runny when you bottle it. It will thicken overnight—in fact to the point where a vigorous shake is necessary before using.

It is recommended that this mask be warmed before using..

HAIR PRODUCTS

It is certainly true that hair plays a very important part in a persons appearance, but TV commercials have made it the embodiment of everything that is good, beautiful, and sexy. Hair products are thus some of the most marketable products you will make. This section gives a variety of formulas for all types of hair products, shampoos, rinses, conditioners, colorants, etc., permitting you, the amateur chemist, to make an abundance of these popular products.

SHAMPOOS

Shampoos are cleansing agents used for the hair and scalp. Soap is being replaced by synthetic detergents plus various additives to leave the hair so that it may be easily combed and shaped and will be soft, nonsticky, and glossy. Presented are formulas for lotions, creams, and paste shampoos, and some that contain egg, protein, color and anti-dandruff preparations.

Clear Liquid Shampoo

I

		%/wt.
A	Diethanolamine Lauryl Sulfate (35% active)	34.3
	Sodium Chloride	1.0
	Armid C	2.0
B	Water	62.7

Clear yellow liquid. Viscosity of 1500 cp. @ 75°F, pH of 5.5*
Armid builds viscosity, increases and stabilizes foam.

Proc.: Mix ingredients in A together. Heat B then add A to B.

II

	%/wt.
Sodium Lauryl Sulfate or Sodium Lauryl Ether Sulfate	25
CLINDROL 200-L	5
Water	70
Preservative	0.15
Sodium Chloride	0.1–1.0
Acid	to pH 7.5–8.0*

Proc.: Perfume and color may be added, as desired, to the above formulation. The acid in the above formulation may be citric, hydroxyacetic or phosphoric, the latter being extensively used.

Liquid Cream Shampoo

		pts./wt.
A	Water	50
B	Sodium Lauryl Sulfate (28%)	40
	CLINDROL Superamid 100L	5
	CLINDROL SEG	1.5
	Lanolin	1
C	Perfume	q.s. to 100
	Preservative	q.s. to 100
	Hydroxyacetic or Citric Acid	to pH 7.5

High Viscosity Liquid Shampoo

		pts./wt.
A	Water	69
B	Sodium Lauryl Sulfate WA Paste (28%)	25
	CLINDROL Superamide 100L	5
	Preservative	q.s.

C   Perfume   q.s.
    Phosphoric Acid   to pH 7.5–8.0*

Proc.: Heat A. Heat B. Add B to A. Add C.

Pearlescent Shampoo

	pts./wt.
Sodium Lauryl Sulfate	20
Water	q.s. to 100
CLINDROL Superamid 100CG	4
CLINDROL SEG	1.5
Preservative	0.15
Citric or Phosphoric Acid	to pH 7.5-8
Sodium Chloride	q.s. 0-1.0
Color	q.s.
Perfume	q.s.

Proc.: The first 4 ingredients are heated at 160°F. until clear and allowed to cool with slow agitation to develop pearl. The preservative is added at 125°F. Neutralization with acid is conducted below 100°F. Salt may be used to adjust viscosity and it is added as a solution using a portion of the water called for in the formula. CLINDROL Superamid 100CG may be used to solubilize the perfume.

Should a clear shampoo be desired, CLINDROL SEG is omitted from the above formula.

Protein Shampoo

	pts./wt.
Water	39.56
Na Lauryl Ether Sulfate (28%)	42.5
Lauryl Diethanolamide	5.0
WSP X-250	5.0
Citric Acid Monohydrate	0.14
TEGOSEPT M	0.2
TEGOSEPT P	0.05
TEGOBETAINE C	7.5
BRONOPOL	0.05

Proc.: Weigh the ingredients together excluding the Tegobetaine C and Bronopol. With agitation heat the mixture enough to melt the lauryl diethanolamide and dissolve the other materials. Cool the resulting solution to room temperature. Add the Tegobetaine C with the Bronopol dissolved in it. Considerable thickening will occur upon addition of the Tegobetaine C.

Non-Irritating Lanolin Enriched
All Family Shampoo

	%/wt.
Water	65.6
Sodium Chloride	1.0
Ethoxylated Lanolin (50%)	1.0
70% Lauric Diethanolamide	
(High Activity)	2.0
Standapol T	15.0
Standapol SH-300	
Preservative	0.2
Perfume	0.2

Proc.: Use the above order of addition. No heat required except to melt the Lauric Myristic Superamide. Viscosity can be modified by variation in NaCl content. Adjust pH to 6.5—7.0 with 50% Citric Acid.

Cream Lotion Shampoo

	pts./wt.
"Ultrawet 60L" (anionic)	33
Glyceryl monostearate	3
Water	64
"Carbopol 934"	1
Triethanolamine	1
Perfume	q.s.

Proc.: Carefully disperse "Carbopol" in water which has been heated to 70°C. Add "Ultrawet" and triethanolamine with gentle stirring to avoid foaming. Add glyceryl monostearate as a melt and carefully disperse this in the mix, cool the cream to 30°C. Add per me at 45-50°C.

Simple Oil Shampoo

Shampoo advertisements have conditioned us to expect that the excellence of a shampoo is directly proportional to the amount of suds it forms. Suds, television commercials notwithstanding, are not the whole show.

The thing to remember about a shampoo such as this one is that you need to rinse well. It is made primarily of sulfonated oils which are effective detergents, and which can act as conditioning agents.

Sulfonated Olive Oil	1½ qts.
Sulfonated Castor Oil	1½ qts.
Distilled Water	6 qts.
Color (food coloring), as desired	
Perfume, as desired	

Proc.: Stir together until well blended. Bottle.

"Natural" Coconut Oil Shampoo

		pts./wt.
A	Neo-Fat 265	13.96
	LANTROL	0.50
	Coconut Oil	0.50
	Cerasynt 840	5.00
	Vitamin E (Alpha Tocopherol)	0.02
B	Triethanolamine	2.19
	Glycerine	5.00
	Potassium Hydroxide Pellets	4.21
	Deionized Water	38.64
	Methocel 60 Hg 4,000 cps (370 aqueous solution)	30.00

Proc.: Add Methocel solution to water first to form thick solution. Add rest of B into solution. Heat both A and B to 80°C. and add A to B with stirring. Cool to secure heavy viscosity.

"Natural" Olive Oil Shampoo

		pts./wt.
A	Neo-Fat 265	13.96
	LANTROL	0.50
	Olive Oil	0.50
	Cerasynt 840	5.00
	Vitamin E (Alpha Tocopherol)	
B	Triethanolamine	2.19
	Glycerine	5.00
	Potassium Hydroxide Pellets	4.21
	Deionized Water	38.64
	Methocel 60 Hg 4,000 cps (3% aqueous solution)	30.00

Proc.: Add Methocel solution to water first to form thick solution. Add rest of B into solution. Heat both A and B to 80°C. and add A to B with stirring. Cool to secure heavy viscosity.

Egg Shampoo

		pts./wt.
A	Water	73
B	Sodium Lauryl Sulfate (30%)	20
	CLINDROL Superamide 100Cg	4
	CLINDROL SEG	1
	Preservative	q.s.
	Phosphoric Acid	to pH 7.5–8*
	Sodium Chloride	q.s.
	Yellow Dye	q.s.

Egg – whole or powdered	2
Perfume	q.s.

Proc.: Heat A, Heat B, Add B to A. Add C.

The above formula may also be used for a pearlescent shampoo by omitting the egg and yellow dye. The color depends on the shade desired. Perfume depends on the strength of the smell.

Paste Shampoo

I

A soapless paste shampoo may include the following ingredients:

	%/wt.
"Standapol" SHC-101	80
"Standamid" LP	10
90/95 Oleyl/Cetyl Alcohol	5
Ethylene Glycol Monostearate	5

Proc.: Heat "Standapol" SHC-101 to 175°F. Under constant agitation, add chunks of "Standamid" LP slowly or premelt it at 130°F. Add the fatty alcohol. Add the flakes of EGMS slowly while maintaining agitation and temperature. When cooled to 130°-140°F., pour into containers.

II

A paste shampoo suitable for packaging in jars or collapsible tubes is suggested:

	%/wt.
"Maprofix" WAC	40.00
"Synotol" LM 90	4.00
"Neobee" M5	1.00
"Wecoline" 1892	5.00
Sodium Hydroxide	0.75
"Veegum" HV	0.50
Ethylene Glycol Monostearate	3.00
Water	45.75

Proc.: Disperse "Veegum" thoroughly (45 minutes) in water at 180°F. Add the "Maprofix" WAC, followed by the sodium hydroxide and the remaining ingredients. Cool with stirring and add the perfume at 120°F. Continue stirring and cool to room temperature.

Cream Shampoo

I

	pts./wt.
White Mineral Oil (0.860)	43.0
White Beeswax	3.0
Stearic Acid	2.4

Triethanolamine	1.2
Glyceryl Monostearate	0.2
Cologne Compound (33% in *Carbitol*)	1.5
Distilled Water	48.7

Proc.: The first three substances and the glyceryl stearate are heated to about 80°C. and the triethanolamine and water, at the same temperature, are then poured in with moderate stirring, which is continued until the batch cools. The perfume dissolved in a little *Carbitol* (diethylene glycol monoethyl ether) is then stirred in. This formula gives a cream stable to a wide range of temperature changes, and does not call for excessively vigorous stirring or homogenization.

II

	pts./wt.
Stearamide	1.5
Soap Chips (Castile)	5.0
Mineral Oil (0.860)	26.0
White Beeswax	1.5
Distilled Water	66.0

Proc.: The ingredients of this emulsion are heated together, stirred until cool, and then preferably homogenized. Apart from the additional stability that stearamide imparts, it also serves to facilitate subsequent shampooing, owing to its stability to reemulsify mineral oil residues if still present in sufficient amount.

Dry Shampoo Powder

	%/wt.
Cocoanut Oil Soap Powder	30
Sodium Carbonate Monohydrated	45
Borax	25
Henna Leaves Powder	trace
Aniline Yellow	trace
Perfume	to suit

Proc.: Mix together and sift. Keep in closed containers.

Dandruff Control Shampoo

I

		%/wt.
A	Loramine SBU 185	2.0
	Lanogel 41	1.0
	Orvus K Liquid	30.0
	Perfume	0.5
B	Allantoin	0.2
	Distilled Water	66.3

Proc.: Heat A to 40°C, heat B to 40°C also. Add A to B and mix well.

II

		%/wt.
A	Dermodor Floranol 9235	1.0
	Loramine DU. 185	1.0
	Synotol LM 60	2.0
	Duponol XL	40.0
	Triton X100	1.0
B	Allantoin	0.2
	Water	54.8

Proc.: Heat A and B to 40°C and with gentle stirring add B to A.

Dandruff Remover and Scalp Stimulant

Absorption Base (Falba)	3 oz.
Pilocarpine Hydrochloride	20 gr.
Colloidal Sulfur	132 gr.
Water	½ oz.

Proc.: Melt the absorption base. Dissolve the pilocarpine hydrochloride in hot water, and pour it slowly, with constant stirring, into the molten base. Stir in the colloidal sulfur, and continue to stir until the mass is cold.

Temporary Color Shampoo

		pts./wt.
A	Maprofix TLS 500	80.0
	Superamid L9	10.0
	Loramine CJA	2.0
	Tween 20	6.0
	Potassium Sorbistate	0.4
B	Water	97.6
C	Perfume Oil	4.0

Proc.: Heat A. Heat B. Add B to A. Add C.

This shampoo contains an antiseboric agent and is effective in the reduction of itching, scaling and dandruff. This shampoo, made with a mild neutral detergent, has a hair conditioning effect and hair combing ability.

The following temporary colors may be obtained by adding 1—2% of these formulas to the above shampoo:

Blond

	%
FD & C Yellow 5	40

| FD & C Yellow 6 | 40 |
| D & C Brown 1 | 20 |

Mid-Brown

	%
FD & C Blue 1	35
Blond Mixture	25
D & C Orange 4	20
FD & C Red 2	20

Black

	%
D & C Black 1	65
FD & C Red 4	15
FD & C Brown 1	10
FD & C Violet 1	10

HAIR CONDITIONERS

Hair conditioners are used after shampooing to deposit a film of conditioning agents on the hair fibers. They are applied, allowed to remain on the hair for a short period of time and then rinsed out with warm water. Conditioners help restore elasticity to the hair and prevents dryness and brittleness of hair fibers.

Hair Conditioner

	%/wt.
Orcol 2000 Mono-Oleate	25.00
Alcohol #40	54.74
Water	20.00
Sod. Carbonate	0.01
S-7	.25
Perfume	.50

Proc.: S-7 is a fungicide and germicide and is bis (2 hydroxy-5 chloro-phenyl-sulfide).

Orcol, is mono-ester of oleic acid and propylene glycol.

Heat Orcol and water, sodium carbonate and S-7. Then mix with alcohol. Add the perfume, and filter.

Protein Hair Conditioner

A typical protein-type hair conditioning formula is as follows:

	%/wt.
Gum Karaya	2.0
Protein Hydrolysate	50.0
Multi Sterol Emulsifier	5.0
Color, Perfume and Preservative	q.s.
Water	43.0

Proc.: Heat ingredients together. Then add color and perfume.

The above product is diluted with water before use, in the ratio of 1:4 or 1:8.

Rinses

Protein Cream Hair Rinse

		%/wt.
A	Miranol S H D Conc.	4.5
	Cerasynt 945	2.0
B	Polypeptide L.S.M. (40%)	10.0
	Tegosept P	0.2
	Water	83.3

Proc.: Melt Miranol S.H.P. and Cerasynt together to 85°C. Separately combine B ingredients and heat to 85°C, add B to A, slowly at first until a smooth white dispersion forms, then more rapidly. Cool to 30°C with sweep agitation.

Cream Rinse for Hair

	pts./wt.
Triton X-400	16.0
Diglycol Laurate	0.64
Acetic Acid	0.64
Water	110.72
Perfume and Color	to suit

Proc.: Mix the acid with the water and heat to 65°C. Disperse the Triton X-400 together with the diglycol laurate in the acidulated water, with constant stirring. Stir in slowly the perfume and color and stir until cold.

HAIR SPRAYS

Hair Sprays

I

	pts./wt.
Luviskol VA 37 or P.V.P./V.A. 1335	6
Aerosol Perfume	0.3
Methylene Chloride	30
Isopropyl Alcohol	63.7

	A.	B.
Base as above	50	30
Propellant 11/12 (25/75)	50	
Propellant 11/12 (50/50)		70

Proc.: The use of isopropyl alcohol and methylene chloride will be noted. The film-forming agent is dissolved in the alcohol and the other ingredients are then added.

II

	pts./wt.
Plastocrex D.M.H.F. Resin	5
Lanolin, Alcohol-Soluble	0.5
Plasticiser (e.g., Glycerin,	
Propylene or Triethylene Glycol)	0.5
Perfume	1
Alcohol	93

Charge for Aerosol Pack

40% Concentrate (as above)
60% Propellant

Proc.: Dissolve the Plastocrex in the alcohol and then add all other ingredients except the propellant. Warm slightly, allow to stand overnight and filter. Add propellant.

Lanolin Hair Sprays

I

	pts./wt.
Shellac, bleached, dewaxed, rosin-free	1.50
Polyethyleneglycol 400 Monolaurate	0.30
Lanolin (anhydrous)	0.25
Perfume	0.45
Alcohol, anhydrous	37.50
Propellant	40.00

The shellac need not be bleached but must be completely de-waxed for reasons of solubility.

II

	pts./wt.
PVP	5.0
Polyethyleneglycol 400 Monolaurate	0.5
Liquid Lanolin	0.5
Perfume	1.0
Alcohol, anhydrous	93.0
Propellant	150.0

Hair Spray for Men

	%/wt.
PVP/VA E 735 Copolymer 50% EtOH	
solution	3.0
SOLAN	0.5
Isopropyl Lanolate	0.3

Ucon 50HB 660	1.5
Anhydrous Ethanol	24.4
Perfume	0.2
Propellant 11	45.5
Propellant 12	24.5

Proc.: Mix all ingredients together. Add the perfume. Charge with the propellant.

Hair Lacquer (Non-Aerosol)

	pts./wt.
Refined (wax free) Bleached Shellac	15.00
Borax	3.45
Water	81.00

Proc.: Heat water to 145°F and add the borax. Add shellac, dissolving it at 145°F using a high speed stirrer. When the shellac is dissolved, cool and filter. Adjust pH with ammonia to about 8.5. This basic formula is compounded or reduced to the desired solids to give either an all water hair lacquer or water alcoholic hair sets.

Hair Lacquer Water Based Non-Aerosol

	pts./wt.
Basic Formula (see above)	80.0
Citroflex A-2	1.0
Perfume	0.2
Water	18.8

Proc.: Mix all ingredients together.

Alcohol-Based Hair Set

	%/wt.
Refined Wax-Free Bleached Shellac	8
PVP/VA E-735 (50% solids)	2
Plasticizers	0.6
Lanolin or Lanolin Derivatives	
(Ethylan Ethoxylan 100,etc.)	0.2
Perfume	0.1
Alcohol SD #40	q.s.

Proc.: Dissolve shellac in alcohol using high speed stirrer. Add PVP/VA, then lanolin and plasticizers. Add perfume last.

NOTE: Solids can be either increased for stiff spray or lowered for soft type.

Clear Gel Hairdressing

I

		%/wt.
A	Mineral Oil, Marcol 70	13.7
	BRIJ 97	15.5

ARLATONE G	15.5
Propylene Glycol	8.6
SORBO	6.9
B Water	39.8
C Perfume	q.s.

Proc.: Heat A to 90°C and B to 95°C. Add B to A with moderate agitation. Add C at 70°C. Pour at 60°C. A clear gel forms on cooling.

II

	%/wt.
A Mineral Oil	10.0
BRIJ 97	20.7
ARLATONE G	10.3
Propylene Glycol	8.6
SORBO	6.9
B Water	43.5

Proc.: Heat (A) to 90°C and (B) to about 95°C. Add (B) to (A) with moderate agitation. Pour at 60°C. A clear gel forms on cooling.

Cream Hair Dressing O/W

I

	%/wt.
Oil Phase:	
Surface Active Extract of	
Lanolin Alcohols (18)	6.0
Mineral Oil, 70 vis	25.0
Petrolatum USP White	5.0
Microcystalline Wax 170°F mp	5.0
Aluminum Stearate	1.0
Sorbitan Sesquioleate	0.2
Water Phase:	
Glycerin	5.0
Water	52.8
Perfume and Preservative	q.s.

Proc.: Add the aluminum stearate slowly to the mineral oil while mixing to prevent lump formation. Continue stirring and heat until clear and stringy. Add the remaining oil phase ingredients and maintain heat until wax is melted. Stir intermittently while cooling to 55°C. Add the water phase at 55°C to the oil phase at 55°C with moderate stirring. Mix slowly while cooling to below 30°C. Remix one day later.

II

	%/wt.
Oil Phase:	
Lanolin Alcohols Ethoxylate (16 mo EO)	3.0
Surface Active Extract of Lanolin Alcohols	5.0
Acetylated Lanolin Alcohols (liquid fraction)	2.0
Glyceryl Monostearate SE	5.0
Water Phase:	
PVP Polymer	3.0
Glycerin	1.0
Purified Water	48.5
Carboxyvinyl Polymer 3.3% slurry	30.0
Protein Derivative	1.5
Triethanolamine	1.0
Preservative	q.s.

Proc.: Thoroughly mix the carboxyvinyl polymer in water at room temperature to make a 3.3% slurry. Continue mixing until all particles are gone, to form a uniform cloudy dispersion. Heat the oil phase and water phase separately to 70°C. Add the water to the oil and mix with moderate agitation. When all of the water phase has been added, add the triethanolamine. Continue mixing while cooling. Perfume below 35°C.

Alcoholic Hair Dressing

	oz.
Odorless Castor Oil	23.04
Deltyl Extra	8.96
Alcohol	96.00
Perfume	to suit

Proc.: Dissolve the castor oil in the alcohol, add the Deltyl extra and the perfume, and shake well. Allow the mixture to stand at least 24 hours.

Hair Groom

I

	%/wt.
Mineral Oil	23.0
Ethoxylated Hydrogenated Lanolin (20 Ethylene Oxide Units Per Mole)	14.0
Ethoxylated Oleyl (7:3) Cetyl Alcohol 4 Ethylene Oxide units per mole	10.0
2-Ethyl-1, 3-Hexanediol	1.0

Preservatives, Perfume, color and
distilled water   q.s.

Proc.: Heat first 4 ingredients to 170°F. Add color and water at same temperature. Cool to 140°F, add perfume, stir and pour.

II

Flaxseed (Whole)	1 lb.
Boric Acid	2 oz.
Glycerin	12 oz.
Water	1 gal.
Color and perfume to suit.	

Boil the flaxseed with the water until syrupy; filter by squeezing through a linen bag. Discard the residue. Add the boric acid and glycerin (above) to the liquid.

Hairdressing & Conditioner

		%/wt.
A	Solulan 98	2.0
	Amerchol L101	4.0
	Stearic Acid XXX	4.0
	Span 60	1.0
	Neobee M5	5.0
B	Propylene Glycol	5.0
	Triethanolamine	1.5
	Water	76.8
	Allantoin	0.2
	Methyl Paraben	0.2
C	Carlin H-8004	0.3

Proc.: Heat A and B to 75°C. With rapid agitation add B to A. Add C at 40°C. Stir again the next day.

Clear Hair Stick

		%/wt
A	Polyamide Resin S43	18.0
	Castor Oil, cosmetic grade	72.3
	Cocoanut Diethanolamide	2.0
	"Volpo" 3	1.0
	Hydrogenated Lanolin	0.7
B	Isopropyl Myristate	3.0
	Perfume	3.0
	Anti-Oxidants, Dyes, Pearl Essence, etc.	q.s.

Proc.: Melt A at 110-120°C. while stirring. Combine B ingredients, heat to 80-90°C., then add to A. Pour into molds when clear and when clear and uniform, then cool. After cooling, remove sticks from mold and set aside at room temperature for about an hour.

The stick should be flamed to make it clearer. This preparation is also quite adaptable as a treatment or a fragrance vehicle.

Brilliantine

I
Oil

	%/wt.
Liquid Petrolatum	32–40
Sesame Oil	10–32
Isopropyl Alcohol	q.s.

Proc.: Mix, perfume, and color as desired.

II
Blue

	%/wt.
Mineral Oil (Water-White, 0.880)	75
Deodorized Kerosene	25
Oil Blue (Dye)	q.s.

Proc.: This type of brilliantine can give an excellent burnish and corrective sheen to ash-blonde and yellowish-gray hair.

III
Jelly

	pts./wt.
Sesame Oil	90
Lanolin	10
Spermaceti	42
Paraffin Wax	6
Beeswax	18
Mineral Oil	300
Cholesterol	6

Proc.: Melt all ingredients together. This may be perfumed and colored with oil-soluble green.

Pomade

		%/wt.
A	Amerchol CAB	15.0
	Microcrystalline Wax	15.0
	Mineral Oil, Heavy	66.0
	Isopropyl Myristate	2.0
	Castor Oil	2.0
B	Color	to suit

C Bergamot Oil   q.s.

Proc.: Heat ingredients in A. Grind color (B) and add to A. Add C.

Hair "Tonics"

I

	%/wt.
Isopropyl Alcohol	70.0
Propylene Glycol	5.0
Eau de Cologne	5.0
Cholesterol	0.5
Perfume	0.5
Distilled Water	19.0

Proc.: Dissolve all the ingredients except water in isopropyl alcohol, add the water last.

II

Gum Benzoin	2 dr.
Castor Oil	4 oz.
Alcohol	1 qt.

Shake well together, then add

Lavender Oil	1 dr.
Bergamot Oil	1 dr.
Clove Oil	30 drops
Rosemary Oil	30 drops
Lemon Oil	30 drops
Neroli Oil	30 drops
Tincture of Cantharides	½ oz.

Proc.: Shake well to cut the oils.

Hair Tonic—Dry Scalp

Castor Oil	1 gal.
Crude Carbolic 30%	8 oz.
Cresol U.S.P.	3 oz.
Lignol	1 gal.
Soya Bean Oil	2 gal.
Precipitated Sulphur	2 oz.

Proc.: Mix the soya bean oil, the castor oil heat to 100°F. and add the lignol. Take a small quantity of this mixture and rub up precipitated sulphur into a smooth paste. Mix with rest of oils. Add carbolic and cresol.

Dry scalp is often a diseased condition, accompanied by dandruff. Often it is caused by poor circulation of blood. Above preparation should be rubbed into scalp at night, and, because odor is obnoxious, shampooed out in morning. Label should contain a statement to the effect that the longer the preparation is left on the better the results will be.

Hair Straighteners
Caustic Type Straightener

	pts./wt.
Gum Tragacanth	2
Boric Acid	1
Water	40

Make a uniform paste and stir in previously dissolved:

Sodium Carbonate	1
Potassium Hydroxide	1
Glycerin	2
Water	8

Proc.: Apply the paste to hair with combing. Allow to remain for ½ hour. Then wash well with water to remove all paste from the hair.

"Thio type" straighteners may contain up to about 9% thioglycolic acid at pH 9.5 to 9.6[*]

Hair Straightening Cream

	%/wt.
Palm Oil	5.0
Sodium Carbonate	3.0
Sodium Hydroxide	6.0
Starch	3.0
Water	83.0

Proc.: Mix ingredients until well blended.

Hair Pressing Oil
I

		%/wt.
A	Mineral Oil, Heavy	15.0
	Modulan	5.0
	Petrolatum	80.0
C	Perfume	q.s.
B	Color	To suit

II

		%/wt.
A	Mineral Oil, Heavy	20.0
	Modulan	5.0
	Amerlate P	5.0
	PEG 400 Monostearate	5.0

Silicone 556 Fluid	3.0
Petrolatum	62.0
B Color	To suit
C Perfume	q.s.

Proc.: Heat ingredients in A. Grind B and add to A. Add C.

PERMANENTS

Cold Wave Lotion

	pts./wt.
A Polyoxyethylene Cetyl Alcohol	
(Texofor A6)	7.5
Mineral Oil, White, sp. gr. 0.83	3.0
B Ammonia Water	q.s. (3–4.5)
Water	105.0
Ammonium thioglycolate 35%	30.0

Proc.: Heat A to 65°C. Heat B to the same temperature and add it to A with constant stirring. When the emulsion has cooled, stir in the thioglycolate, and continue stirring until it is cold.

Permanent Wave Solution

	oz.
Triethanolamine	14.72
Sodium Sulfite	1.92
Borax	5.12
Sodium Carbonate	2.56
Water	103.68

Proc.: Mix the triethanolamine with the water, add the powdered chemicals, and shake well until solution is complete. Perfume and color may be added if so desired.

Permanent-Wave Fluids

I

Sodium Carbonate (powdered)	8 oz.
Sodium Hyposulfite (powdered)	6 oz.
Water (distilled)	to make 1 gal.
Morpholine	6 fl. oz.
Castor Oil (sulfonated)	1 fl. oz.
Perfume Oil and Color	to suit

Proc.: Dissolve the sodium carbonate and sodium hyposulfite separately, each in 3 pt. of water. Add the morpholine to the sodium carbonate solution. Then mix the two solutions and stir in the sulfonated castor oil, which has been previously mixed with the perfume oil. Color to suit with a trace of caramel or certified cosmetic color. Do not overcolor.

This waving fluid is suitable for both spiral and croquignole waves. Contains no caustic alkalies and may be safely used on all colors and qualities of hair.

II

Triethanolamine	16 fl. oz.
Sodium Hyposulfite	6 oz.
Water	to make 1 gal.
Perfume	to suit

Proc.: Dissolve the sodium hyposulfite in ½ gal. of water. Add the triethanolamine and then make up to 1 gal. with water. Add any desired perfuming oils.

This is an ideal product where a soft deep wave instead of a tight curl is wanted.

HAIR COLORANTS

Considered a concomitant of old age, grey hair is a liability to the men in business. Statistics indicate an increasing number of men using hair coloring. Metallic dyes, so-called color restorers, account for the major portion of men's hair colorants on the market.

Hair Darkener

I

	pts./wt.
Precipitated Sulfur	20
Glycerin	40
Lead Acetate	15
Perfume Oil	5.0
Water q.s.	1000

Proc.: Dissolve the perfume oil in glycerin. Dissolve lead acetate in water; add sulfur and perfume oil-glycerin. Mixture: The mixture must be well shaken before bottling and use. After application, clean hands.

II

	pts./wt.
Lead Acetate C.P.	6
Glycerin	50
Sodium Thiosulphate (Sodium Hyposulphite)	18
Alcohol #40	60
Perfume Oil	5
Water (dist.) q.s.	1000

Proc.: Sodium thiosulphate is dissolved in about 300 cc of water.

The lead acetate is dissolved in about 300 cc of water mixed with glycerin and this solution is poured slowly, with stirring, into the solution of sodium hyposulphite. The alcohol and the perfume oil are then added and the mixture is filled up with water to the required volume.

Similar ingredients are used in the manufacture of hair darkening creams, the lead acetate and sulfur being incorporated in a suitable emulsion of mineral oils.

Hair Bleach

	pts./wt.
Hydrogen Peroxide Solution, 3% U.S.P.	124.29
Alkyl Ethyl Morpholinium Ethosulfate	1.92
Adipic Acid	1.03
Sodium Stannate	0.76

Proc.: Mix thoroughly until solution is complete. This acid bleach is good for partially decolorizing the hair. The pH value of this lotion is 4–4.5.

"Henna" Shampoos

Soapy Base

Transparent Soft Soap	60 g.
Potassium Carbonate	3 g.
Distilled Water	450 g.
Alcohol	50 cc.

Dyeing Shampoo
Chestnut

Toulyene Solution	7 g.
Pyrogallol Solution	2 g.
Soapy Base as above	15 g.
Logwood Solution	2

Developer for 25 g. of liquid dye: 3 tablets of urea peroxide.

Dyeing Shampoo in the Blond–Light Chestnut range

Toluylene Solution	5 g.
Ammonia (0.97)	2 g.
Pyrogallol Solution	4 g.
Soapy Base	15

Developer for 25 g. of liquid dye: 1 to 2 tablets.

Proc.: Mix all the ingredients together until uniform and add the hair color.

Hair Dyes

I
(Jet Black)

No. 1 (Color)
Toluylene solution   20 cc.
No. 2 (Developer)
6 tablets of 0.5 g. urea peroxide
Contact   1 hour

II
Spanish Black

No. 1 (Color)
Toluylene solution   10 cc.
No. 2 (Developer)
5 tablets dissolved in:
   Water   10 cc.
Contact   40 minutes

III
Brown

No. 1
Toluylene Solution   5 cc.
Water   5
No. 2 (Developer)
5 tablets dissolved in:
   Water   10 cm
Contact   30/40 minutes

IV
Chestnut

No. 1
Toluylene Solution   5 cc.
Pyrogallol Solution   2.5 cc.
Ammonia (0.97)   2.5 cc.
No. 2
4 tablets dissolved in:
   Water 10 cc.   10 cc.
Contact 30 minutes   30 min.

V
Gold Chestnut

No. 1
Toluylene Solution   5 cc.
Logwood Solution   5 cc.

No. 2
3 tablets dissolved in:
Water   10 cc.
Contact   ½ hour

VI
Dark Auburn

No. 1
Toluylene solution   10 cc.
No. 2
Permanganate solution   10
    Contact   40 minutes

VII
Gold Auburn

No. 1
Toluylene Solution   5 cc.
Pyrogallol Solution   2.5 cc.
Ammonia (0.97)   2.5 cc.
No. 2
Permanganate Solution   10 cc.
    Contact   30/40 minutes

VIII
Blond A

No. 1
Toluylene Solution   2 cc.
Pyrogallol Solution   2 cc.
Ammonia (0.97)   2 cc.
Water   4 cc.
No. 2
2 tablets, dissolved in:
Water   10 cc.
Contact   20 minutes

IX
Blond B

Toluylene Solution   3 cc.
Pyrogallol Solution   3
Ammonia (0.97)   2
Water   2
No. 2
2 tablets, dissolved in:
Water   10 cc.
Contact   ½ hour

Proc. I–IX: Mix ingredients together to form cold solutions. Add colors as necessary for desired results.

FACE MAKE-UP PREPARATIONS

Lipstick

I

	%/wt.
Beeswax, White	33.0
Cetyl Alcohol	12.0
Sesame Oil	20.0
Castor Oil	29.0
Perfume Oil	2.0
Tetrabromfluorescein	4.0

Proc.: Dissolve the tetrabromfluorescein in castor oil; Melt beeswax, cetyl alcohol, sesame oil together. Mix with the bromo/castor oil solution. Add the perfume.

II

	%/wt.
Cocoa Butter	7.0
Cetyl Alcohol	3.0
Paraffin	9.0
White Beeswax	22.0
Cholesterin Absorption Base	22.5
Petrolatum	18.0
Perfume	0.9
Preservative*	0.1
Bromo Acid	2.5
Butyl Stearate	5.0
Color	10.0

Proc.: Heat the butyl stearate to 160 F. and dissolve the bromo acid in it. Melt the waxes and fats, mix in the bromo acid solution. Then incorporate the color pigment, preservative and perfume. Mix, mill and case into molds.*

*Paraminobenzoaic Acid

Super Moisturizing Lipstick Base

	%/wt.
Candelilla Wax	5.00
Yellow Ozokerite	3.50
Carnauba Wax	2.00
LANFRAX	2.00
Paraffin Wax	2.00
Beeswax	1.00
Arlacel 20	1.00
HYDROXYOL	3.00
Tenox 2	0.10

Myristyl Lactate	2.00
DISTILLED ISOPROPYL LANOLATE	5.00
Oleyl Alcohol, Deodorized	12.00
Castor Oil	61.40

Proc.: Heat until all ingredients are melted. Mix and pour into molds.

ǀ

The Brush-on Lipstick which follows is a modification of a basic creamy lipstick softened for application to the lips by brush. It is generally marketed in the form of patties which are hot poured into metal pans or directly into small styrene, nylon, tyril or polypropylene compacts. The formulation shown is a heavily pearled, rich, emollient cream.

Frosted Brush-On Lipstick

	pts./wt.
CRODACOL C	3.5
POLYCHOL 5	10.0
NOVOL	8.0
LIQUID BASE	15.0
Candelilla wax	7.5
Carnauba wax	3.0
Neobee M-20	26.3
Natural pearl essence in castor oil	20.0
D&C Red #21-Bromo acid	1.5
D&C Red #27-Lake #3127	4.2
Perfume	1.0

Proc.: Dissolve the bromo acid in the POLYCHOL and NOVOL then disperse the Red #27 Lake and mill until finely dispersed. Combine with the remaining ingredients at about 90-95°C and stir slowly, avoiding entrapping air, until all the ingredients are molten and uniform. Begin cooling mass to about 70°C and add perfume. At 60-70°C, or just before the point at which the mass viscosity becomes too thick to handle, pour into pans or compacts.

Lip Gloss

	%/wt.
Carnauba Wax	3.60
LANTROL	8.00
Propyl Paraben	0.10
ACETOL	3.00
Ozokerite, 170°F	7.00
Candelilla Wax	6.00
Camphor	0.10
Castor Oil	72.20

Proc.: Heat to 70-80°C. Mix slowly and pour into molds.

Lip Pomade

	%/wt
Spermaceti	7.0
Soft White Petrolatum	40.0
Glyceryl Monostearate N.F.	26.0
Noebee M5	14.8
Robane	10.0
Allantoin	0.2
Perfume oil	2.0

Proc.: Melt the first five ingredients. Then add the Allantoin and mix well. Add perfume and pour into molds.

ROUGE

Paste Rouge

I

	%/wt.
Cetyl Alcohol	3.0
Coca Butter	3.0
White Beeswax	3.0
Spermaceti	5.0
Petrolatum, Short Fiber, White	77.0
Cosmetic Lake Color	8.0
Preservative	0.1
Perfume	0.9

Proc.: Melt, mix and mill, adding perfume last.

II

	%/wt.
White Beeswax	12.5
Benzionated Lard	10.0
Petrolatum	68.29
Amiline Violet	0.2
Cosmetic Lake Color	8.0
Perfume	1.0
Preservative	0.01

Proc.: Melt, mix and mill, adding perfume last.

Gel Cheek Rouge

	%/wt.
1. Deionized Water	33.74
2. Carbopol 934 (2.5% solution)	40.00
3. Sequestrene NA4[22]	0.01

4. FD & C Red #2 (2.5% solution)	5.00
5. Glycerine	16.00
6. ETHOXYOL 40	4.00
7. Methyl Paraben	0.15
8. Propyl Paraben	0.10
9. Triethanolamine 85-87%	1.00

Proc.: Dissolve 2 and 3 into 1 and add 4. Mix well. Dissolve 6, 7 and 8 into 5. Add this to previous mixture at room temperature. Slowly add 9 while gently mixing.

MAKE-UP

Liquid Make-up

I

	%/wt.
A 585 Cosmetic Liquid	5.00
Marcol 72	4.00
Propylene Glycol Monostearate	0.50
Stearic Acid, triple pressed	2.00
Triethanolamine	1.00
Propyl p-hydroxybenzoate	0.10
Lanolin USP	3.40
B Titanium Dioxide	3.00
Talc	5.00
Iron Oxide to shade	1.25
C Water, Deionized	67.50
Veegum HV	0.55
D Propylene Glycol USP	5.00
Sodium Lauryl Sulfate	
(28% in Water)	1.00
CMC 7HF	0.25
Methyl p-Hydroxybenzoate	0.20
E Perfume Oil	0.25

Proc.: Pre-blend B. Heat A to 70°-75°C. Stir C for 15 to 20 minutes. Add D to C. With stirring, heat to 60°-65°C. Add B to Parts C and D with stirring. Heat to 65°-60°C with stirring. Add A with stirring. Cool to 45°-50°C with stirring. Add E with stirring. Cool to 25°-30°C with stirring. Homogenize and package.

II

	pts./wt.
Talcum Powder	50.0
Cellocize 10% Sol.	5.0

Glycerine	2.5
Water	25.0
Perfume	q.s.

Proc.: Mix all ingredients together then add perfume.

Lotion for Blotched Skin

Orange Water (or Rose Water	2 fl. oz.
Extract of Witch Hazel	1 fl. oz.
Magnesium Sulphate	½ oz.
Borax	¼ oz.

Proc.: Mix all ingredients together until uniform.

(Wash face with soap and hot water, then apply lotion at bedtime.)

Cover Paint for Skin Discolorations

So-called cover paints used to conceal and blend the white, pigment-free skin spots characteristic of vitigilo (acquired leukoderma) can also be used to temporarily cover other smooth skin blemishes, scars or marks. Moreover, such products, if properly made, can also be useful to soften and blend the often sharp sunburn lines so frequently seen at the end of the summer season.

Zinc Oxide	45 g.
Prepared Calamine	45 g.
Glycerin	4–16 cc.
Rose Water, To make	500 cc.

To this is added, drop by drop, sufficient ˈchthyol to cause the paint to match or blend with the surrounding skin. Usually from 10 to 60 drops are needed, and the addition should be made carefully, since success of the paint depends on the closeness of the match obtained. Face powder may be applied after using this concealing paint.

Pigmented Make-up Base

	%/wt.
Gel Base	32.0
Arlacel 186 (1), 1 part	
Sorbo (2) 9 parts	
Mineral Oil	10.0
Beeswax	1.5
Ceresin Wax	1.0
Titanium Dioxide (pigment)	20.0
Water	35.5

Proc.: Prepare intermediate gel. Blend and heat together to 70 C. all ingredients except the pigment and water. Mix to a uniform paste. Blend in the pigment* and mix until uniformly dispersed.

Heat the water to 72 C. and mix into the blend thoroughly. Stir until cool.

Preparation of gel base: Add small amounts of Sorbo to the arlacel 186 and form a thick slurry by using mechanical agitation. Then add remainder of Sorbo slowly with agitation.

*Instead of titanium dioxide, Lake colors may be substituted.

Creamy Make-up

		%/wt.
A	585 Cosmetic Liquid (a)	24.25
	Parmo 10 (a)	6.00
	White Microcrystalline Wax (m.p. 158°F)	4.00
	Carnauba Wax	3.50
	Lanolin USP	9.00
	Deca-Glycerol Deca-Oleate	2.65
	Tenox II (c)	0.06
	Propyl p-Hydroxybenzoate	0.10
B	Zinc Oxide USP	20.00
	Titanium Dioxide	19.84
	Kaolin	5.80
	Iron Oxide to shade	2.75
	Calcium Carbonate	1.80
C	Perfume Oil	0.25

Proc.: Pre-blend Part B. Heat Part A to 80°-85°C. Stir Part B into Part A. Roller mill* the preparation. Reheat to 70°-75°C with stirring. Add Part C with stirring. Pour at 60°-68°C (just above solidification point). Allow to set overnight.

Make-up Cover Stick

	%/wt.
ACETOL	10.00
ETHOXYOL-5	5.80
Paraffin Wax, 143-150°F.	7.00
Ozokerite, 170°F.	9.00
Beeswax	3.50
Candelilla Wax	1.50
White Petrolatum	16.00
Mineral Oil, 70 visc.	10.00
Oleyl Alcohol	1.00
Methyl Paraben	0.10
Propyl Paraben	0.10
Titanium Dioxide	19.67
C33-8075 (Cosmetic Russet Oxide)	3.95
C33-8073 (Cosmetic Yellow Oxide)	12.38

Proc.: Heat the above with mixing to 85°C. Mix slowly, deaerate and pour into molds at 65°C.

EYE MAKE-UP

Powder Eye Shadows

	I %/wt.	II %/wt.
VEEGUM F	7	7
SNOW GOOSE 400S	60	—
NYTAL 400	—	50
Titanium Dioxide	5	—
Zinc Oxide	—	4
Zinc Stearate	8	11
Kaolin	8	10
Pigments	12	18

Proc.: Mix the ingredients together.

Dip finger in eye shadow or apply with a brush.

NOTE: These eye shadows may be packaged as a loose powder or a pressed cake.

Pressed Powder Pearlescent Eye Shadow

		pts./wt.
A	VEEGUM F	5
	Mearlin AC	35
	NYTAL 400	29
	Zinc Stearate	8
	Magnesium Carbonate	1
B	Acetol	3
	Polysorbate 20	9
	Water	10
	Preservative	q.s.

Proc.: Micropulverize A. Mix B and add to A. Continue to blend in mixer. Screen through a No. 16 sieve. Compress.

Wet brush and rub on cake surface. Apply to eye lid, and allow to dry.

Cream Eye Shadow

		%/wt.
A	CMC P75-M	0.15
	Veegum HV	2.50
	Propylene Glycol	7.00
	Water, Dist. or D.I.	69.65
	Methyl Paraben	0.20
	Duponol C	0.30
	Triethanolamine	0.50
B	Glycerol Monostearate	3.50
	Stearic Acid	1.70

Lanolin, Anhydrous	5.70
Sesame Oil	2.90
Isopropyl Myristate	4.30
Mineral Oil, Lt.	1.40
Propyl Paraben	0.20
Base (above)	70.00
Pearl-Glo pearlescent pigment	10.00
Bi-Lite 20 pearlescent pigment	10.00
Chroma-Lite pearlescent color	10.00

Proc.: Disperse the Veegum and CMC in the propylene glycol. Add the water. Stir to complete dispersion of solids. Add rest of A. Heat to 65°C. Weigh out Part B and heat to 65°C. with agitation. Stir B into A with rapid agitation, to emulsify. Remove from heat. Gently paddle-stir to room temperature.

Allow to set overnight so viscosity may reach equilibrium. May be stored in closed vessel at cool temperatures for 30 days with no adverse changes. Weigh out base. Amount depends on concentration of pearl desired. Slowly sift in pearlescent pigments with rapid agitation, at room temperature. When pigments are completely dispersed, product is ready to be packaged.

Frosted Cream Eye Shadow

		%/wt.
A	VEEGUM	4.3
	Water	63.7
B	Propylene Glycol	1.7
	Modulan[3]	1.7
	Amerchol L-101[3]	5.1
	Petrolatum	8.5
C	Mearlin AC[25]	15.0
	Preservative	q.s.

Proc.: Add the VEEGUM to the water slowly, agitating continually until smooth. Add B to A and heat to 70°C. Mix until uniform. Add C. Mix until pigment is thoroughly dispersed.

Liquid Eye Liner

		%/wt.
A	VEEGUM	2.5
	Water	75.5
B	Polyvinylpyrrolidone	2.0
	Water	10.0
C	Pigment	10.0
	Preservative	q.s.

Proc.: Add the VEEGUM to the water slowly, agitating continually until smooth. Dissolve the polyvinylpyrrolidone in water, using
a little heat. Add B to A and then add C. Mix well.

NOTE: For a product with a soft, cream like viscosity that can be
applied with a brush, increase the VEEGUM content to 3.5%.

Eye Liner

	%/wt.
Carboset 514 (30%)	93.3
Pigment	6.2
Carbopol 961	0.5
Antifoam and preservative	q.s.

Proc.: Mix all the ingredients except the Carbopol; grind with a
mortar & pestle to desired particle size; slowly add Carbopol with
good agitation. The water-resistant film may be removed by peeling
or with soapy water.

Eyebrow Pencil Base

	%/wt.
Soft Paraffin	40.0
Hard Paraffin	42.0
Color, as desired	18.0

Proc.: Melt together all ingredients except perfume. Stir in perfume and package in a suitable container.

Cream Mascara

I

		%/wt.
A	VEEGUM	2.0
	Sodium carboxymethylcellulose	
	(low visc.)	0.1
	Water	47.9
B	DARVAN No. 1	0.2
	Propylene Glycol	1.5
	Water	30.9
C	Beeswax	6.5
	Light Mineral Oil	3.5
	Channel Black	1.0
D	Stearic Acid	1.0
	Carnauba Wax	5.0
E	Morpholine	0.4
	Preservative	q.s.

Proc.: Dry blend the VEEGUM and CMC and add to the water slowly, agitating continually until smooth. Mix A and B and heat to 65-70°C. Mill C, and add to Part D and heat to 70°C. Add Part E to the A and B mixture and then immediately emulsify by adding the C and D mixture, constantly mixing until cool.

This cream mascara formula is designed to be water resistant. It should be applied with a brush and taken off with eye makeup remover.

II

Tegin	36 oz.
Petrolatum, White	13½ oz.
Isopropyl Linoleate	4½ oz.
Oleyl Alcohol	3 oz.
Methyl p-Hydroxybenzoate	131 gr.
Water	75 oz.
Earth Pigment*	15 oz.
Veegum	3 oz.
Perfume oil	to suit

*Only mineral colors or lakes may be used around the eyes; see note under Cake Mascara.

Proc.: Melt the first four components. Dissolve the preservative in the water and heat to the same temperature. Mix the two solutions with constant stirring. Add the Veegum.

Eye Make-up Remover Stick

	%/wt.
LANTROL	30.00
Isopropyl Myristate	26.00
White Petrolatum	27.70
Ozokerite, 170°F.	4.00
Candelilla Wax	10.00
Carnauba Wax	2.00
Propyl Paraben	0.10
Butylated Hydroxy Anisole[2 7]	0.20

Proc.: Heat all ingredients to 85°C. Mix thoroughly and pour into molds at 70°C.

SHAVING PREPARATIONS

Aerosol Shave Cream

I

		%/wt.
A	Carbopol 941 Resin	0.06
	Natrosol 250 HR	0.06

Glycerine	4.00
Triethanolamine	2.90
Water	82.38
Preservative	q.s.
B Amerlate P	0.80
Amerlate LFA	0.80
Solulan 98	1.00
Wecoline 1260A	2.00
Stearic Acid, T.P.	5.50
Monamid 150 LW	0.50
C Perfume	q.s.

	%/wt.
Concentrate	92.0
Propellant	8.0
12/114=40/60	

Proc.: Prepare A by dispersing the Carbopol resin thoroughly in the water using high-speed agitation. Add the Natrosol when the Carbopol slurry is homogeneous. Add the glycerine, preservative and triethanolamine when the Natrosol has dissolved. Heat to 85°C. Add B to A at 85°C while mixing. Continue mixing while cooling to room temperature. Add the perfume and mix well.

Packaging Instructions: Use a Precision 0.018 x 0.080 valve, a Precision foam spout, and a side-seamed, lined can.

	II %/wt.	III %/wt.
A Carbopol 934 resin	0.2	0.1
Water (deionized)	83.05	83.4
Triethanolamine	0.25	0.25
B Stearic acid, T.P.	6.0	7.75
Pegosperse 100S	1.0	1.0
Isopropyl myristate	1.25	—
Mineral oil, light	—	1.25
Polyethylene glycol 600 monostearate	2.0	—
Sorbo (70% solution)	3.0	3.0
C Triethanolamine	3.0	3.0
D Coconut Oil	0.25	0.25
Perfume	q.s.	q.s.

Packaging:

Concentrate	93.0
Propellant	7.0
Genetron 12/114/152A=38/54/8	

Proc.: Prepare A by dispersing the Carbopol resin in water then neutralizing with the amine when slurry is homogeneous. Prepare B and add it to A. Heat to 82°C. Heat Part C to 82°C and add it to blend of A and B. Cool to room temperature, then add Part D. Use continuous agitation.

Concentrate

	pts./wt.	pts./wt.
A	G-E SF-96 (350) Silicone Fluid	0.75
	Palmitic Acid	1.00
	Stearic Acid	3.00
	Cetyl Alcohol	0.25
	TEA Lauryl Sulfate (40%)	3.00
	Lanolin	0.50
	Hexachlorophene	0.25
B	Sorbitol (70%)	2.50
	Triethanolamine	2.00
	Part I Water (Sterile)	76.35
C	Part II Water (Sterile)	10.00
	Aromox DMCD	0.40
	Perfume	q.s.
	Propyl Parasept	q.s.

Proc.: Heat A and B separately to 85°C (185°F). Add B to A slowly and continue mixing until mixture reaches room temperature. Mix ingredients of C together and add to mixture A-B with mild agitation.

Aerosol Loading: Concentrate 96.5%; Isobutane 3.5%.

Aerosol Shaving Cream With Protein

		%w/wt.
A	Carbopol #394	0.15
	Water	80.98
	Triethanolamine	0.25
	Methyl Paraben	0.10
	Butyl Paraben	0.02
	Germall	0.50
B	Stearic acid T.P.	7.00
	PEG Monostearate (600)	2.00
	Sorbo 70%	3.00
C	Polypeptides #37-S	2.00
	Maypon SK	1.00
	Triethanolamine	3.00
D	Perfume M-497	0.50

Heat A and B separately to 80° C, mixing each until all components are dissolved. Add A slowly to B with good agitation, maintaining the temperature. Add C and cool to 40°C. Add D and mix until cool.

Aerosol Formulation: Propellant 114/12 43/57, 8.0%; Above concentrate, 92.0%.

Hot Shave Co-Dispensed Aerosol Lather

	%w/wt.
CRODAFOS N3 NEUTRAL	1.0
FLUILANOL	1.0
HARTOLAN	1.0
LIQUID BASE	0.5
POLAWAX A31	1.0
SKLIRO	1.0
Stearic Acid	4.5
Behenic Acid	1.5
Lauric Acid	1.0
Sorbo 70% (23)	4.0
Sodium Alginate (34)	0.25
Diethanolamine (DEA)	2.0
KOH to pH 9 . . . approximately	1.0
Distilled Water	71.25
Potassium Sulfite	9.0

Proc.: Mix all except the DEA, KOH, water and potassium sulphite and heat to 80°C. Dissolve the DEA and KOH in a little of the water and add to the warm solution previously prepared. When homogeneous add the remainder of the water and cool to 40°C followed by the addition of the potassium sulphite. Fill into the codispensing package 80% above concentrate 20% of a 7 soln of H_2O_2. Pressurize with 5% Isobutane/Propane (84:16) propellent mix.

Lathering Shave Cream

I

A shaving lather may be composed of these ingredients:

		pts./wt.
A	Stearic Acid	32.50
	Coconut Oil fatty acid	3.50
	Peanut Oil	3.26
B	Triethanolamine Stearate	4.00
C	Caustic Potash solution (45%)	18.10
	Caustic Soda solution (33%)	0.40
	Potassium Carbonate	2.90
	Glycerin	11.60
	Propylene Glycol	2.85

Cellulose Glycolate (4%) aqueous sol.	11.20
Water	54.09
Boric Acid	1.20
D Stearic acid for fitting	4.00

Proc.: Heat ingredients of A to melting point and pass through a cambric strainer into a jacketed kettle, heated by indirect steam. B is made from

	pts./wt.
Stearic acid	2.75
Triethanolamine	1.25

The stearic acid is heated to 75°C., then the triethanolamine is heated to the same temperature and stirred into the stearic acid. A together with B is heated in the jacketed kettle to 85°C. until the triethanolamine stearate is melted.

C is heated separately to 90°C. and then stirred into the mixture. (Add in small portions since the batch will rise in the kettle.) After saponification is complete, permit to stand for two hours for after saponification. Test for alkalinity and neutralize with stearic acid (D); adjust to about 2 per cent excess stearic acid. Perfume by adding about one per cent perfume oil (such as eau de cologne) at a temperature of 40°C.

II

	%/wt.
Coconut Oil (Cochin)	10.0
Stearic Acid XXX	35.0
Caustic Potash (as 100%)	6.8
Caustic Soda (as 100%)	1.5
Glycerin	18.0
Perfume	0.5
Water	28.2

The caustic solution is run with agitation into the heated coconut oil and glycerin (165°-170°F); when saponification is complete, it is followed by the melted stearic acid and the balance of the water. It is imperative that the coconut oil be saponified before the stearic acid is added, in order to ensure that the remaining unsaponified material present in the finished cream is actually stearic acid.

Electric Pre-shave Lotion

I

When using an electric razor it is well to prepare the skin before beginning to shave with this:

Alcohol	3 qt.

Sorbo	2 cup
Menthol	1 tablespoon
Water	3 qt.
Perfume, as desired	
Color (food coloring), as desired	

Proc.: Dissolve menthol in alcohol, then add other ingredients and stir until clear. Add perfume and food coloring as desired.

II

	%/wt.
Alcohol #40	60.0
Silicon S.F. 1075	0.5
Isopropylmyristate	2.5
Color	q.s.
Perfume Oil	q.s.
Water q.s.	100.0

Proc.: Add perfume oil to the alcohol and mix. Then add silicon, isopropylmyristate, and water. Cool to $32°$F. and filter. After filtration add the color.

III

	%/wt.
CRODAMOL	15.0
Germicide	0.1
SD 39C Alcohol (95%)	74.9
Perfume	q.s.

Proc.: Blend all ingredients and filter.

IV

	%/wt.
ISOPAR COSMETIC GRADE	81.0
Isopropyl Myristate	19.0
FDC Yellow No. 4 (0.2% in oil)	q.s.
Perfume	q.s.

Proc.: Mix all ingredients together until uniform.

Swedish Face Tonic
(After Shave Lotion)

1. Zinc Phenolsulfonate	½ oz.
2. Witch Hazel	15 oz.
3. Isohol	10 oz.
4. Glycerine	1 oz.
5. Balsam Peru	¼ oz.
6. Lavender Oil	10 gm.

Proc.: Dissolve Nos. 1 and 2 and then dissolve Nos. 4, 5, and 6 in

No. 3. Mix both solutions and stir thoroughly. Allow to stand overnight and filter.

Mild After Shaving Balm

	%/wt.
POLYCHOL 40 or SOLAN regular	0.5
VOLPO 10	0.5
LIQUID BASE	1.0
Perfume	1.0
SD 40 Alcohol	15.0
Water, Distilled	71.5
Glycerin	1.0
10% Aq. Soln. Diisopropanolamine	1.5
1% Carbopol 941 Acid form stock solution	8.0

Proc.: Combine Carbopol stock solution with the water and add to the emulsifier/ oil phase at 50°C. Dissolve perfume in the alcohol and add to the emulsion at 45°C then continue cooling and neutralize with the diisopropanolamine solution to complete the batch.

After-Shave Lotion

I

		%/wt.
A	Ethyl Alcohol 96%, Denatured	55.5
	Perfume Oil H + R	0.5
B	Distilled Water	36.6
	Allantoin	0.2
	1.2-propylene Glycol twice dist.	2.0
	Hamamelis Distillate H + R	5.0
	Lactol 1	0.2

Proc.: Mix A and B separately, add B to A, cool to 0°C and filter.
Filter: Sometimes materials are not suluble. To get a clear solution pour through funnels and filter paper to trap solid particles.

II

		%/wt.
A	Ethanol SDA-40	52.0
	Ethomeen T/15	2.0
	Allantoin	0.1
	Menthol	0.1
	Arquad C/50	0.2
B	Glycerine	2.5
	Water	43.1

Proc.: Heat A and B separately. Add A to B and mix until uniform. Ethomeen provides lubrication and emollient feel at lot concentrations. pH of 7.2 and no apparent viscosity. Clear solution with Gardner 1 color.

Pearly After-Shave Lotion

		%/wt.
A	Carbopol 941 resin (1% soln in ethanol SD 40)	30.00
	Menthol	0.10
	Ceraphyl 41	2.50
	Ceraphyl 230	2.00
	Water, deionized	20.00
B	Ethanol SD 40	20.00
	Diethanolamine	0.30
	Water, deionized	24.60
	Mearlin, Pearlwhite	0.30
	Allantoin	0.20
	Color	q.s.

Proc.: Add B to A very slowly with high-speed agitation.*
Note: B itself must be stirred continuously during the addition to prevent settling of the pearl pigment.

BATH PREPARATIONS

Bubble Bath

	pts./wt.
Water	q.s. 100
Sodium Dodecyl Benzene Sulfonate (60%)	20
CLINDROL Superamid 100 L	1—5
Perfume	1—3
Preservative	q.s.
Phosphoric Acid	to pH 6.7-6.9

Proc.: The order of addition of ingredients is that in which they are listed above. Where higher concentrations are desired, small amounts of isopropanol may be added to obtain a pourable composition.

Bubble Bath Gelee

	%/wt.
Maprofix TLS-500	30.00
Maprofix MB	20.00
Monamid 150 LMW-C	20.00
Phosphoric Acid, 85%	0.50
Arlatone T	4.50
Sesame Oil	3.00
LANTROL	5.00
Avocado Oil	0.50
Apricot Kernel Oil	0.60
Tween 80	7.00

Vitamin A	0.25
Vitamin D2	0.25
Vitamin E	0.25
Formalin	1.00
Perfume	5.00

Proc. Mix and heat all ingredients to 80°C. Cool to 50°C. Add perfume and formalin. Cool to 30°C with careful stirring.

Bubble Bath Powder

Powdered Sodium Lauryl Sulfate	15.0 oz.
Powdered Sodium Sesquicarbonate	27.0 oz.
Powdered Sodium alginate	1.0 oz.
Perfume and Color	q.s.

Proc.: Mix the powders thoroughly.

Milk Bubble Bath

I

		ptsl/wt.
	Whole milk solids	4
	Water	55
A	Sodium Lauryl Sulfate (28%)	36
B	CLINDROL Superamid 100CG	5
C	E-284 Latex	2
	Preservative	0.1
	Phosphoric Acid	to pH 6.5
	Color	q.s.
	Perfume	q.s.

Proc.: Dissolve A, B, & C and mix with the milk and water under heat.

II

		pts./wt.
	Whole Milk Solids	4
	Water	55
A	Sodium Lauryl Sulfate	36
B	CLINDROL 200-L	5
C	E-295 Latex	2
	Preservative	0.1
	Phosphoric Acid	to pH 6.5
	Color	q.s.
	Perfume	q.s.

Proc.: The milk solids are dissolved in 45 parts water. The lauryl sulfate is added and dissolved followed by CLINDROL 200-L. The

E-295 Latex is dispersed in 10 parts water and added to the mixture with stirring.

Vitamin Bubble Bath

	%/wt.
Texapon Q	90.0
Comperlan KM	3.0
Soluvit-Complex	4.0
Essential Oil	3.0
Blue Dye	q.s.

Proc.: Comperlan KM is dissolved in Texapon Q at approximately 100°F. or without heating over night. The essential oil is compounded with the vitamins and added to the above mixture.

Bath Oil

Bath oils are among the oldest cosmetics. Using them creates the most luxurious, relaxing moments during a grooming routine. This is an extremely simple, and very pleasing, preparation.

I

Tween 20	3 pt.
Distilled Water	6 qt.
Perfume, as desired	up to 1 cup

Proc.: Stir together all ingredients until well blended.

II

	pts./wt.
Mineral Oil	60.0
Isopropyl Myristate	25.0
PEG-400 Dilaurate	1.5
Diglycol Laurate	3.5
Aromox DM18DW	6.0
Perfume 5864 K (IFF)	4.0

Proc.: Mix all materials until solution is complete.

Egg-Cream Bath Oil

Whole Egg	1
Sesame Oil (or substitute safflower oil)	½ cup
Liquid Dishwashing Detergent	1 tsp.
70% Isopropyl Rubbing Alcohol	2 tbsp.
Milk	¼ cup
Orange or Lemon Extract or Perfume of your choice	¼ tsp.

Proc.: Mix all of the ingredients together until uniform.

Floating Bath Oil

	%/wt.
Mineral Oil Med. Vis	46.00
Isopropyl Myristate	48.00
Perfume Oil	5.00
Arlaton T	1.00

Proc.: Mix Arlaton T with the perfume oil, and add this mixture to the cosmetic oils with agitation.

Bath oils have increased in popularity during recent years. The floating type of bath oil has won exceptional popularity because it not only gives a pleasant scent to the bath, but may also help prevent skin dryness, through spreading a thin layer of oil over the body.

Dusting Powder

	pts./wt.
Talc	50
Precipitated Chalk	25
Thixcin R	4
Perfume	1
Boric Acid	5
Colloidal Kaolin	15

Proc.: Mix all ingredients together until uniform.

Talcum Powder

Talc	71 g.
Precipitated Chalk	20 g.
Zinc Stearate	3 g.
Boric Acid	5 g.
Perfume	1 g.

Proc.: Mix all ingredients together until uniform.

DEODORANTS AND ANTIPERSPIRANTS

Deodorant Powder

I

	pts./wt.
Talcum	105
Sodium Bicarbonate	60
Magnesium Oxide	15
Starch	5
Perfume	To suit

Proc.: Mix well in a high-speed mixer. This powder is very effective, will not injure clothing, and is nonirritating.

II

Talcum Powder	11 lb.
Cornstarch, Powdered	1.40 lb.
Aluminum Sodiumsulfate, Powdered	4.50 lb.
Salicylic Acid, Powdered	11.25 oz.
Boric Acid, Powdered	11.25 oz.

Proc.: Blend the powders thoroughly, and put the mixture through silk of at least 120 mesh. Use a modern mixing and sifting apparatus for this operation.

Antiperspirant Cream

	I %/wt.	II %/wt.
Sodium Aluminum Chlorhydroxy complex, 40% Solution	50.0	50.0
Specially Denatured Alcohol	42.0	42.0
Propylene Glycol	3.0	3.0
Sodium Stearate	6.0	
Sodium Hydroxide		0.75
Stearic Acid		5.25
Water		2.0
Perfume	q.s.	q.s.

Proc.: In the first formula, the aqueous aluminum salt solution is heated to 60 to 65°C. and the alcohol and propylene glycol slowly added with agitation and added heat to maintain the temperature. The sodium stearate is added with stirring until the soap dissolves. The perfume is finally added and, after thorough mixing, the solution is poured into molds.

In the second mixture, with sodium stearate formed in situ, the stearic acid is dissolved in the warmed alcohol, the sodium hydroxide is dissolved in the water and the alkali solution is added to the aluminum complex solution. The aqueous solution is heated to 65 to 70°C. and the alcoholic solution is about the same temperature is added with stirring. The formation of the soap takes place rapidly and after slight cooling the perfume is mixed in and the solution poured into molds. There is little or no difference in result if the alcoholic solution is added to the aqueous or vice versa, but stearic acid must be dissolved in the alcohol and the alkali in water.

Antiperspirant Gel

	%/wt.
A Brij 56	15.0
Brij 98	10.0
Mineral Oil	5.0

B Chlorhydrol 50% W/W Solution	40.0
C Aluminum Chlorhydroxy Allantoinate	0.3
Distilled Water	29.2
D Dermodor 3066 Bis Perfume Oil	0.5

Proc.: Slowly stir B at 80°C into A at 80°C. Add C heated to 70°C. and at 50°C. add D.

Antiperspirant Lotion

I

A water/oil antiperspirant lotion may be formulated from the following ingredients:

	%/wt.
A "Volpo" 3	3.75
"Polychol" 5	1.25
Mineral Oil, Heavy	5.00
B "Robane"	8.00
Glycerine	4.00
C Water, Deionized	38.00
D "Chlorhydrol" (50% w/w solution)	40.00

Proc.: Melt A at 40°C. Cool A to 30°C, and add B. Add C dropwise to AB with good agitation. Next, add D in the same manner. Add perfume. Reheis Chemical warns that, in packaging, all contact between "chlorhydrol" solutions and iron, copper, brass or aluminum should be avoided

II

	%/wt.
A Glycerol Monostearate Pure	3.5
Mineral Oil	1.5
Armeen DM18D (Stearyl Dimethyl Amine)	0.3
Lactic Acid (80% edible)	0.1
Ethomeen 18-60 (Ethoxylated Fatty Amine)	1.0
B Water	53.6
Dridol (50% Aluminum Chlorhydroxide)	40.0

Proc.: Heat A, then combine with B and mix until smooth. White roll-on lotion. Cationic system reduces tackiness of deposited film. pH of 3.5-4.0*.

Anti-Perspirant Roll-On

	%/wt.
"Volpo" 3	4.0
Polychol 5	1.0
Mineral Oil 70 visc.	5.0
"Robane," Perhydrosqualene	8.0
Glycerin	4.0
Water	38.0
Chlorhydrol, 50% alum.	
Chlorhydroxide Solution	40.0

Proc.: Dissolve the "Polychol" in the "Volpo" and the other oils, then add the water to the oils with vigorous agitation. When the emulsion is formed and all the water is added, add the chlorhydrol solution.

II

		%/wt.
A	Hexachlorphene	0.10
	Cerasynt 1000 D	2.00
	Emulsynt 2400	6.25
	Solulan 24	1.50
	Ceraphyl 31	1.50
B	Propylene Glycol	3.00
	Distilled Water	43.95
	Aluminum Chlorhydroxy Allantoinate	0.20
	Veegum	1.00
C	Chlorhydrol (50%)	40.00
D	Dermodor 4775 Perfume Oil	0.50

Proc.: Mix Veegum in the water with rapid mixing for about 45 min. Then add propylene glycol and Aluminum Chlorhydroxy Allantoinate and heat to 70°C. Heat A to 70°C. Stir A into B until 45°C is reached. Add C, mix well and add D.

Quick Breaking Foam Anti-Perspirant

	%/wt.
SOLE-MULSE A-66	3.0
Anhydrous Alcohol	42.4
Water	26.0
Aluminum Chlorhydroxide	18.0
Quarternary Germicide	0.1
Perfume	0.5
Propellent 12/114 (40:60 ratio)	10.0

Proc: Heat the water/alcohol solution to about 110°C to dissolve the waxes and other active ingredients. Add the warm solution to the containers in one step.

Deodorant Aerosol

	%/wt.
Diaphene	0.1
Aluminum Chlorhydroxy Allantoinate	0.2
Propylene Glycol	1.0
S.D. Alcohol #40 90 proof	40.7
Aluminum Sulphocarbolate	2.0
Fiorodor 50504	1.0
Freon 12/114 40/60	50.0

Proc.: Dissolve the Aluminum Chlorhydroxy Allantoinate in water with heat. Add the S.D. Alcohol, then dissolve the other ingredients with heat. q.s. to 50 volume with S.D. Alcohol. Package.

DEPILATORIES

Depilatory Paste

Strontium Hydrate	50	g.
Calcium Oxide	12	g.
Colloidal Clay	102	g.
Methyl Cellulose	11	g.
Thioglycollic Acid	12	cc.
Water	300	cc.
Perfume	0.8	cc.

Proc. Mix together until uniform.

Depilatory Cream

	kg.
Oil phase:	
Lauric Monoethanolamide (m.p. 89-93°C.)	3.250
Stearic Monoethanolamide (m.p. 69-73°C.)	3.000
Diethanolamine Fatty Acid Condensate (2:1) (coconut/oleic 60/40)	2.250
Sorbitan Trioleate	1.750
Bithionol	0.500
Aqueous phase:	
Water, Distilled	7.500
Sodium Hydroxide Pellets	0.004
Sodium Sulphite Crystals	1.500
Alkali-Thioglycollate suspension:	
Water, Distilled	7.500
Calcium Thioglycollate	4.200
Calcium Hydroxide (USP)	1.000
Strontium Hydroxide, Milled	2.000

Perfume   0.250

Proc.: Heat water phase. Heat oil phase. Mix together until uniform.

Depilatory Wax

	pts./wt.
Rosin	1700
Vegetable Oil	900
Triethanolamine	100
Benzoin	10
Balsam Tolu	10
Lemongrass Bouquet	5
Butyl p.aminobenzoate	10
Alcohol	5

Proc.: This wax is spread on the rough side of a strip of thin kraft paper, the smooth side of which is silicone-treated.

PERFUMES AND COLOGNES

Cologne

I

Alcohol # 40	105 fl. oz.
Perfume Oil	4 fl. oz.
Water q.s.	128 fl. oz. (1 gal.)

Proc.: Mix all ingredients until uniform. Color solution as desired.

Eau De Cologne

I

(First quality)

Alcohol	3 l.
Oil Bergamot	7 g.
Oil Lemon	17 g.
Oil Neroli	20 g.
Oil Sweet Orange	20 g.
Oil Rosemary	7 g.

Proc. Mix all ingredients together until uniform.

II

(German)

Alcohol	2.00 l.
Oil Bergamot	10.00 g.
Oil Lemon	8.00 g.
Oil Petitgrain	5.00 g.
Oil Neroli	5.00 g.

Oil Lavender	0.30 g.
Oil Jasmin	0.02 g.

Proc.: Mix all ingredients together until uniform.

Moisturizing Cologne Aerosol

		%/wt.
A	CRODAFOS N3 NEUTRAL	1.5
	POLYCHOL 5	3.0
	POLAWAX A-31	1.25
	NOVOL	2.0
B	SD 40 Alcohol (95%)	46.0
	Water	44.75
	Perfume	1.5-5.0
	Fill:	
	Above Concentrate	95%
	Propellent 12/114 (30:70)	5%

Proc.: Heat A until dissolved. Heat B and add to A. Add C and then charge with propellant.

Note: The water alcohol ratio is critical and may require adjustment depending upon perfume and concentration.

Pearlized Cologne

		%/wt.
A	Allantoin	0.2
	Carbopol 941	0.3
	Distilled Water	30.0
B	S.D. Alcohol 39C	50.0
	Dulchena S-2025	2.0
C	Triethanolamine	0.3
	Distilled Water	5.0
D	Mearlmaid OL 6419 (Pearl essence)	1.5
	Solulan 98	3.0
	Distilled Water	7.7

Proc.: Dissolve the Allantoin in water A then sprinkle the Carbopol 941 onto this solution and mix with rapid agitation. Add phase B and continue mixing. Stop mixing to allow the escape of air. Then add phase C and mix again. Add phase D and mix well.

Formula For Solid Deodorant Cologne

		pts./wt.
A	Stearic Acid T.P.	5.
	Alcohol #40	80.
	Propylene Glycol	3.

B Sodium Hydroxide 0.75
   Water 8.

C G-11 0.25

D Perfume Oil 3.

Proc.: Weigh the ingredients of A into a container provided with a source of heat and agitation and heat to 50 degrees C.

Weigh the ingredients of B into a second container and warm gently to a temperature of about 50°C.

When both temperatures are alike, add B to A, stirring constantly. Continue to heat and stir until the temperature reaches 70°C., at which point the source of heat is removed, but stirring is continued. Add the G-11 and the Perfume Oil when the temperature has reduced to 60°C., then pour into molds.

Cyclamen Flower Oil
(Synthetic)

p Isopropyl Alpha-methyl hydrocinnamic Aldehyde (Alpine Violet)	1.5
Hydroxycitronellol	3.0
Hydroxycitronellal Dimethylacetal	1.0
Linñol, Extra	1.5
Benzyl Acetate Coeur	1.0
Terpineol, Extra	1.0
Rhodinol, Extra	0.8
Lemon Oil, Italian	0.6
Geranium Oil, Algerian	0.5
Linalyl Acetate, Extra, 96-98%	0.5
Methyl Ionone	0.5
Tolyl Alcohol	0.5
Ylang-Ylang Oil, Bourbon Premier	0.5
Musk Ketone	0.4
Alpha Ionone	0.4
Cinnamic Alcohol, Pure	0.4
Petitgrain Oil, S.A.	0.4
Phenylacetaldehyde Dimethylacetal	0.4
Hydratropyl Alcohol	0.3
Cumin Ketone	0.2
Amyl Cinnamic Aldehyde	0.2
Aldehyde C_{14}, 10% in dimethyl phthalate	0.1
Benzyl Propionate	0.2
Coumarin, N.F.	0.1
Phenylethyl Propionate	0.1
Benzyl Alcohol, N.F.F.F.C.	1.4

Proc.: Mix well and allow to stand a week or more with frequent shaking.

Neroli Synthetic

	g.
Petitgrain	50
Neroli petals	5
Methylanthranilate	5
Benzyl Benzoate	40

Proc.: Combine all ingredients.

Clover Flower Oil, Synthetic

	pts./wt.
Orchid oil, artificial	25.0
Coumarin, N.F.	1.6
Ylang-Ylang oil (Bourbon Premier)	1.5
Amyl Salicylate, Extra	1.0
Bergamot Oil, Extra, N.F.	0.7
Musk Ketone	0.6
Heliotropine, recrystallized	0.5
Benzyl Acetate Coeur	0.3
Isobutyl Phenylacetate	0.3
Dimethyl Hydroquinone	0.3
Lavender Oil, U.S.P., Barreme, 42%	0.3
Palatone, 5% in phenylethyl Alcohol	0.3
Acetophenone	0.2
Oakmoss Resin	0.2
Tuberyl Acetate	0.2
Vanillin, U.S.P.(from Eugenol)	0.2
Benzyl Alcohol, N.F.F.F.C.	1.3

Proc.: Mix well and allow to stand a week or more with frequent shaking.

Amber Oil

	pts./wt.
Methyl Ionone	4
Labdanum Absolute	2
Rose Oil	2
Vanillin	2
Musk Ambrette	1
Musk Ketone	1
Benzyl Salicylate	1

Proc.: Mix well, and allow to stand a week or more with frequent shaking.

Powder Perfumes

I

Balsamic Base (for powder)

Phenyl ethyl cinnamate	300

Amyl salicylate	300
Ylang	150
Heliotropin	100
Musk ketone	50
Oakmoss resinoid	20
Vanillin	10
Civet absolute 10% in alcohol	30

Proc.: Mix all ingredients together until uniform. If a less expensive base is desired, ylang ylang may be replaced partially or in its entirety with synthetic ylang.

II
Talcum Powder Perfume

	pts./wt.
Bergamot	200
Sandalwood E.I.	320
Ionone	100
Amyl salicylate	120
Geranium Bourbon	100
Oakmoss resinoid	150
Benzyl acetate	80
Phenyl ethyl alcohol	80
Heliotropin	60
Coumarin	50
Musk xylol	50
Musk ambrette	50
Vanillin	45
Neroli synthetic	30
Lavender oil	20

Proc.: Mix all ingredients together until uniform. Bergamot, sandalwood and geranium oils may be substituted partially with their synthetic counterparts for less expensive variation of this formula.

Perfume Foams
I

		%/wt.
A	Myristic acid	1.5
	Glycerin	3.0
	Triethanolamine	3.0
B	Water	75.9
	Bactericide	1.0
C	Perfume oil	6.0
	Filling:	%/wt.
	Emulsion	90
	Propellant 12/114-40:60	10
	Use foam valve.	

Proc.: Heat A and B separately to 75°C. Add B to A at this temperature and stir until cool. Add perfume oil at 45°C. and pass through homogenizer.*

II

To yield a quick breaking foam, the emulsion is formulated with high alcohol and low fat content. A quick breaking perfume foam is composed of:

A	"Croda" wax GP 200	1.7
	Distilled water	39.2
B	Ethyl alcohol	50.0
	Isopropyl adipate	1.0
C	Glyceryl monostearate	5.0
	Perfume oil	4.0
	Filling: Solution of active substances	93.0
	Propellant 12/114-40:60	7.0

Proc.: Melt part A by heating gently, then add B and then separately C, which has been mixed previously.

Instability of the foam is due to the alcohol content. By boosting the water component at the expense of the alcohol, stability may be increased.

Lotion Sachet

A medium viscosity, high gloss vehicle for prestige fragrances. This rich lotion vanishes on the skin with just a touch, leaves no objectionable greasy or sticky aspect.

Oil Phase	%/wt.
Cetyl Alcohol, N.F.	1.50
Glyceryl Monosterate	7.00
Mineral Oil, "Carnation"	2.00
Water Phase	
Fungitex R	0.05
Glycerin, U.S.P.	5.00
Lactic Acid, U.S.P., 85%	0.34
Methocel 90 HG, 15,000 cps	0.50
Sapamine COB-ST	1.20
Water, Distilled	76.41
Perfume, 9449	6.00
Certified dye	q.s.

Proc.: Heat oil phase and water phase separately. Add water phase to oil phase. Mix together. Add Perfume. Mix until smooth.

Imitation Parfum Joy

Heliotropin	0.886 g.
Rose Oil	3.434 g.
Bergamot Oil	1.772 g.
Moskus	0.25 mg.
Gray Amber	0.177 g.
Jasmin Autf.	0.354 g.
Neroli Oil	0.25 mg.
Angelica Oil	0.50 mg.
Vetivert Oil	0.50 mg.
Alcohol No. 40	85.05 g.

Proc.: Mix ingredients together until uniform.

NAILS

Nail Lacquer

A nail lacquer, with water as the primary solvent, is suggested:

	%/wt.
Carboset 525 (1)	13.7
Water	75.1
Ammonium Hydroxide 28%	1.1
Pigment	9.1
Wetting Agent	1.0
Antifoam and Preservative	q.s.

Proc.: This formulation is resistant to water on drying and is rapidly removed by means of soap and water. It may be made considerably more resistant to soap by the addition of varying proportions of a zinc-complex of the following formula:

	pts./wt.
Water	71.4
Ammonium Hydroxide 28%	8.4
Ammonium Carbonate	12.7
Zinc Oxide	7.2

Combine in the order listed to form a clear solution.

Nail Enamel

	pts./wt.
Dibutyl Phthalate	20.25
D & C Red #19	0.15
D & C Red #31	0.30
Oil Pink	0.15
Nitrocellulose	17.10
Ethyl Acetate	47.25
Butyl Acetate	45.00
Tri-o-cresyl Phosphate	12.45
Alcohol, Anhydrous	7.35

Proc.: Mix the colors in the dibutyl phthalate, and grind the mixture in a pebble or ball mill.* Dissolve the nitrocellulose in the ethyl acetate, and add the butyl acetate to this mixture. Incorporate slowly the color dispersion in the nitrocellulose solution and mix with thorough stirring. Finally, stir in the rest of the ingredients, and continue to stir until the product is homogeneous. Add some perfume oil if so desired. For best results fill the enamel at once.

Cuticle Remover

	%/wt.
Trisodium Phosphate	8
Glycerin	20
Rose Water	50
Distilled Water	22

Proc.: Mix ingredients until uniform.

Cuticle Softner

	pts./wt.
Caustic Potash	2
Alcohol	25
Water	75

Proc.: Mix ingredients until uniform.

Nail Polish Remover

I

This cleanser for the nails consists of substances that dissolve nail lacquer, the job which all polish removers must accomplish without harming the surface of the nails and without leaving the surface dry and brittle. This is an efficient, satisfactory cleanser for a specific grooming step.

Ethyleneglycol Monoethyl Ether	1/4 cup
Butyl Acetate	2 tablespoons
Propylene Glycol	1 teaspoon
Castor Oil	1 teaspoon

Proc.: Stir together until well blended, then bottle.

II

	%/wt.
CRODAMOL DA (diisopropyl adipate)	2.5
LANEXOL (water & alcohol soluble liq. lanolin)	2.5
Methyl Cellosolve	q.s. to 100%
Perfume	0.5-2.0

Proc.: Blend all ingredients, filter* and fill.

MOUTHWASH PREPARATIONS

Mouthwash

I

(similar to Cepacol)

	%/wt.
Quaternary Ammonium Salt	9.5
(Funtigtex, Ciba or Cetol, Fine Organic Co.)	
Sorbo 70	30.00
Cinnamon Oil	0.05
Peppermint Oil	0.10
Citric Acid	0.10
Tween 60	0.30
Ethyl Alcohol	10.00
Water	q.s.
Color	q.s.

Proc.: Mix all ingredients together until uniform.

Alcohol in a mouthwash serves the following purposes:

(1) Assists in solubilizing flavor oils (2) It reduces surface tension (3) Assists in helping in the active ingredients to penetrate and to become effective (4) Alcohol serves as a mild astringent.

Mouthspray Refresher

	%/wt.
Fungitex R	0.10
Saccharin	0.50
Pepperming Oil	0.50
Cinnamon Oil	0.25
Water	5.00
Glycerine	10.00
Alcohol #38/B	83.65

Proc.: Mix all ingredients together until uniform. Fill spray-type bottles with this solution.

Zinc Chloride Mouth Wash

	pts./wt.
Zinc Chloride	1.00
Menthol	0.50
Oil of Cinnamon	1.30
Oil of Cloves	0.50
Solution of Formaldehyde	0.50
Saccharin	0.40
Naphthol Yellow S	0.24
Alcohol	45.00

Talc.	15.00
Diluted Hydrochloric Acid	1.00
Distilled Water, to make	1000.00

Proc.: Dissolve the zinc chloride in a solution of the dilute hydrochloric acid and 25 cc of distilled water, and add this to 800 cc of distilled water. Dissolve the oil of cinnamon, oil of cloves, menthol, solution of formaldehyde and saccharin in 45 cc of alcohol, and mix this with 15 Gm of talc. Add to this, with constant stirring, the previously prepared zinc chloride solution, shake resulting mixture occasionally during 24 hours, and filter. Add distilled water to make one liter, and add a suitable color.

DENTIFRICES

Tube Toothpaste

The following formulations are recommended.

I

	pts./wt.
Sorbo (70% Sorbitol in water)	1000
Sodium Saccharine	6
Sodium Lauryl Sulfate	70
Carbopol 934 Dispersion (6% in water)	600
Water	484
Oil of Spearmint	40
Sodium Hydroxide (50% in water)	31
Dibasic Calcium Phosphate Dihydrate	1800

Proc.: Dissolve the sodium lauryl sulfate in one third of the water. Mix the Sorbo, sodium saccharine and 6% Carbopol solution with the rest of the water. This will be a thin solution. Blend the two solutions carefully to avoid foaming. Slowly add the sodium hydroxide to neutralize the Carbopol 934 while gently stirring. The solution will become thick as the neutralization progresses. Slowly mix in the calcium phosphate. Stir until the particles are wetted out and all the lumps are broken up. Mill* and deaerate the batch on conventional equipment. Add the flavoring during the last stages of milling to avoid evaporation loss. Tube the product.

II

	pts./wt.
Lathanol LAL-70 Powder	3.0
Tricalcium Phosphate	26.6
Gum Tragacanth	1.0
Glycerin	45.5
Saccharin	0.2

Water	23.1
Flavor	0.6

Proc.: Solution I. Disperse the gum in 25 parts of the glycerin.

Dissolve the saccharin in the water. Slowly add to gum/glycerin mixture, agitating continually until smooth.

Solution II. Mix Lathanol LAL-70 with remaining glycerin (use mortar, mixer, blender, or similar equipment).

Add tricalcium phosphate to Lathanol LAL-70 and glycerin. Mix until a thick paste is obtained.

Add Solution I to Solution II and continue mixing until uniform. Add flavor.

Note: Tricalcium phosphate may be increased or decreased to obtain desired consistency.

Clear Gel Toothpaste

		pts./wt.
A	Glycerine	40
	Water	282
	"Carbopol" 940	0.6
	Sodium Saccharine	6.6
B	"Duponol" C (Sodium Lauryl Sulfate U.S.P.)	40
	Sodium Hydroxide (10% solution)	28
	Peppermint Oil	4 drops

Proc.: Heat A until dissolved. Add each of the remaining elements of B separately stirring until each is dissolved before adding the next.

Tooth Powder

This old-fashioned tooth powder is a simple to make, efficient cleanser.

Precipitated Chalk (Calcium Carbonate)	1¾ cup
Magnesium Carbonate	1 cup
Powdered Sugar	¾ cup
Sodium Perborate	½ cup
Powdered Neutral White Soap	¼ cup
Flavor, as desired	

Proc.: Put all ingredients except flavoring into mortar and rub down well. Add flavor slowly and continue to rub well until evenly mixed.

Denture Cleaners
I

		pts./wt.
A	"Veegum WG"	5
	Sodium Perborate	13

Tetrasodium Pyrophospate Anhydrous	25
Sodium Chloride	13
Tartaric Acid	9
Sodium Phosphate Dibasic	12
Citric Acid	7
Sodium Carbonate	16
B Isopropyl Alcohol	25
C "Veegum WG"	2

Proc.: Granulate A with B. Pass through a No. 10 mesh screen. Dry the granulation one hour at $105°C$. and pass through a No. 16 mesh screen. Dry blend with C and compress.

Directions for use: Add one tablet to a glass of hot water. Soak dentures in solution one-half hour or overnight. Rinse in fresh water.

Although the regular grade of "Veegum" is used as a binding agent, the WG grade finds use as a tablet disintegrator. This property is utilized in the above formula.

II

	%/wt.
A Veegum	1.00
Sodium carboxymethylcellulose (med. visc.)	0.50
Water	28.90
B Saccharin	0.10
Sodium benzoate	1.00
C Sorbitol (70% soln.)	9.00
Glycerin	9.00
D Dicalcium Phosphate Dihydrate	36.00
Dicalcium Phosphate Anhydrous	12.00
E Flavor	0.50
F Sodium lauryl sulfate	2.00

Proc.: I. Dry blend the Veegum and the CMC. Add to the water slowly, continually agitating until smooth. II. Add B to I. and mix. Blend C and add to II. IV. Blend D and add to III. V. Add E and F, one at a time, to IV and mix until uniform.

FOOT PREPARATIONS

The feet are probably the most neglected part of the human body. They take the whole weight of the body and, in so doing frequently have to cope with highly restrictive, undersized shoes equipped with narrow toes and spike heels. Thus, no air can circulate; carbon dioxide is entrapped and oxygen cannot reach the surface of the foot. Numerous sweat glands pour their secretions into wrappings of nylon, wool or cotton, and much of this sweat is taken up in turn by the shoe leather.

Foot baths have a soothing effect on tired and aching feet. Hot water soaks and baths increase the local blood supply and promote circulation. Alkali helps to soften the outer horny layer of skin. Stale secretions, debris and possible sources of odor are removed. The bacterial count is temporarily reduced. The whole effect is refreshing, cleansing and tonic.

Foot powders, creams and antiperspirants have become prevalent as deterrents to the above conditions. Aerosols, the newest way to apply foot preparations, are an ideal method of application.

Foot Bath Salt

I

	pts./wt.
Borax	10.0
Sodium Bicarbonate	35.0
Sodium Carbonate, Dried	55.0

II

	pts./wt.
Extract of Cannabis Indica	1
Alcohol	20
Ether	32
Collodion, enough to make	100

Proc.: Incorporate the acids and extract of cannabis in the alcohol and ether and add collodion.

Recipes for corns depend, in great measure, on salicylic acid. It should be noted that the concentration of the acid in collodion, ointment, etc., is more than six per cent. The reason is that salicylic acid is one of the chemicals the action of which on the unruptured horny epidermis varies with the concentration below and above six per cent. The lower concentrations are keratoplastic; they cause the horny layer to become thicker. Concentrations above six per cent cause keratolysis: the horny layer is lifted off.

Salicylic Corn Collodion

Salicylic Acid	10 g.
Lactic Acid	10 g.
Flexible collodion, enough to make	100 g.

Proc.: Dissolve the acids in the collodion.

Corn and Callus Remover

Salicylic Acid	10 g.
Lactic Acid	10 g.

Foot Cream

		%/wt.
A	Ottasept Extra	1.0
	Tegacid	16.00
	Cetyl Alcohol	1.00
	Amerchol L101	3.00
	Solulan 98	0.50
	Mineral Oil	3.00
B	Aluminum Chlorhydroxy Allantoinate	0.25
	Propylene Glycol	5.00
	Water (Distilled)	69.95
	Perfume Oil	0.30

Proc. Heat A and B to 75°C. With rapid agitation add B to A. Add Perfume Oil at Cream point.

Aerosol Foot Powder

Concentrate:	%/wt.
"Nytal 400" (fine talc)	86.0
Magnesium Carbonate	5.0
Kaolin	5.0
Zinc Stearate	3.0
"Methyl Tuads" (Tetramethylthiuram Disulfide)	0.5
Isopropyl Myristate	0.5
Packaging:	
Concentrate	10
Propellant 12/11 (50/50)	90

Proc.: Add to the talc each component in order listed with high speed mixing after each addition.

Antiperspirant Foot Spray

		pts/wt.
A	Talc USP	33
	Cab-O-Sil	2
	MICRO-DRY Ultra fine impalpable Aluminum Chlorhydrate	5
	Tribromosalicylanilide	1
	PROCETYL AWS	4
	PVP-VA E735	3
B	Menthol	0.5
	Alcohol SD40	51.5

Above concentrate	15
C Propellant 114	35
Propellant 12	50

Proc.: Heat A until smooth. Mix B together than add to A, Charge with propellent C.

Note: for anti-fungal purposes undecylenic acid or zinc unde-cylenate may be added to this system.

BABY PREPARATIONS

Baby Oil

I

	oz.
White Mineral Oil (65/75 Viscosity)	105.00
Isopropyl Myristate	3.50
Liquid Lanolin	3.50
Ethyl Stearate	4.50
Methyl p-Hydroxybenzoate	0.20

Proc.: Mix all ingredients together until homogeneous.

II

	%/wt.
"Prodipate"	10.0
"Lanogene"	10.0
Sorbitan Trioleate	2.0
Bactericide	0.2
Mineral Oil (80 vis.)	77.8

Proc.: Mix all ingredients together.

Baby Powder

A soothing powder for baby's tender skin.

I

Talc	3 qts.
Kaolin	1 pt.
Boric Acid	3 cups
Calcium Carbonate (Precipitated Chalk)	1 cup
Zinc Stearate	1 pt.
Perfume, as desired	
Color, as desired	

Proc.: Put all ingredients except for color and perfume into mortar. Rub down well, then add perfume. A few drops of food coloring may be used to tint the powder a delicate color.

Baby Lotion

	oz.
White Mineral Oil (65/75 Viscosity)	7.500
Silicone Oil	1.125
Glyceryl Monostearate S	1.500
Span 80 (Sorbitan Mono-oleate)	0.450
Benzalkonium Chloride	0.075
Glycerin	2.625
Water	61.500
Perfume Oil	0.225

Proc.: Heat the first five ingredients to 65°C. Dissolve the glycerin in the water and heat solution to the same temperture. Mix the two solutions, allowing the emulsion to cool. When the mass is at 35°C., stir in the perfume oil and continue stirring until cold.

Diaper Rash Ointment

	oz.
Balsam Peru, U.S.P. XI	1.0
Castor Oil, U.S.P. XI	1.0
Petrolatum, White, U.S.P. XI	86.0
Boric Acid, Powdered, U.S.P. XI	2.0
Zinc Oxide, U.S.P. XI	5.0
Zinc Stearate, U.S.P. XI	5.0

Proc.: Make a mixture of the castor oil and the Peruvian balsam; stir this mixture into the petrolatum, mix thoroughly, and stir in the powered components. Mix carefully, and put the mass through an mill* until it is perfectly smooth.

Diaper Rash Cream

		%/wt.
A	Tegacid	16.0
	Neobee- M.5	5.0
	Cetyl Alcohol	1.0
	Amerchol C.A.B.	5.0
	Solulan 98	1.0
B	Allantoin N-Acetyl DL-Methionine	0.2
	Propylene Glycol	5.0
	Distilled Water	66.5
C	Perfume Oil	0.3

Proc.: Heat A and B to 75°C. and with rapid agitation add B. Add C at the point that the cream begins to form.

Baby Shampoo
I

	%/wt.
MIRANOL 2MCAS MODIFIED	35.0
Super Amide L-9	4.0
Tween 20	7.0
Water	54.0

Proc.: pH* adjusted to 7.1 with Muriatic Acid

II

	%/wt.
MIRANOL 2 MCAS MODIFIED	30.0
Super Amide L-9	2.0
Propylene Glycol	2.0
Perfume (about 1/8%)	
Tween 20	1.0
Water	65.0

Proc.: Mix all ingredients together until uniform.

SUNTAN AND SUNSCREEN PREPARATIONS

Suntan Cream

		pts./wt.
A	Silicone Fluid SF-96 (350)	2.0
	Isopropyl Myristate	18.5
	Cetyl Alcohol	0.2
	Stearic Acid	4.0
	Giv Tan F	1.5
B	Water (Sterile)	67.5
	Triethanolamine	1.2
	1% Carbopol 934 Solution	5.0
	Cinaryl 200	0.1
C	Perfume	q.s.

Proc.: Heat A and B to 70°C (158°F). Add B slowly to A with good mixing. Continue slow mixing until the temperature reaches 30°C (86°F), then blend in C.

"Natural" Suntan Cream

		%/wt.
A	Cocoa Butter	20.00
	Cetyl Alcohol	1.00

	Emersol 132	2.00
	LANTROL	10.00
	Vitamin E (Alpha Tocopherol)	0.10
B	Veegum R	2.00
	Triethanolamine	0.75
	Methyl Paraben	0.15
	Propyl Paraben	0.10
	Deionized Water	q.s.

Proc.: Heat A to 80°C. Add Veegum to water and add rest of B (disperse Veegum with Lightning Mixer*.) Heat B to 75°C. Add A to B with handstirring. Cool to 40°C and perfume. Cool to room temperature.

Suntan Lotion

		%/wt.
A	Carbopol 940	0.5
	Water, deionized	60.8
B	Amerchol L-101	5.0
	Isopropyl Linoleate	1.0
	Cerasynt 840	3.0
C	Alcohol SD #40	28.0
	Escalol 506	1.2
D	Triethanolamine	0.5
E	Perfume	q.s.

Proc.: Heat and stir A until homogeneous and no lumps remain (70-75°C). Heat B to 70-75°C. When both A and B are about the same temperature, add B to A with agitation. Continue agitation and cool to 50°C, then add C. When completely dissolved, add D whereupon immediate thickening will take place. At 45°C, add E and continue stirring until thoroughly dispersed. Package.

Quick Suntan Stick

		%/wt.
A	Alcohol #40	79.0
	Water	3.0
B	Propylene Glycol	8.0
	Sodium Stearate	6.0
	D.H.A. (Dihydroxyaceton)	2.5
	Amino Benzoic Acid	0.5
C	Perfume Oil	1.0

Proc.: Heat A and B separately. Add A to B stirring until uniform. Add C.

Sun Screening Gel

	pts./wt.
Ethanol (SD-40)	48.0
"Carbopol" 940	1.0
"Escalol" 106 (Glyceryl-p-amino-benzoate)	3.0
Monoisopropanolamine	0.09
Water	47.91
Perfume	9.5

Proc.: Disperse the "Carbopol" 940 in the alcohol, and then dissolve the "Escalol" 106 (Van Dyk & Co.) in the dispersion. Slowly add the monoisopropanolamine. This will give a thin but very cloudy solution. Slowly add the water and stir carefully to avoid air entrapment; the solution will clear up and thicken. Add perfume and color as desired.

Sunscreen Cream

	%/wt.
Vaseline	100
Lanolin	100
Zinc Oxide	30
Lycopodium Dust	30
Salicylic Acid	3
Sweet Almond Oil	50
Spermaceti	50
Virgin Wax	10
Rose Water	5

Proc.: Weigh the ingredients and heat in a double boiler until smooth.

Sun-Screen Lotions
I

		%/wt.
A	Glycery Mono Stearate, Pure	1.0
	Oleyl Polypeptide	0.4
	Spermaceti U.S.P.	1.0
	Ninol 2012 Extra	0.25
B	Methyl Paraben	0.2
	Butyl Paraben	0.02
	Germall #115	0.3
	Polypeptide LSN (40%)	3.0
	Maypon 4CT	1.0
	Water	88.03
	Natrosol 250 HR	0.8
C	Isopropyl Myristate	2.0
	Giv Tan F	1.5

D Perfume: Spring Blossom M-1227   0.50

Color DC Red #19 0.05%
4-5 drops/100 gms.

Proc.: Heat A and B separately to 80°C. Add B to A, agitating and maintaining the temperature. Prepare C and add slowly to the A-B mixture. Cool to 40°C, add D and continue the agitation until cool. Adjust pH to 6.8 with Triethanolamine.*

II

	pts./wt.
Methyl Cellulose	0.64
Glycerin	2.56
Alcohol	13.44
Water	102.40
Menthyl Salicylate	8.96

Proc.: Wet the cellulose with the alcohol, add the glycerin and water, and stir until solution is complete. Add the menthyl salicylate and mix thoroughly.

Freckle Preventive Treatment
I

Sulphur lotions should not be used as these tend to increase the pigmentation, neither should tar preparations, ichthyol, or resorcin be ingredients of the lotions. Before the hot weather comes the following ointment should be used:

	pts./wt.
Quinine Bisulphate	1.5
Aesculin	1.0
Simple Ointment	27.5

II

The following ointment, to be applied twice a day, left on for thirty mintues, and then wiped off with a cleansing tissue, is recommended by Continental dermatologists:

	pts./wt.
Zinc Peroxide	20
White Soft Paraffin	70
Anhydrous Lanolin	10

Proc.: The affected parts are then powdered with a powder composed of magnesium peroxide 30 parts, talc 50 parts, zinc oxide 20 parts.

4. Drugs

This chapter has been prepared with several purposes in mind. First and of paramount importance has been to present formulas and procedures that will be easily understood by the layman. The reader can therefore approach any formula in this chapter with the confidence that he will be able to come out with a successful product at the end of his labor.

Secondly, you are given as wide a variety of formulas as possible—ranging from skin preparations for all parts of the body to cold preventatives and cures. Not only are modern formulas given, but folk and herbal remedies have been included, too. More and more, people are recognizing that modern remedies are not the sole way of treating maladies, and that many older approaches are valid. In some cases more than one formula is given for one type of preparation. The reader can then decide which formula he feels more comfortable with, or he can make sample batches of each, and make his own choice.

The formulas herein are presented with an eye towards public appeal and saleability. Careful work, coupled with ambition can result in a successful small business for you the reader.

It is important to remember that when working with chemicals every precaution should be taken. Whenever possible, do not touch the chemicals with bare hands unless you know the raw ingredients to be harmless. Adhere to the amounts called for in the formulas until you have gained familiarity with the product. Be sure to double check your weights and measurements carefully. These remedies are not offered as a substitute for the care of a physician.

HEALTH

Tonic (Health Restorer)

In order to return the body to a state of good health after a long illness or a generally run down or weakened condition, it is best to be supplied with an extra nutrient booster. This energy fortifier, when consumed daily, in addition to a normal diet, results in the speedy return to good health.

Combine and blend with an electric mixer:

Whole Eggs	2 tablespoons
Lechithin	1 tablespoon
Vegetable Oil	1 tablespoon
Calcium Lactate	1 1/2 teaspoon
Magnesium Oxide	1/2 teaspoon
Yogurt	1/4 cup

When these ingredients have been thoroughly blended, add:

Skim Milk	1 1/2 cups
Yeast (fortified with calcium)	1/3 cup
Non-Instant Powdered Milk	1/3 cup
Soy Flour	1/4 cup
Pure Vanilla	1 teaspoon
Frozen, Undiluted Orange Juice	1/4 cup

Proc.: Pour into a jar and then add the rest of the quart of milk. This recipe is sufficient for a one-day supply. Multiply the ingredients according to the quantity desired.

Castor Oil Oral Emulsion
(36.0%)

%/wt.	
Castor Oil, USP	36.00
Vanillin	0.02
Methyl Salicylate	0.13
Saccharin Calcium	0.13
Citric Acid Monohydrate	0.05
Sodium Benzoate	0.20
Propyl Gallate	0.01
Methylcellulose 4000 cps.	0.75
Water, Distilled	q.s.

Proc.: Slurry the methylcellulose in the castor oil. Add the propyl gallate and methyl salicylate. Dissolve the vanillin, saccharin, citric acid, and sodium benzoate in water. Add the aqueous solution to the oil slurry with rapid agitation. Stir until viscous. Bring to full volume.

With today's cultural emphasis on the natural organic products, people are turning to the old-fashioned, folk medicine, tonics, cold preventatives and cures. Many find that what has worked for thousands of years, still works. The following formulas are old folk medicinals that could have a popular market today.

Cherry Tonic

Proc.: 1 ounce of the stalks of any kind of cherry. Add 1 quart of cold water. Simmer in a saucepan for an hour then strain. Dosage: 1 tablespoon 3 times a day.

Hay Fever Aid

To 1 quart of cold water, add:
½ oz. centaury
½ oz. horehound
½ oz. red sage
½ oz. vervain
½ oz. yarrow

Proc.: Bring to boil, then simmer 15 minutes. Strain. Dosage: ½ cup, 4 times a day., children ¼ cup, 4 times a day.

Asthma Aid

The following is smoked in a pipe or as a cigarette.

Powdered Grindelia Robusta	240 g.
Powdered Jaborandi Leaves.	240 g.
Powdered Eucalyptus Leaves	120 g.
Powdered Cubeb	120 g.
Powdered Stramonium Leaves	450 g.
Powdered Potassium Nitrate	360 g.
Powdered Cascarilla Bark	30 gr.

Proc.: Mix until a uniform powder is obtained.

Asthma Relief
Cigarette Tobacco

	pts./wt.
Stramonium Leaves	65
Henbane Leaves	20
Wild Mint Leaves	10
Corn Silk	5
Potassium Nitrate	2½

Proc.: Crush the leaves and mix all the ingredients together. Then use it like a tobacco. Roll into cigarettes, or use in a pipe.

Nasal Drops

Menthol	2.5 g.
Camphor	2.5 g.
Eucalyptus Oil	6.0 cc.
Mucilage of Methyl Cellulose	60.0 cc.
Chloretone	5.0 g.
Dextrose	45.0 g.
Distilled Water q.s.	1,000.0 cc.

Proc.: Liquefy the menthol, camphor, and eucalyptus oil by trituration in a glass mortar. Add the mucilage under constant stirring until the oily drops disappear. Dissolve the dextrose and chloretone in the boiling water. When cool, mix the liquids, add water to make 1 l., and shake. Label: Shake the bottle.

Cough Syrup

	pts./wt.
Dihydrocodeinone Bitartrate	0.32
Pyrilamine Maleate	2.40
Potassium Guaiacol Sulfonate	32.00
Sodium Citrate	16.00
Citric Acid	16.00
Sorbic Acid	0.50
Water	160.00
Honey	1150.00

Proc.: The solid ingredients are dissolved in about 140 ml of water with the aid of gentle heat. The solution is filtered*; the filter is washed with the remainder of the water, and the filtrate and washings are mixed with the honey. To produce this formula one must have a license.

Old-Fashioned Cough Candy

Canada Snake-Root	1 oz.
Pectoral Species, N.F.	2 oz.
Sugar	12 lb.
Molasses	8 oz.
Oil of Wintergreen	10 minims
Oil of Sassafras	10 minims
Oil of Anise	10 minims
Water	Sufficient

Proc.: Make a decoction of the herbs with 4 pints of water, and strain. Cook the sugar, the molasses and a little water until it forms a homogeneous mass; slowly stir in the decoction, and cook the batch to 310°F. Pour the mass on an oiled slab; and as it cools incorporate the oils. Finally cut into drops.

Menthol Cough Drops

Gelatin	1.0 oz.
Glycerin (by wt.)	2.5 oz.
Orange Flower Water	2.5 oz.
Menthol	5 grains
Rectified Spirits	1 drachm

Proc.: Soak gelatin for 2 hours in water. Heat until dissolved. Add 1.5 oz. of glycerin. Dissolve the menthol in the spirit, mix with the balance of the glycerin, and add to gelatin mixture. Pour into a well greased shallow tray. When cold divide into 120 lozengers.

BURNS

Burn Dressing

I

	pts./wt.
Petrolatum	500
Sorbitol Monöoleate (oil soluble)	20
Sorbitol Monöoleate (water soluble)	3
Sulfathiazole Powder	50
Distilled Water	500

Proc.: The petrolatum and the sorbitols are mixed and heat-sterilized. The sulfathiazole is sterilized and dissolved in the sterile distilled water. The solution is then emulsified with the petrolatum mixture, with aseptic precautions throughout the process. A quantity of 3-by-12-in. strips of fine mesh gauze are placed in a sterile porcelain tray, and on top of these is placed a quantity of the ointment. The tray is covered and heat treated below 210°F, just long enough for thorough permeation of the ointment into the gauze. The use of such impregnated gauze strips makes possible a more uniform dressing and facilitates its application.

II

		pts./wt.
A	Glycerol	18.0
	Procaine HCl	1.2
	Magnesium Sulfate Heptahydrate	20.0
	Magnesium Oxide	6.0
B	Water-Soluble Base	54.8

Proc.: Heat all ingredients in A and B separately. Then mix A into B.

Sunburn

The best way to deal with sunburns is to avoid overexposure to the sun. However, this is no comfort to one who is afflicted with a painful sunburn. A lecture will not help his suffering, but one of these formulas certainly will.

Sunburn Ointment

	pts./wt.
Petrolatum	36
Stearyl Alcohol	3.5
Mineral Oil	15
Sesame Oil	2
Calcium Stearate	10
Kaolin	30

Proc.: Mix all ingredients together until smooth.

Sunburn Lotion
I

	%/wt.
Boric Acid	4
Acetic Acid	2
Citric Acid	1
Alcohol	10
Glycerin	5
Water	78

Proc.: Heat part of the water and dissolve the boric acid in it. Dissolve the citric acid in the alcohol and the acetic acid in the remainder of the water. Add the glycerin, boric acid solution, the citric acid solution, mix and filter.

II

	%/wt.
Picric Acid Solution (20%)	1.5
Alcohol (90%)	10
Water	88
Perfume	.5

Proc.: Dissolve the perfume in alcohol, add to the water and then dissolve the picric acid in the solution. This is excellent for sunburn but as it is will stain the skin yellow, it may be found objectionable for facial sun-quickly or subjected to percussion, it is safer to purchase the 20%.

Wound-Burn Antiseptic

"Cetrimide"	1 g.
β-Phenoxyethyl Alcohol	2 cc.
Water	To make 100 cc.

Proc.: Stir all ingredients until uniform.

Ear

The ear is a very sensitive part of the anatomy. In most instances "home remedies" are not recommended. Accumulation of ear wax is common in some people, and those of us who swim in pools are ripe for fungus infections in the ear. Simple remedies follow.

Ear-Wax Softener

	pts./wt.
Sodium Bicarbonate	0.6
Glycerin	15.0
Water	To Make 30.0

Proc.: Mix all ingredients until sodium bicarbonate is dissolved. Apply with an ear dropper.

Treatment for "Swimming Pool Ear"

Boric Acid	2 g.
Mercury Bichloride (1:1000)	8 cc.
Alcohol, To Make	30 cc.

Proc.: Mix all ingredients until the powdered ingredients are dissolved. A few drops in the ear will help prevent fungus infection.

Anti-Infective-Analgesic Ear Drop

Tyrothricin (0.1% w/v)	10 g.
Benzocaine (5% w/v)	500 g.
Chlorobutanol (0.5% w/v)	50 g.
Polyethylene Glycol-400, q.s.	10 l.

Proc.: Triturate the tyrothricin in portions with small quantities of the polyethylene glycol-400. Add more polyethylene glycol-400 and heat to 50°C. Dissolve with stirring the benzocaine and chlorobutanol. Bring up to volume with polyethylene glycol-400 and filter solution through kieselguhr.

Earache Oil

Oil Thyme	2 oz.
Oil Cajeput	2 oz.
Ethyl Amino Benzoate	3 oz.
Oil Apricot Kernel	93 oz.

Proc.: Dissolve Ethyl Amino Benzoate in Apricot Kernel Oil by gently heating. When completely dissolved, allow to cool and add Oil of Thyme and Oil of Cajeput. Finally add sufficient Oil Soluble Chlorophyll to make it a light green color.

EYE

Irritated eyes is a common disorder. The following are very soothing:

Eye Lotions
I

Sodium Chloride	5 g.
Borax	4 g.
Sodium Bicarbonate	2 g.
Distilled Water, To Make	1 fl. oz.

II

Mercuric Oxycyanide	1 g.
Distilled Water	6 fl. oz.

Proc.: Dilute with equal amount of warm water for use.

MOUTH CARE

Since there is no substitute for your dentist, we can only try to give you in this book items that will promote good dental hygiene. Keeping the mouth and gums properly cleansed, together with good dietetics, should hold your dental expenses to a minimum. Experiment with the formulas. The ones you find most successful should be saleable.

We start out with a cavity preventive.

Dental Caries Preventive
I
(Rinse)

Dibasic Ammonium Phosphate	50.0 g.
Urea	30.0 g.
Glycerin	100.0 cc.
Alcohol	40.0 cc.
Saccharin Sodium	1.0 g.
Menthol	0.4 g.
Amaranth Solution (U.S.P.)	2.0 g.
Sodium Benzoate	1.0 g.
Distilled Water To Make	1,000.0 cc.

Proc.: Mix until all ingredients are dissolved.

II
(Powder)

Dibasic Ammonium Phosphate	5.0 g.
Urea	3.0 g.
Bentonite	5.0 g.

Precipitated Calcium Carbonate	86.6 g.
Saccharin	.2 g.
Menthol	.2 g.
Peppermint Oil	.2 cc.
Cinnamon Oil	.1 cc.

Proc.: Mix all ingredients until smooth.

Gargle, Sore Throat

The preparation is used, diluted with two or three parts of water, either from a spray applied to nose and throat, or as a nasal douche from a nasal irrigator or syringe. Habitual users commonly inhale the solution into the nostrils from the palm of the hand. The preparation is also a most useful gargle for sore throats:

	%/wt.
Sodium Bicarbonate	1.00
Borax	2.00
Sodium Benzoate	0.80
Sodium Salicylate	0.52
Menthol	0.03
Thymol	0.05
Eucalyptol	0.13
Oil of Pumilio Pine	0.05
Oil of Wintergreen	0.03
Alcohol (90 per cent.)	2.50
Glycerin	10.00
Solution of Carmine	0.52
Talc or Kaolin	sufficient
Distilled Water	q.s.

Proc.: The salts are dissolved in 80 of the water and the glycerin added. The other ingredients are dissolved in the alcohol and the alcoholic solution is triturated with the talc (about 5 per cent.), and the mixture added to the salt solution. The solution of carmine is added and the whole is filtered, distilled water being passed through the filter to produce the required volume. Filtration through talc or kaolin is essential to the production of a clear and bright solution.

Antiseptic Mouth Wash

Boric Acid	50.000 g.
Benzoic Acid	1.000 g.
Eucalyptol	0.125 cc.
Oil of Peppermint	0.500 cc.
Oil of Wintergreen	0.250 cc.
Oil of Thyme	0.100 cc.
Grain Alcohol	250.000 cc.

Water to make up to	1000.000 cc.
Caramel	To Color

Proc.: The boric acid is dissolved in the water or about 700 cc. of same. All the other products are dissolved in alcohol and the two solutions mixed and colored to a very pale straw. The above product must be labeled 25% grain alcohol for commercial use.

Mentholated Throat and Mouth Wash

A	Alcohol	4 3/4 gal.
	Ethyl Amino Benzoate	12 oz. 350 gr.
	Thymol	1 oz. 120 gr.
	Eucalyptol	1 oz.
	Oil Wintergreen	3/4 oz.
	Menthol	100 gr.
B	Boric Acid	3 lb.
	Distilled Water	5 1/4 gal.

Proc.: Dissolve ethyl amino benzoate, thymol, eucalyptol, oil wintergreen and menthol in alcohol. Dissolve boric acid in hot distilled water, cool and filter. Add B to A while stirring and then filter.

When confronted with a mouth infection, and until you can get an appointment with your dentist, the following will be very effec‑tive.

Dental Desensitizer (Pain Killer)

I

Thymol	4 dr.
Benzocaine	28 gr.
Acetone	52 min.

Proc.: Stir until dissolved and keep in stoppered brown bottle.

II

Camphor	2
Clove Oil	2
Cajeput Oil	2
Peppermint Oil	1
Menthol	1
Hops Oil	2
Alcohol	15
Ether	10

Proc.: Dissolve the oils in the alcohol and add the ether.

Toothache Oil

	pts./wt.
Clove Oil 12	12
Camphor	6

Chloroform	6
Phenol	2
Menthol	2
Cinnamon Oil	3

Proc.: Mix all the ingredients together until smooth.

HAIR TREATMENT

Falling Hair Treatment

To discover that one's hair is falling out can be a distressing experience. However, there is more than can be done than just looking and worrying.

A powdered mixture of the following can be prepared. This is a daily dose. For larger quantities multiply accordingly. Most people prefer taking in the powdered form, but the powder could be put into a gelatin capsule. Nutritionists recommend taking daily until all signs of the condition disappear.

Inositol	1/2 tsp.
Folic acid	5 mg.
Biotin	50 mg.
PABA	300 mg.

Dilute 1/2 oz. of pearl ash and 1 gill of onion juice with 1 pint of water. Massage into the scalp daily.

"Falling Hair" Ointment

Salicylic Acid	1.0
Resorcinol Monoacetate	1.5
Precipiated Sulfur	1.5
Ointment to Rose Water to make	30

Proc.: Mix salicylic acid, resorcinol, sulfur and add to the warm rose water ointment:

Rose Water Ointment:	
Soap powder	10 gr.
Alcohol	20 cc.
Glycerin	150 cc
Rose Perfume	q.s.
Water	1000 cc.

Proc.: Dissolve the soap in the alcohol. Heat the boric acid with the glycerin until dissolved. Add hot water and mix until uniform. Add rose perfume.

Rub into the scalp vigorously once a day.

Dandruff Ointment

	oz.
Lanolin	12

Water	15
Silver Lactate	3
Tincture Fish Berries	5
Sulfur Iodide	3
Balsam of Peru	15
Cocoa Butter	20
Petrolatum	60
Glycerin	10
Perfume to suit.	

Proc.: Dissolve the silver lactate in water and the sulphur iodide in glycerin. Melt the petrolatum, the lanolin and the cocoa butter, stir in the silver lactate solution, add the sulphur iodide solution and finally the balsam of Peru and the fish berries.

Dandruff Remover

I

	pts./wt.
Swift's "Solar 25 CG," Neutral	25.00
Germicide	0.10
Water	74.90

Proc.: Place Swift's "Solar 25 CG" in a vessel, equipped with agitator. Add germicide. Warm with agitation to dissolve. Next add this solution to the water, preferably warmed, and agitate for approximately 15-30 minutes to assure homogeneity.

II

Mercury Bichloride	0.5 g.
Resorcinol	5 g.
Alcohol	125 cc.
Water	125 cc.

Proc.: Dissolve the bichloride and the resorcinol in the water. Then add alcohol. Apply on the dry scalp and rub thoroughly—then shampoo the hair. One treatment a week is usually sufficient for a complete absence of dandruff.

Crab Louse Ointment

	pts./wt.
y-Benzenehexachloride	1
Stearyl alcohol	3
Beeswax	8
Petrolatum, white	82

Proc.: Warm and mix until uniform.

SKIN

The following is a group of ointments, creams, lotions and salves that have been tested and found successful in aiding the healing of irritations, preventing infections in wounds, and the treatment of minor skin irritations and infections.

Experiment with small batches. Find the ones that suit your means of manufacture best. Test their effectiveness, and then consider marketing them. Everyone has a favorite home remedy. Perhaps yours will make your fortune.

Wrinkle Aid

Wrinkling is caused by loss of elasticity of the skin through degeneration of its fibers. No treatment can cure it or restore to the skin its youthful look. Slight inflammation causes swelling which temporarily smooths out the wrinkles. Such an inflammation, followed by skin peeling, can be brought about by ultraviolet radiation, by freezing with crushed dry ice or by chemical irritants. The following is an irritant type formula:

	pts./wt.
Betanaphthol	10
Precipitated Sulfur	40
Soft Soap, U.S.P.	25
Petrolatum	25

Proc.: Combine all ingredients until a homogeneous paste is formed. This paste may darken, but it will not lose its effectiveness. Caution: Must not be used where a history of kidney disease is known.

Area should be washed with ether or benzine, and the paste should be spread thickly and allowed to remain 20 or 30 minutes. Then remove thoroughly. Soon after, a slight burning is felt, but this will stop shortly. The skin will remain red for a few hours. Treatment should be repeated each day until a tightening sensation is felt, or pealing shows that treatment has been sufficient. Five days is usually enough. During period of treatment no soap or water should be used on treated areas. They may be cleaned with a solution of 0.5% salicylic acid in alcohol.

Until recently, the value of Vitamin E as a skin cream has been underestimated. Research has revealed that the vitamin can prevent scarring, and in some cases remove old scars. Used on wounds and severe burns it promotes the process of healing and reduces the chances of scar tissue formation. When Vitamin E Ointment has been used for skin ulcerations, the healing time has been reduced to a fraction.

Vitamin E Ointment

		pts./wt.
A	Petrolatum	28.0
	Vitamin E	4-6
	(Depending on Strength of Ointment desired)	
	"ARLACEL" 83	2.0
	"TWEEN" 80	1.0
B	Water	q.s. ad 100.0
	Preservative	q.s.

Proc.: Heat A to 85°C., B to 87°C. Add B to A slowly with moderate agitation. Stir until cool.

Patients who are confined to bed whether in a hospital or at home are subject to special skin problems. The following are useful. The chronically ill are particularly bothered by bed sores.

Acne

Acne is known to be the blight of most adolescents. A product that can alleviate the effects of this condition will have a large public appeal. What follows is a group of formulas that successfully treats this unhappy condition.

Acne Lotion, Peeling

A keratolytic ointment recommended as part of a carefully developed routine for treating acne, particularly for adolescents. Unlike proprietary products, it is a flexible formulation in which the active ingredients can be adjusted to give an optimum peeling effect without inflammation.

	pts./wt.
Resorcin	2.4- 9.6
Precipitated Sulfur	2.4-12.0
Zinc Oxide	15.0
Talc	15.0
Glycerin	10.0

Proc.: Mix all ingredients. Stir until homogeneous.

Acne Cream
I

		pts./wt.
A	"Veegum"	1.75
	CMC (7 MSP)	0.40
	Water	34.60
B	Glycerin	5.00
	Allantoin	0.25
	Resorcinol	3.00

"Triton" X-100	0.20
Water	29.70
C "Nytal" 300	16.00
Titanium Dioxide	2.90
Iron Oxides	1.10
Sulfur	5.00
D "Vancide" BN	0.10

Proc.: Dry blend "Veegum" and CMC and add to the water slowly, agitating continually until smooth. Add B to A. Pulverize C and add to B. Incorporate D by making up concentrate in portion of finished cream, triturating thoroughly. Add this to product, milling or homogenizing finished cream.

II

	pts./wt.
Methyl Paraben	0.1
Propyl Paraben	0.05
De-ionized Water q.s.	100.00
Perfume, color to suit.	

Proc.: Melt all ingredients together, excluding the water, to about 80°C. Add water at 85°C. with agitation and stir until cool. Color and perfume to suit.

Acne Lotion
I

Rose Water	2½ gal.
Alcohol	1 gal.
Glycerin	½ pt.
Menthol	½ oz.
Phenol	1 oz.
Methyl Salicylate	½ oz.
Benzaldehyde F. F. C.	¼ lb.
Zinc Oxide	1¼ lb.
Calamine	1¼ lb.
Boric Acid	5/8 lb.

Proc.: Mix all ingredients together until a homogeneous lotion is formed.

II

	pts./wt.
Calamine	4
Zinc Oxide	8
Phenol	2
Glycerin	8
Spirit of Camphor	4

Distilled Water To make   240

For patients who have a dark complexion and for those whose skin is quite oily, a more "drying" lotion is prescribed:

Blackhead and Whitehead Lotion

	pts./wt.
Pepsin, of 4000 activity	1.28
Aluminum Potassium Sulfate, U.S.P.	2.56
Lactic Acid, U.S.P.	1.28
Boric Acid, U.S.P.	1.28
Distilled Water	121.60

Proc.: Mix and dissolve all ingredients in the water. Shake well, and allow solution to stand overnight. Filter if necessary.

Apply this lotion in the form of wet compresses, or dab it on with cotton. This lotion digests and dissolves blackheads and whiteheads. It serves remarkably to clarify the skin and give it a clean, fresh appearance.

Impetigo can cause a great deal of discomfort. It can flare up on all parts of the body unless checked. Here are two formulas that control this skin disease.

Impetigo Jelly

Methyl Cellulose (15 Centipoises)	10.0
Sulfathiazole	1.0
Carbowax 1500	3.0
Glycerin	1.0
Alcohol (70%) To make	100.0 cc.

Proc.: Mix all ingredients together until uniform. The local application of this preparation has the distinct advantage of drying rapidly and not requiring a bandage to retain the therapeutic agent in contact with the lesion.

Impetigo Lotion

Boric Acid	4 gr.
Zinc Sulfate	4 gr.
Saturated solution of Sulfanilamide in freshly distilled water, to make	1 fl. oz.

Proc.: Combine all ingredients. Bottle and cork securely.

Ringworm is a common, annoying skin ailment which can be quickly cleared up by any of the following preparations.

Ointment for Ringworm
I

Salicylic Acid	0.5
Iodine	0.5

Sulfur, Precipitated	5.0
Triethanolamine	6.0
Zinc Undecylenate	20.0
Undecylenic Acid	5.0
Washable base	63.0

Proc.: Dissolve the iodine in the triethanolamine. Triturate the powders with a small amount of the washable base and finally combine and mix all ingredients. Mill if necessary to make smooth.

II

Precipitated Sulfur	1.5 g. - 6 g.
Petrolatum	30.0 g.

Proc.: Mix until uniform cream is formed. Rub in gently once or twice a day.

Lassar Paste

	pts./wt.
Lard	50
Zinc Oxide	25
Starch	25
Salicylic Acid	3

Proc.: Melt together and mix well.

Medicated Cream Base

	pts./wt.
Cetyl Alcohol	6.4
Stearyl Alcohol	6.4
Sodium Lauryl Sulfate	1.5
White Petrolatum	14.3
Mineral Oil	21.4
Water	50.0

Proc.: Warm the fatty alcohols on a water bath to 65° C., and stir in the sodium lauryl sulfate. Add the mineral oil and petrolatum. Cool to room temperature and add the water slowly, with constant stirring.

Antiseptic Ointment

The value of ointments seems to depend as much on the type of base used as on the antiseptic constituent according to recent test.

	%/wt.
Petrolatum	73
Anhydrous Lanolin	25
Phenol U.S.P.	2

Proc.: Variations of more than 1/2% will materially lower the antiseptic value. Grind together.

Ointment for Boils

	pts./wt.
Tetracycline	2.4
Zinc Oxide and Talc	36.0
Glycerol	24.0
Bentonite	12.0
Water To make	240.0

Proc.: Mix with mortar and pestles all of the ingredients together until uniform. Such products should never be considered medical prescriptions.

Wart Treatment

Of the non-specific methods of treating warts probably paints of various kinds are most commonly employed. A well-known one is 10 per cent salicylic acid in flexible collodion. Owing to the horny surface of a wart it is essential to use a keratolytic substance, and caustic potash, phenol, and glacial acetic acid belong to this category. A useful combination is the following:

	pts./wt.
Phenol	10
Glacial Acetic Acid	10
Salicylic Acid	10
Tincture of Iodine	20
Alcohol	100

Success may often be obtained by constant daily application of such remedies.

Corn Removers
I
(pts./wt.)

Solution of monochloroacetic acid in ratio of 1:2 is suitable, but stronger solutions should not be used as they irritate skin. Another preparation contains 10 parts salicylic acid and 90 parts glacial acetic acid. This is thickened with mucilage containing 0.5 part gum tragacanth, 3 parts pectin, 3 parts glycerin and 43.5 parts water. About 5 parts of this mixture is used for thickening the preparation. Another composition contains 1 part glacial acetic acid, 8 parts lactic acid, 3 parts dried salicylic acid crystals and 8 parts of aforementioned thickener. Formic acid and carbolic acid, thickened with same thickener, may also be used.

II

	pts./wt.
A Salicylic Acid	30
B Ammonium Chloride	30
C Acetic Acid (90%)	8
D Lanolin	30
E White Beeswax	30
F Lard	40
G Water	40

Proc.: Melt together, D, E, and F. Add to the warm mass A, and, after cooling, mix into it B dissolved in G, and finally add C. Place in tin tubes or jars.

Prickly Heat Preparations

I

Menthol	2 gr.
Camphor	10 gr.
Eucalyptus Oil	3 min.
Paraffin Wax To Make	1 oz.

II

	pts./wt.
Sublimed Sulfur	10
Zinc Oxide	10
Boric Acid	10
Starch	10

Insect bites, mosquito, fly, chigger, bee, etc., all aid and abet the summer miseries. Lotions, creams and salves help to remove the irritations and prevent further infections. Once again may we say that many people have their favorites. It is up to you to come up with new ones. Use the formulas given as "starters." Once you have these down pat start your experimenting.

Insect Bite Lotion

I

Ethylaminobenzoate	1.0 g.
Camphor	1.0 g.
Liquified Phenol	0.3 cc.
Alcohol (70%)	30.0 cc.

Proc.: Mix and dissolve all chemicals.

II

Ethylaminobenzoate	1.0 g.
Camphor	1.0 g.

Liquefied Phenol	0.3 cc.
Alcohol (70%)	30.0 cc.

Proc.: Mix and dissolve all chemicals.

Insect Bite Cream

I

	%/wt.
Vanishing Cream	90.0
Benzene	6.0
Naphthalene	3.0
Oil of Pennyroyal	1.0

Proc.: Mix all ingredients together until smooth using heat to about 120°F. using double-boiler type vessel.

II

	%/wt.
Glyceryl Monostearate	20.00
Propylene Glycol	5.00
Spermaceti	5.00
Water	67.90
Camphor	0.50
Menthol	0.50
Phenol	0.50
Clove Oil	0.25
Eucalyptol	0.25

Proc.: Heat the first four ingredients together to boiling, remove from the heat, and stir until cool. Mix the last five ingredients together and stir or grind in a mortar until a clear mixture results. When the emulsion is cool, add the second mixture to it slowly, with stirring, and mix thoroughly.

Calamine Lotion
I

Calamine	80 g.
Zinc Oxide	80 g.
Glycerol	20 cc.
Calcium Hydroxide (Saturated Solution)	1000 cc.
Phenol	10 g.

Proc.: Mix ingredients together until smooth.

II

Calamine	80.00 g.
Zinc Oxide	80.00 g.
Medium-Viscosity Sodium Carboxymethyl Cellulose	20.00 g.

Dioctyl Sodium Sulfosuccinate	0.65 g.
Glycerin	30.00 cc.
Water	960.00 cc.

Proc.: Dissolve the cellulose in 600 cc. of the water with vigorous stirring; then, with slower stirring, add 285 cc. of the water and then add the sulfosuccinate previously dissolved in 65 cc. of water. Rub the calamine and the zinc oxide to a smooth paste with the glycerin and about 40 cc. of the prepared suspension base; then add the remainder of the base in portions with trituration after each addition, or slowly with continuous trituration.

This lotion pours easily from small dispensing bottles without clogging, spreads smoothly on the skin, and dries to a flexible film which does not rub off easily on fabrics. It is compatible with liquefied phenol and can be used to prepare neocalamine lotions which are blended to match skin colors.

Poison Ivy Remedies
I
Dusting Powder

	pts./wt.
Colloidal Sulfur	50
Sodium Hyposulphite (Powder)	5
Lycopodium (or Colloidal Clay)	35
Calcium Stearate	10

Proc.: Mix. Particularly useful when lesions are weeping profusely.

Poison Ivy Lotion
II

		pts./wt.
A	"Veegum"	1.50
	Water	49.40
B	Zirconium Oxide	4.00
	Propylene Glycol	5.00
	"Vancide" BL	0.25
C	Isopropanol	8.00
	Benzocaine	1.50
	Menthol	0.05
D	Sodium Carboxymethylcellulose (med. misc.)	0.30
	Water	30.00

Proc.: Add the "Veegum" to the water slowly, agitating continually until smooth. Add B to A. Add C with rapid agitation. Add the CMC to the water slowly, mixing until smooth. Add D and mix well.

Protective Cream
(Poison Ivy—Poison Oak)
I

	pts./wt.
Glyceryl Monostearate (Aldo 28)	7.50
Lanolin Alcohol	4.50
Triethanolamine Ricinoleate	0.75
Glycerin	4.50
Distilled Water	50.85
Methanol	4.50
Oleyl Alcohol, N.F.	2.25
Chlorocresol	0.15
Zirconium Oxide (this can be varied)	3.00

Proc.: Melt the first 3 ingredients. Dissolve the glycerin in the water and heat, in a separate container, to the same temperature. Pour the water solution slowly into the oily mixture with constant stirring. While it is cooling, stir in the methanol, oleyl alcohol, and chlorocresol. Finally, stir in the zirconium oxide, and continue stirring until the mass is cold.

II

	pts./wt.
Monostearin, Self-emulsifying	10.0
Glycerol	6.0
Triethanolamine Ricinoleate	1.0
Wool Alcohols	6.0
Oleyl Alcohol	3.0
Chlorocresol	0.2
Zirconium Oxide	4.0
Distilled Water	To make 100.0

Proc.: Mix all ingredients together until uniform.

If you have an itch, instead of scratching, try these formulas.

Itch Ointment
I

	pts./wt.
Dyclonine Hydrochloride	1.0
Stearic Acid	18.0
Polyethyleneglycol 1000 Monostearate	8.0
Polyethyleneglycol 400 Monostearate	5.0
Beeswax (White)	2.0
Hydrous Chlorobutanol	0.3
Distilled Water	To make 100.0

Proc.: Mix all ingredients until homogeneous ointment is reached.

II

This is a doughy non-greasy cakelike material, which is applied by hand in a thick layer over the itching regions, but not rubbed in.

N-Butyl-p-Amino-Benzoate	100.0 g.
Benzyl Alcohol	170.0 c.c.
Anhydrous Lanolin	20.0 c.c.
Cornstarch	640.0 g.
Sodium Lauryl Sulfate	64.0 g.

Proc.: More benzyl alcohol is added in mixing the other ingredients, according to the following directions:

Warm the benzyl alcohol, and dissolve it in the n-butyl-p-aminobenzoate making approximately saturated solution.

Add melted lanolin, keeping the mixture warm and stirred until as much of the lanolin as will dissolve is in solution.

Mix well the cornstarch and sodium lauryl sulfonate, and add slowly to this powder, a little at a time, the warm lanolin mixture. Knead this mixture to distribute the liquid evenly through the powder.

Add benzyl alcohol about a tenth of that already used to produce a doughy, non-greasy, cakelike ointment that can be packed in ointment jars or other suitable containers. It is better not to use containers made of metal.

Infant's Wound Ointment

Infants and young children have particularly delicate skin, and so their ailments must be handled with particular care. The following preparations are geared to the young or those with young and tender skin.

	pts./wt.
Zinc Oxide	10
Talc	10
Balsam Peru	3
"Emulgade" F	12
"Cetiol" V	6
White Petrolatum	18
Mineral Oil	6
Water To Make	100
Preservative	To Suit

Proc.: "Emulgade" F, "Cetiol" V, the white petrolatum, and the mineral oil are melted in a double boiler at 158 to 176°F. The water, in which the preservative has been dissolved, is heated to the same temperature and then stirred into the fatty mixture. The zinc oxide is mixed with the talc and then blended with the balsam of Peru. This mixture is stirred into the ointment base at about 86°F.

Baby-Skin Lotion

Oxyquinoline Sulfate	0.1
Cholesterol Absorption Base	10.0
Mineral Oil	65.0
Water	24.9

Proc.: Mix all the ingredients together. Heat to 100°F., and stir until dissolved.

Frost Bite Prevention Cream

Lanolin	10	g.
Cocoa Butter	15	g.
Olive Oil	25	g.
White Beeswax	22	g.
Water	25	g.
Borax	1.5	g.
Benzoic Acid	.5	g.
Perfume	1	g.

Proc.: Mix the fat oil and wax at about 49°C., and the solution of borax and water which should be about 10 degrees warmer. The whole mixture should be stirred until the temperature has dropped to 45°C. The last ingredients are the perfume and the benzoic acid which should be dissolved in a little alcohol to make a more "elegant" preparation.

Antibiotic Cream
I
(Germicidal)

	pts./wt.
Glyceryl Monostearate	15.00
Spermaceti, U.S.P.	5.00
White Mineral Oil, Heavy	4.00
Glycerin	6.00
n-Butyl-p-hydroxybenzoate (Butoben)	0.10
Dioctyl Sodium Sulfosuccinate (wetting agent)	0.02
Distilled Water	69.88

Proc.: Place the monostearate, spermaceti, mineral oil, glycerin, and Butoben in a suitable container, and heat to 75°C. Place the wetting agent in the water and heat to the same temperature. Pour the water solution into the melted fats, stirring at a medium rate of speed. Continue to stir until the cream is cold.

II

	pts./wt.
Glyceryl Monostearate SE	15.00
Glycerin	6.00

Spermaceti	5.00
Liquid Petrolatum, Heavy	4.00
"Butoben"	0.10
"Aerosol" OT	0.02
Distilled Water, q.s. ad	100.0

Proc.: The antibiotic is dissolved in a small quantity of sterile saline and incorporated with a sufficient amount of the prepared base. Dermatologic antibiotic cream should be kept in the refrigerator to help preserve potency.

Antiseptic Cream

Carbolic Acid	5 gr.
Camphor	10 gr.
Lanolin (Anhydrous)	1 oz.
Paraffin Wax (Soft)	1½ oz.
Cocoa Butter	½ oz.

Proc.: Melt the lanolin, paraffin, and cocoa butter together. Rub together the carbolic acid and camphor, and when they are liquid, add to the first mixture. Stir until cold.

Healing Ointment

	%/wt.
Camphor	5
Menthol	4
Turpentine	1
Eucalyptus Oil	1
Cedar-Leaf Oil	1
Nutmeg Oil	1
Thymol Oil	1
Wintergreen Oil	1
Petroleum Jelly	85

Proc.: Melt together and mix well.

Analgesic Balm Ointment

	pts./wt.
Menthol	10.0
Menthyl Salicylate	10.0
White Wax	5.0
Petrolatum	10.0
"Arlacel" 80	2.5
Lanolin, Anhydrous	45.0
Water Distilled q.s.	100.0

Proc.: Fuse wax, petrolatum, lanolin, and "Arlacel" at 70°C. Add menthol and methyl salicylate. Heat water to 72°C. and add to oils with agitation.

Photography has become a popular hobby among many novices. Working with photographic developers can cause a skin rash called metol poisoning. Although prevention is the best cure, there is a formula which is successful in the treatment of this dermatitis.

Metol Poisoning Salve

Wash developer from hands with dilute acetic acid (1%). The following salve has been found helpful for dermatitis resulting from metol.

	pts./wt.
Ichthyol	10
Lanolin	40
Petrolatum	30
Powdered Boric Acid	25

Proc.: Mix until uniform.

Here are some popular formulas for liniments. They can be liberally used to relieve muscle aches and pains.

Liniment

I

Wormwood Oil	5	c.c.
Menthol	1	g.
Thymol	0.50	g.
Iodine Crystals	0.25	g.
Cajeput Oil	2	c.c.
Tyrolese Pine Oil	2	c.c.
Ethanol	25	c.c.

II

	%/wt.
Spirits of Camphor	20
Spirits of Ammonia	5
Wintergreen Oil	3
Rubbing Alcohol	72

Stimulating Liniment

I

Tincture of Iodine, 7%	12.0 oz.
Turpentine Oil	3.0 oz.
Glycerin	3.0 oz.
Croton Oil	25.0 min.

Proc.: Mix the iodine with the glycerin, then add the croton oil and turpentine oil, and mix well. Apply daily without any covering.

II

	oz.
Tincture of Iodine, 7%	9.0
Spirit of Camphor (1 oz. camphor in 9 oz. alcohol)	9.0
Glycerin	9.0
Isopropyl Alcohol	23.0

Proc.: Mix thoroughly. Apply to affected surface, and rub well. Cover with cotton and bandage.

Rheumatism Liniment

Menthol	1 lb.
Camphor Liniment	6 lb.
Olive Oil	5 lb.
Methyl Salicylate	20 lb.
Alcohol	4 lb.

Proc.: Dissolve the menthol in the camphor liniment. Add oil, methyl and alcohol. Filter.

Analgesic Back Rub

	pts./wt.
Methyl Salicylate	40
Methol	5
Oil Turpentine	20
Oil Clove	5
Propylene Glycol	5
Oleic Acid	3
Triethanolamine	5
Water	17

Proc.: Mix the first six ingredients together. Mix the last two ingredients together. Then blend all with stirring.

Analgesic Ointment for Arthritic Pain

	%/wt.
Methyl Salicylate	40.0
Menthol	10.0
Eucalyptol	3.0
Camphor	1.0
White Petrolatum	46.0

This is *extra strong*. No rubbing needed.

Proc.: Dissolve menthol and camphor in the menthyl salicylate and add the eucalyptol. Then mix with the white petrolatum until smooth.

Many people find relief from the discomfort of the congestion that accompanies a cold by the use of a chest rub salve.

Chest Rub

	%/wt.
Nicotinic Acid Methyl Ester	0.5
Rectified Oil of Turpentine	5.0
Pine Needle Oil	5.0
Oil of Thyme	1.0
Rosemary Oil	1.0
Eucalyptus Oil	1.0
Oil of Juniper	0.5
Oil of Sage	0.5
Flowers of Camphor	3.0
"Emulgade" F	18.0
"Cetiol" V	5.0
Liquid Paraffin	5.0
Peanut Oil, Hydrogenated	5.0
Dist. Water To Make	100.0
Preservative	q.s.

Proc.: Mix all ingredients until a uniform solution is obtained. Bottle and store in a cool place.

Hospital Massage Lotion
I

	pts/wt.
Olive oil	560.0
Stearic acid	600.0
Cetyl alcohol	300.0
"Aquaphor"	375.0
Aluminum dihydroxy allantoinate	30.0

II

Triethanolamine	250.0
Sorbitol solution (70%)	1.0
Perfume	
Ethyl alcohol (95%)	100.0
Purified water	11.4

Proc.: Levigate allantoinate with olive oil. Heat Group 1 ingredients to 70°C. Place triethanolamine in water and heat to 70°C. Slowly add melted Group 1 ingredients to aqueous solution. Add sorbitol solution. Mix perfume with alcohol. Add alcoholic mixture to lotion when it has reached room temperature and stir well.

Antiseptic Massage Lotion

	%/wt.
Germicide	0.5
Diethylene Glycol Monostearate "C"	2.0
Stearic Acid T.P.	2.0
Cetyl Alcohol N.F.	0.5
Deltyl Extra	10.0
Lanolin U.S.P.	1.0
"Trisamine"	1.0
Water	82.5
Perfume Oil	0.5

Proc.: Mix all ingredients until uniform. The results of tests show that the cream and lotion containing Germicide are highly effective against the antibiotic-resistant staphylococci.

Massage Cream

	g.
"Polawax"	4.0
Super Harton Lanolin Alcohol, B.P.	30.0
Olive Oil	32.0
Nolasia or Perfume Oil	3.0
"Velvasil 300" Silicone Fluid	5.0
Witch Hazel Solution	260.0
Distilled Water	227.0
Preservative	9.5

Proc.: Melt together and mix well.

Athletes' Rub

Alcohol	100 oz.
Witch Hazel	50 oz.
Methyl Salicylate	2 oz.

Proc.: Mix and filter.

STOMACH REMEDIES

Liquid Antacid Drug

I

		pts./wt.
A	"Veegum"	1.0
	"CMC" (type 7 MSP)	0.5
	Water	71.4
B	Saccharin	0.1
	Water	6.0
C	Aluminum Hydroxide gel F2000	12.0

D Magnesium Trisilicate   9.0
   Preservative   q.s.

Proc.: Dry-blend A adding to the water slowly, agitating continually until smooth. Mix B together and add to A. Add small portions of A-B to C, triturating until smooth. Add D and mix until smooth.

Liquid antacids are particularly difficult to formulate as they tend to settle with hard packing, or gel on aging. This formula is very stable on aging and is easily restored to its original viscosity by shaking.

II
Calcium Carbonate

	pts./wt.
Calcium Carbonate pptd	0.3300
Mannitol	0.3300
Saccharin, Soluble	0.0007
10% Starch	0.0240
Paste, *ca.* (as starch) 0.24 g	
Peppermint Oil	0.0007
Talc	0.0076
Magnesium Stearate	0.0070

Proc.: Blend calcium carbonate and mannitol. Dissolve saccharin in starch paste and use this to granulate the medicament-mannitol blend. Dry at 140°F. and screen through 16-mesh screen. Incorporate flavor, talc, and magnesium stearate; mix well. Age at least 24 hours and compress into tablets.

Indigestion Mixture

Bismuth Subcarbonate	1 lb.
Powdered Rhubarb	1 lb.
Sodium Bicarbonate	4 lb.
Magnesium Carbonate	2 lb.
Peppermint Oil	2 oz.

Proc.: Mix the peppermint oil with the magnesium carbonate. Add other powders and mix in ball mill* for one hour. Take one-half teaspoonfull in one-half glass water.

Diarrhea Relief
I

Whiskey	½ pt.
Powdered Ginger Root	1 oz.
Powdered or Grated Nutmeg	1 oz.

Proc.: Mix well by shaking thoroughly. The dose is 1 teaspoonful every 2 to 3 hours until relieved.

II

Calcium Carbonate, Precipitated, U.S.P.	1 oz.
Tincture of Gambir	3 dr.
Tincture of Opium	30 min.
Oil of Cinnamon, U.S.P.	3 min.
Distilled Water, enough to make	7 oz.

Proc.: Mix the tinctures and dissolve the oil of cinnamon in the liquid. Stir in the carbonate, then the water, and shake well. The dosage is 1 ounce.

Constipation Relief

	%/wt.
Magnesium Sulphate	6.5
Sodium Sulphate	6.5
Potassium Sulphate	.06
Sodium Bicarbonate	.52
Sodium Chloride	.42
Water	36
Honey	50

Proc.: Dissolve the several ingredients in the water, add the honey.

For Haemorrhoids

	pts./wt.
"Dehydag" Wax SX	10.0
"Cetiol" V	6.0
Lanolin	20.0
Petrolatum, White	20.0
Procaine Hydrochloride	1.0
Menthol	0.2
Bismuth Subgallate, Basic	5.0
Zinc Oxide	5.0
Rice Starch	5.0
Witch Hazel Extract, dist.	10.0
Water	17.8
Preservatives. As desired	

Proc.: Heat "Dehydag" Wax SX, "Cetiol" V, lanolin, petrolatum, and bismuth gallate on the water bath to 70-80°C. (158-176°F.). Dissolve procaine, witch hazel extract, and preservatives in the water, likewise heated to 70-80°C. (158-176°F.), and stir these into the molten fats.

Emulsification takes place while stirring until cold. Into resultant ointment incorporate the zinc oxide and rice starch at room temperature. Finally pass the finished ointment through a roller mill and fill into tubes or other suitable containers.

Suppositories

Here is a simple formula for glycerin suppositories.

Mix:

Sodium Bicarbonate	1 g.
Glycerin	33 g.

Proc.: Heat for 20 minutes over moderate flame. Add: 3 grams of stearin. Stir mixture over low flame until it becomes homogeneous. Allow to cool and form into molds.

FEET

Foot Powder

The ordinary old-time foot powder is composed principally of some such base as talc and starch, together with a little boric or salicylic acid. Modifications of this old formula are as follows:

I

Salicylic Acid	6 dr.
Boric Acid	3 oz.
Powdered Elm Bark	1 oz.
Powdered Orris	1 oz.
Talc	36 oz.

II

Salicylic Acid	1 dr.
Powdered Zinc Oleate	1½ oz.
Starch Powder	3 oz.

Proc.: Mix together until uniform.

Foot Relief

Tannic acid, 30 grams; powdered alum, 60 grams; ticture iodine, 2 cubic centimeters; grain alcohol, 60 cubic centimeters; water, to make 500 cubic centimeters. Dissolve first two in water, cool and add rest. Let stand 1 day, then filter.

Use 1 tablespoonful per quart of water. Bathe the feet every night in this warm solution. Gently massage the feet while in the water. Do not use any soap. Let the feet dry without wiping.

Athlete's Foot Powder

Athlete's foot can be nasty. Fortunately we have the following:

A	Undecylenic Acid	2.0
	"Cabosil" M-5	0.5

Mix and Add

B Zinc Undecylenate 20.0
  Magnesium Stearate 2.0
  Talc USP 75.5

Proc.: Mix the ingredients in A. Add B to A with stirring.

Athlete's Foot Control Cream

	pts./wt.
Undecylenic Acid	5.00
Triethanolamine, enough to buffer to*	pH 6.50
Zinc Undecylenate	20.00
Hydrophyllic Ointment Base USP	75.00

Proc.: Rub up zinc undecylenate with ointment base until smooth and add the buffered acid. Mix slowly.

Foot Corn Ointment

Corns can usually be prevented by wearing well-fitted shoes. Some how or other most of us manage to develop corns now and then.

	%/wt.
Benzoic Acid	1.0
Lanolin	44.5
Salicylic Acid	5.0
Sulfathiazole	5.0
Vanishing Cream	44.5

Proc.: Warm the lanolin and vanishing cream together and beat in a finely pulverized mixture of the benzoic acid, salicylic acid, and sulfathiazole. When all the ingredients have been added, continue to beat until a good, fluffy cream is obtained. This cream is particularly useful for soft corns between the toes and similar foot irritations.

MISCELLANEOUS

Vaginitis Suppository

Sulfanilamide	1.0 g.
Sulfathiazole	1.0 g.
Zinc Peroxide	0.5 g.
Sodium Tetradecyl Sulfate	0.2 cc.
Cocoa Butter To make	4.0 g.

Proc.: Mix all ingredients together under heat, and pour into molds. Note: One must have a license to make this medicine.

Spray-On Bandage

	%/wt.
PVP/VA (35/65)	10.0
Dibutyl Phthalate	1.0
Acetone	20.0

Alcohol Anhydrous 69.0

Charge:

35% Concentrate

65% Propellant 12/11 (70/30)

Proc.: Mix ingredients together and charge with the propellant. See Aerosol section for instructions.

Styptic Pencil

	pts/wt.
Ammonium Chloride	6
Copper Sulfate	6
Zinc Sulfate	6
Ferric Sulfate	6
Alum	To make 100

Proc.: Heat and mold.

Smelling Salts

	%/wt.
Phenol	1
Menthol	1
Camphor	2
Weak Solution of Iodine (2.5 per cent. v/v)	1
Oil of Pumilio Pine	1
Oil of Eucalyptus	1
Strong Solution of Ammonia	3
Ammonium Carbonate	90

Proc.: The ammonium carbonate should be packed into the bottle, the strong solution of ammonia added, then the other ingredients, previously mixed. Sodium sesquicarbonate is sometimes substituted for ammonium carbonate.

5. Farm, Garden and Home Specialties

Whether home is a city apartment, a house with a small plot of land, or a farm of many acres, the care of plants and animals is a part of our lives.

This chapter covers all types of insecticides, whether they be for roaches in a city apartment, thrips on the gladiolas in the garden, or fleas on your favorite cat.

Care of crops or small gardens include the use of fertilizers, herbicides, insecticides, and pesticides. Formulas for all these products are included.

The use of insect repellents has grown tremendously since World War II, and a complete selection is included for your experimenting.

Before preparing any of the "killers", be sure to check the ingredients with the local U.S.D.A. authorities. Everyday new list of forbidden ingredients are issued. Before actually manufacturing, formulas must be submitted to your branch office of the U.S.D.A. who will check out the formula, and advise you of label requirements.

Algae Control

The treatment consists of two phases, algaecidal and algistatic. Generally, the addition of 1 gallon of a 10% "Hyamine" 2389 solution to each 10,000 gallons of water, providing a concentration of 10 ppm, is adequate for algaecidal treatment. This may be added with other water treatment chemicals over a period of several hours to several days. After the algaecidal treatment is complete, dead growth should be removed or the water changed. One gallon of 10% "Hyamine" 2389 solution is added to each 40,000 to 50,000 gallons of make-up water, providing 2-3 ppm "Hyamine" 2389 to prevent further growth. In certain cases, periodic shock treatment may be preferable.

Pond Larvicide

	pts/wt.
Light Fuel Oil	95
Concentrated Pyrethrum Extract (Containing 2.5g. Pyrethrins per cc.)	5
Coconut Oil-Potash Liquid Soap	5
Water	45

Proc.: The soap is agitated with water until it foams. The pyrethrum is thoroughly mixed with the oil. This oil mixture is then gradually added to the soap solution, with constant stirring and pumping. When all the oil has been mixed, agitation should continue for 5 to 10 minutes or until a homogeneous emulsion results. This forms a stock emulsion which should be freshly prepared and is diluted in the proportion of 1 qt. to 4 gal. water just before use.

Increasing Natural Food for Fish In Ponds

Addition of the following increased fish production from 134—578 lb. per acre.

	pts./wt.
Ammonium Sulphate	40
Superphosphate (16%)	60
Muriate of Potash	30
Basic Slag or Calcium Carbonate	15
per acre	

Proc.: Mix all ingredients together until uniform.

Swimming Pool Algaecide

Used for preventing growth of unsightly scum:

I

	%/wt.
"Variquat" 415	10
Water	90

As in all quaternary formulations, it is advisable to use softened water for making up this formulation. For a 50,000 gallon swimming pool, it is recommended that 1 gallon of product be added at the start, then another quart every few days.

HOUSE AND GARDEN PLANTS

Cut-Flower Preservative

I

(Australian Patent 165,666)

2,2^1-bis (4,6-Dichlorophenol) Sulfide water solution. Use up to 5 p.p.m.

II

(German Patent 952,753)

	pts./wt
Hydrazine Sulfate	12
Sodium Chloride	6
"Pril" (Detergent)	1

Add 1 g. of above to 1 liter of water.

Summer-Foliage Preservative

A solution made from 1 part glycerin and 2 parts water serves a the preservative for magnolia, beech, eucalyptus leaves and many other tree and shrub materials. Stems are split and placed to a depth of 3 to 5 in. in the glycerin-water solution. Allow to stand for approximately 2 weeks or until they have turned a warm brown in color. Foliage preserved in this way is soft, pliable, and attractive, lasts indefinitely, and can be used in either fresh or dry arrangements.

Preserving Color of Leaves

Immerse leaves in	
Glycerin	5 g.
Copper Sulphate	2 g.
Water	93 cc.

Nursery Stock Coating Wax Emulsions

I

	%/wt.
Paraffin Wax	32.3
Drying Oil*	10.8
Trihydroxyethylamine Stearate	8 1
Water	48.8

Proc.: Heat all ingredients and stir until uniform.

II

	%/wt.
Paraffin Wax	4.60
Drying Oil*	1.53
Trihydroxyethylamine Stearate	0.31
Water	92.02

Bentonite	1.23

Proc.: Heat all ingredients and stir until uniform.

Liquid Grafting Wax
I

Rosin	16 lb.
Tallow	2 lb.
Isopropyl Alcohol	3 qt.
Turpentine	1 pt.

Proc.: Melt rosin and tallow together, cool sufficiently, and dilute with alcohol and turpentine. Keep tightly stoppered.

II

	%/wt.
Scale Wax	40
Paraffin Wax (135°F.)	47
Microcrystalline Wax (170°F.)	8
Anhydrous Lanolin	5

A portable lantern is available which will keep a small pot of this wax in a "just melted" condition. The wax must not be so hot as to injure plant tissues. The formula can be hardened or softened by adjusting the ratio of soft paraffin to hard paraffin.

Plant Leaf-Shine Spray
(Aerosol)

	%/wt.
PVP/VA 1-535	3.00 to 5.00
"Carbowax" 1500	0.20 to 0.35
Isopropanol	31.80 to 29.65
Propellents 12/11 (40/60)	65.00

Proc.: Dissove the PVP/VA 1-535 in isopropanol. Add "Carbowax." Fill under pressure.

Antifreeze for Plants

	%/wt.
PVP/VA S-630	1-5
Water, to make	100

Proc.: Mix well. Spray on plants to prevent excessive transpiration and freezing. Will also prevent wiltage of ornamental plants during transplanting.

Plant Aphicide

Tetraethyl Pyrophosphate (40%)	4 oz.
Water	50 gal.

Proc.: Apply by spray. It is very toxic when freshly mixed with

water, but rapidly loses toxicity after mixing so that no harmful
residue remains on the plants after 24 hours.

Artificial Humus

Sugar	10 kg.
Hydrochloric Acid (0.2N)	100 kg.

Proc.: Heat for 3 hours at 100°C.

Plant Growth Accelerator
(British Patent 506,910)

	pts/wt.
Bentonite	1½
Water	50
Ammonium Linoleate	1½
Paraffin Wax	5

Proc.: To the above which has been emulsified by heating with
vigorous stirring add β-Indolylacetic acid to give a concentration of 1
in 300,000 parts by weight.

Root Growth Stimulant
(British Patent 514,250)

β-Indolpropionic Acid	4.000 kg.
Anthraceneacetic Acid	0.005 kg.
Petrolatum	985.000 kg.
Fungicide	1-70.000 kg.
Calcium Chloride	Trace

Proc.: Heat all ingredients and stir until uniform.

Soil Acidifier

	pts./wt.
Flowers of Sulfur	1
Ammonium Sulfate	1
Aluminum Sulfate	1

Proc.: Add to soil until it gives the pH value desired.

Trace Elements for Use on Garden Soils and Plants

		pts./vol.
A	Manganese Sulfate	1
	Magnesium Sulfate	1
	Ferrous Sulfate	1
	Aluminum Sulfate	1
	Copper Sulfate	1
	Zinc Sulfate	1
	Dissolve in 1 gal. water.	
B	Sodium Borate	1

Proc.: Dissolve in 1 gallon water. To apply to plants add 1 teaspoonful of A and B to each gallon water which is then applied to the base of plants or sprinkled over cultivated ground.

Plant Food

	pts./wt.
Trisodium Phosphate	2
Potassium Sulphate	2
Sodium Nitrate	3

Proc.: Grind together and mix well. Only about a half gram of the above mixture should be used per plant every month or two. Caution: Using too much of any plant food is dangerous.

House Plant Food

Potassium Nitrate (Saltpeter)	3
Tribasic Sodium Phosphate	2

Proc.: Mix and dissolve about one tablespoon to the gallon. Of this solution, use one gill for each average size plant, once every two weeks.

Fertilizer for Potted Plants

	%/wt.
Potassium Nitrate	44.00
Ammonium Sulfate	6.00
Urea	10.00
Triple Superphosphate	15.00
Calcium Carbonate	0.04
Inerts (Clay, Etc.) To make	100

Proc.: Mix all ingredients together and stir until uniform.

Fertilizer for Acid-Loving Plants
(Azaleas, Camellias, Gardenias)

	pts./wt.
Ammonium Sulfate	26
Superphosphate	31
Potash	190

Proc.: Mix 1 lb. of this fertilizer with 5 cu. ft. redwood or cypress sawdust. Use about half this material and half garden loam for the planting mixture.

Nutrient for General Vegetable Culture
I

Potassium Nitrate	1.09 oz.
Magnesium Sulfate	1.10 oz.
Calcium Sulfate	2.01 oz.
Monocalcium Phosphate	.57 oz.

Water	10 gal.

Proc.: Mix all ingredients until uniform.

II

Magnesium Sulfate	.73 oz.
Potassium Nitrate	.93 oz.
Calcium Nitrate	1.01 oz.
Monoammonium Phosphate	.17 oz.
Water	10 gal.

Proc.: Mix all ingredients until uniform.

Hydroponic Plant Nutrient
(For tomatoes)

A	Sodium Nitrate	0.05 oz.
	Calcium Superphosphate	0.56 oz.
	Potassium Sulfate	0.49 oz.
	Magnesium Sulfate	0.38 oz.
B	Water	10 gal.

Proc. Heat B to 120°F. Mix A together and add to the hot water under stirring.

Soilless Culture
Hydroponic Plant Nutrient
(For tomatoes)
I

Sodium Nitrate	.05 oz.
Calcium Superphosphate	.56 oz.
Potassium Sulfate	.49 oz.
Magnesium Sulfate	.38 oz.
Water	10 gal.

Proc.: Dissolve all ingredients in water and stir until uniform.

II

Sodium Nitrate	.81 oz.
Monoammonium Phosphate	.13 oz.
Potassium Sulfate	.59 oz.
Calcium Nitrate	.10 oz.
Magnesium Sulfate	.39 oz.
Water	10 gal.

Proc.: Dissolve all ingredients together and stir until uniform.

Repellent Coating for Seeds
I
(Birds)

	%/wt.
"Drinox"	0.5-1

Spreader-Sticker	1
Water, to make	100

Proc.: Heat all ingredients and stir until uniform.

II
(U.S. Patent 2,900,303)
(Birds, Rodents)

	%/wt.
Iron Oxide	33-75
Calcium Carbonate	25-66
Copper Oxalate	2-10

Proc.: Mix all ingredients together until uniform.

Rabbit Deterrent
(Spanish Patent 199,772)

	%/wt.
Coal Tar	60
Nicotine	30
Alcohol	5
Turpentine	5

Proc.: Mix all ingredients together.

A rope is impregnated with this solution and placed around the field about 4 in. above ground.

Protecting Mixture for Young Trees Against Game

	%/wt.
Ceresin (58-60° C.)	20
Spindle Oil, Distilled	60
Dippel's Animal Oil or Carbolineum	20

Proc.: Melt up and stir until cold.

Deer Repellent

	pts./wt.
Bon-Tar Oil	9
Polyglycol 400 Monolaurate	1

Proc.: Emulsify 3 to 4 qt. of this with 100 gal. water and spray on the tree trunks.

Garter Snake Control

Place open pans of dilute nicotine sulfate solution near snake pits.

Bee Repellent

An absorbent pad, impregnated with propionic anhydride is placed in a special fume chamber of the beehive.

This chamber is then fitted over a section of the hive called the *super* from which honey is harvested by beekeepers. Air forced with bellows into the fume chamber drives the bees into other supers, or into the brood chambers below the supers.

When propionic anhydride is used, bees become gentle and easy to work with. They are repelled from sealed honey supers in one to two minutes, in shade as well as in sun. There is no change in flavor or odor of the honey.

Chigger Repellent

	pts./wt.
Benzyl Benzoate	29.7
Butylacetanilide	29.7
2-Butyl-2-Ethyl 1, 3-Propanediol	29.7
Lindane	1.0

Proc.: Mix all ingredients together until uniform.

To Control Land Crabs

	pts./wt.
Rice Bran	10
Lindane (50% wettable Powder)	1
Water, q.s. to make paste.	

This quantity is sufficient for 700 crab holes.

Destroying Man Scent on Animal Traps

	pts./wt.
Anise Oil	1
Rhodium	1
Tincture of Musk Tonquin	1
Terpineol	57

Proc.: Mix all ingredients together until uniform.
A little of this mixture gives a good masking odor.

AGRICULTURAL AIDS

Defoliant and Fertilizer
(U.S. Pat. 3,440,034)

Sodium Arsenite	6.4–12.8 oz.
Nitrogen Fertilizer	48–224 oz.
Water	10 gal.

Proc.: Mix and heat all ingredients together.

Fruit Tree Defoliant

	%/wt.
Magnesium Chlorate	0.25 - 0.5
"Endothal"	0.075 - 0.1
Water	To make 100

Proc.: Dissolve magnesium chlorate and endothal in water.

Protection of Seeds Against Crows
(German Patent 947,210)

	pts./wt.
Diphenylguanidine	70
Oleic Acid	430

Proc.: Add the above to 50 kg. seeds and coat evenly.

Tree Wound Dressing
(British Patent 839,789)

	pts./wt.
Water	131.0
Bentonite	6.5
Potassium Hydroxide	1.8
Ligno Sulfonic Acid	5.0
Bitumen (60-80 penetration)	217.0
Stearic Pitch	17.0
Copper 8-Quinolinolate	5.2

Proc.: Mix all ingredients together (using heat) until dissolved.

Tree-Wound Paint

The addition of ¼% of phenyl mercuric nitrate to an asphalt based paint or varnish prevents fungus growth.

Micronutrient Concentrate

	pts./wt.
"Hamp-OL" 9% iron chelate	55.000
"Hamp-OL" 9% manganese chelate	27.500
"Hamp-OL" 9% copper chelate	14.000
"Hamp-ENE" zinc chelate	9.000
"Solubor"	2.500
Sodium molybdate	0.025

Proc.: Dry-blend ingredients and package in containers that prevent moisture pickup and can be resealed if contents will not be rapidly used up. The micronutrient mix can be applied either bi-monthly by diluting 1/3 oz. (1 level tablespoon) in a gallon of water or mixed with home and garden fertilizers at a rate of 50 to 200 lb. per ton.

Potato Sprouting Stimulant

	pts./wt.
Ethylene Chlorhydrin	7
Ethylene Dichloride	3
Carbon Tetrachloride	1

Proc.: Mix all ingredients together.

Improving Tomato Yield
(Plant Hormone)

	pts./wt.
Naphthylacetic Acid	.3
Cyclohexanone	2.7
Dimethyl Ether	27.0

Proc.: Spray on tomato blossoms.

Orange-Tree Trace-Element
Mixture

	pts./wt.
Nitrogen	4.00
Potash	0.35
Magnesium Oxide	0.2-0.4
Manganese Oxide	0.03
Zinc Sulfate	3.00
(plus lime to neutralize it)	
Water	100.00 gal.
Borax	1.00 lb.

Proc.: Dissolve the borax in the water.

These amounts are necessary per box of bearing trees. The solution should be applied as a spray. When yellow spot is apparent in the grove, 1 to 22 oz. of sodium molybdate should be added to each 100 gal. of spray.

Soil Stabilizer
(U.S. Patent 2,698,252)

	%/wt.
Crude Fly Ash	10-30
Soil	70-90
Lime	2-9

Proc.: Mix all ingredients together.

Soil Sterilization in Field and Garden
I

The stand of such vegetables as peas, spinach and beets can usually be greatly improved by watering, immediately after planting, with a dilute solution of formaldehyde.

	pts./wt.
Formaldehyde (40%)	1
Water	124

Proc.: Use this solution at the rate of 1 gallon for 200 feet of row.

II

	pts./wt.
Formaldedyde (40%)	15
Infusorial Earth	85

Proc.: When infusorial earth is used as a carrier the full strength of the formaldehyde is maintained for a longer time than when other materials, such as charcoal or muck, are employed. Mix thoroughly, taking care to break up lumps. Use 6 oz. of this dust for each bushel of soil, or 1½ oz. per square foot of flat area. Insure that the dust is well mixed with the soil. After placing in flats, sow seed and water immediately.

Activation of Manure Fertilizer
(French Patent 883,190)

The manure is treated with 0.4% potassium permanganate solution which converts the skatole to indole and with 0.1% copper acetate solution that converts the indole to phytohormones.

Fertilizing Compost
(German Patent 933,989)

Penicillin Fermentation Waste (24% Dry Basis)	1-13
Peat (75% Dry Basis)	1

Proc.: Allow to ferment.

Fertilizer for Flowers and Herbs

	pts./wt
Ammonium Nitrate	20
Ammonium Chloride	2½
Ammonium Phosphate	10
Potassium Nitrate	12½
Calcium Sulphate	3
Iron Sulphate	2

Proc.: Thix mixture is employed by dissolving a teaspoonful in a gallon of water, and sprinkling the latter on the plants.

Stump Killers

Effective tree stump killers not only kill the tree but cause its wood and root system to deteriorate. In the old days, holes were drilled into the stump and were filled with potassium nitrate (saltpeter) crystals. In the course of time, the salt permeated the woody structure and rendered it automatically combustible. When the stump had dried out, it was ignited and allowed to burn throughout its root system. The treatment described may be effective for one type of wood (pine) but not for another (eucalyptus). In the writer's experience a large eucalyptus stump is difficult to kill. A stump of this type, 2 ft. in diameter and over 3 ft. high, eventually was killed by using a gallon (about) 6 lb.) of sodium chlorate following treatment with copper sulfate crystals. Sucker growth stopped forming, and the thick bark dried up and fell off. In applying treatment to a stump of this type, it is convenient to introduce it between the bark and the wood. Typical formulas for stump killers are shown below. All chemicals in No. I to IV are in powder form.

The material is placed in holes or troughs, then soaked with water to partially dissolve it.

I

	pts./wt.
Potassium Nitrate Powder	100

II

Sodium Chlorate Powder	100

III

Potassium Nitrate Powder	90
Copper Sulfate Powder	10

Proc.: Mix powder together.

To Suppress Sprouting of Potatoes
(U.S. Patent 2,999,746)
I

In a closed bin introduce 1 pound of diproparyl ether for each 1000 pounds of potatoes.

II

Per bushel of potatoes, 2/3 to 1-1/3 g. of methyl α-naphthalene acetate dust is used.

Potato Sprouting Stimulant

	pts./wt.
Ethylene Chlorhydrin	7
Ethylene Dichloride	3
Carbon Tetrachloride	1

Proc.: Mix all items together.

Grain-Dust Prevention Emulsion
(U.S. Patent 2,585,026)

	%/wt.
Mineral Oil	4.00
Propylene Glycol Monolaurate	0.08
Water	95.92

Proc.: Mix well. Use 0.02 to 0.08% of mineral oil content on the weight of grain.

Preventing Citrus Decay

	%/wt.
"Dowcide" A	2
Hexamine	1
Water	97

Proc.: Dip the fruit in this solution for 2 minutes; then drain and rinse with water.

Disinfecting Vegetable Sacks

"Hyamine" 2389	1 oz.
Water	2 gal.

Proc.: Soak bags for 1 hr. in above solution.

Grain Fumigant
(U.S. Patent 2,895,870)
I

	%/wt.
Carbon Tetrachloride	82-83
Carbon Disulfide	16½-17½
Petroleum Ether	½-1

Proc.: Mix all ingredients together.

II
(Tablet)

	pts./wt.
Aluminum Phosphide	70
Ammonium Carbamate	26
Paraffin Wax	4

Proc.: Heat all ingredients and stir until dissolved.

Nonflammable Fumigant
(U.S. Patent 2,803,581)

	%/wt.
Carbon Bisulfide	13
Ethylene Dibromide	7
Methylene Chloride	2-20
Carbon Tetrachloride	66-78

Proc.: Mix all ingredients together and stir until dissolved.

Increased Flash Point Fumigant
(U.S. Patent 2,824,040)

	%/wt.
Carbon Tetrachloride	79 l.
Carbon Disulfide	21 l.
Benzol	3 l.

Sulfur Dixoide 0.5-1.25
Petroleum Ether 3 l.
Flashpoint 120°F.

Proc.: Mix all ingredients together and stir until uniform.

Household Pest Control Fleas

A dog or cat may be rid of fleas by the application of a level teaspoonful of derris powder to the skin along the back, neck, and the head.

Derris powder, whose effective agent is rotenone, may be diluted with talcum powder to reduce the rotenone content to 1 per cent.

Pyrethrum powder may be used where derris powder is not at hand.

Infested areas in houses and barns should be sprayed with creosote oil. Because of its odor and because it burns plants and animals, creosote oil, however, cannot be used in every case. Where fleas occur in living quarters, scatter flaked naphthalene over the floor of the infested rooms at the rate of 5 pounds per room. The room should be kept closed from 24 to 48 hours.

Sticktight fleas which are troublesome in the South can be controlled by spraying with creosote oil infested chicken houses and areas beneath buildings frequented by poultry or pet animals. The masses of fleas attached to poultry or pet animals may be treated with derris powder or carbolated vaseline.

Space Spray
(For flies, mosquitoes, etc.)

	%/wt.
Velsicol	34
Kerosene	61
Glycox 1300	5

Proc.: Mix all ingredients together.

A knockdown agent, such as pyrethrum, Lethane, 384, or Thanite may be added to the concentrate.

Mosquito Larvicide

	pts./wt.
Cottonseed Pitch	3
Kerosene	1

Proc.: Mix all ingredients together.

Mosquito Spray for Outdoor Gatherings

Kerosene Containing Pyrethrum Extract Equivalent to 1 lb. of Flowers (Analyzing 0.9% Pyrethrins) per Gallon   10 gal.

Water	5 gal.
Sodium Laurel Sulphate (Emulsifier)	1.6 oz.

Proc.: The emulsifier is first mixed with the water and transferred to the mixing tank. The oil is added with agitation until the mixture is thick and homogenous, showing no oil on the surface. Store in tight container.

Directions for Spraying

About half an hour before the gathering takes place the area is completely sprayed with the larvacide diluted 1:10 or 1:12, that is 1 part of larvicide is mixed with 10 or 12 parts of water. The spraying is done with a power sprayer capable of developing a pressure of 100 pounds or more per square inch and equipped with a spray gun. Before mixing with water the concentrated stock larvacide should be well shaken. Also the diluted spray should be frequently stirred or agitated in order to secure uniform distribution throughout the spraying operation. The spray is applied in the form of a fine fog directly to the grass, grounds, tents, trees, shrubs, etc. Then the stream is directed upward so as to saturate the atmosphere with the fog. At no time should a coarse spray be applied, since it is unnecessary and may injure vegatation. The grounds for about 20 feet outside the area should also be thoroughly fogged, especially when tall grass, shrubs, woodland and other vegetation are present offering a hiding place from which adult female mosquitoes may issue suddenly at dusk in large numbers. If the area has been thoroughly fogged one treatment may suffice for two hours or even the rest of the evening. If mosquitoes become bothersome later in the evening, the area on the outside of the "gathering" grounds should again be fogged, directing the stream primarily upward and towards the ground to be protected. This outside fogging may be repeated again if necessary. On small areas, such as back-yards, private lawns, etc., a knapsack sprayer or bucket pump capable of producing a fog spray, of 10 to 15 feet high, can be used.

Ants

Ant control must center on the destruction of the queen and the young in the nest itself. Where the ant colony can be located it may easily be destroyed with a tablespoon or so of carbon disulphide. When the worker ants appear from under the stones or between the bricks of a walk, their colonies may also be destroyed by pouring a tablespoon of carbon disulphide down the crack.

Where the nests are in woodwork, by means of a small syringe inject a tablespoon of carbon disulphide and then close the opening with plastic wood or putty. Where the ant galleries are widely separated it is desirable to make injections at intervals in the wood.

For unlocated ant colonies the use of baits, powder, sprays, or chemical barriers is recommended. No one preparation will do for all kinds of ants. Some ants eat one kind of poison but not another. Some eat only sweet, others eat only meats and grease.

Dusting sodium flouride powder about window sills, drainboards, foundations, and other places where ants crawl sometimes drives them away.

Pyrethrum sprays can be used but they kill only the ants actually hit by the spray.

The use of poison baits is also recommended but the type of bait used depends upon the variety of the ant.

To Kill Ants in Lawns and Gardens

Make a hole about 18 inches deep in the center of the ant hill with an old broom handle and then pour in a solution of poison made by dissolving 1 oz. of sodium or potassium cyanide* in 1 gal. of water. Cover with dirt. If the soil is alkaline use one-half the quantity of water and make another solution of 1 oz. of alum to 2 qt. of water and pour one- half of each in the hole.

*Deadly poison. Do not allow contact with broken skin or cuts.

Ant Poison

Sugar	1 lb.
Water	1 qt.
Arsenate of Soda	125 g.

Proc.: Boil and stir until uniform; strain through muslin; add 1 spoonful of honey.

Ant Control

Use a 2% solution, or emulsion of chlordane. Paint a 6 inch swath along baseboard. Same for long wing American cockroach.

Insecticide for Flies

	%/wt.
Methoxychlor	1.5
Lindane	0.5
Lethane	1.5
Deodorized Naphtha	96.5
Perfume	As desired

Proc.: Mix all ingredients together until dissolved.

Non-Poisonous Fly-Papers

Quassia	16 oz.
Colocynth	2 oz.
Long Pepper	4 oz.
Water	to make 1 gal.

Proc.: Boil until the decoction is reduced to 4 pints; strain;

dissolve in the clear liquid 4 oz. of sugar. Dip the absorbent paper in this solution.

Cobalt Fly-Papers

Dissolve cobalt chloride, 1 oz., and Tartar Emetic, 1 d., in 1 gal. of the Quassia decoction (formula above), and dip the paper in the resulting solution.

Fly Catcher

	pts./wt.
Colophony (Rosin) G	49
Mineral Oil (Viscosity 3½-4° E at 50°C.)	36
Lanolin, Anhydrous	4
Beeswax, Pure	1
Castor Oil	2

Proc.: Heat all ingredients and stir until uniform.

Killing Fly Larvae in Cesspools

Add 0.15% by weight of sodium cyanide to the fecal matter.

Fly Dishes

A	Quassia Wood	500 g.
	Black Pepper	50 g.
	Water	2 l.

Proc.: Extract cold 4 days, then evaporate to 1 liter, filter and add:

B	Sugar	100 g.

Proc.: Color with red or green aniline dye. Impregnate cardboard dishes with solution; dry in air.

Mothproofing Aerosol
I

	%/wt.
p-Chloraniline Oleate	1.84
p-Chloraniline Salicylate	0.16
Petroleum Hydrocarbon Oil	38.00
Propellent ("Freon" 11-12 50/50)	60.00

Proc.: Mix all ingredients together until uniform and add this mixture to aerosol containers and charge with propellent.

Moth Powder

	pts./wt.
Camphor	4
Benzoin	1
Black Pepper	2
Cedar Sawdust	5

Proc.: Mix after reducing the solids to a coarse powder.

Derris Insecticide for Caraway Moth

	pts./wt.
Derris Root Powder (8% Rotenone)	1
Talc	3

Proc.: Mix powders until uniform. Apply at rate of 75 kg. per hectare in two applications.

Clothing Insecticide Impregnant

	%/wt.
Benzyl Benzoate	30
N-Butylacetanilide	30
2-Butyl-2-Ethyl-1,3 Propanediol	30
"Tween" 80	10

Proc.: Mix all ingredients together until uniform.

Cockroaches

Cockroaches destroy food, book bindings and fabrics.

To eliminate cockroaches trade at roach-free stores and watch all boxes or baskets of food and laundry brought into the house.

Fumigation by a professional fumigator is the most effective way of ridding the house of roaches. This method, however, is expensive and in infested areas is apt to give only shortlived relief.

Amateur exterminators should first fill up the cracks which lead to roaches' hiding places with putty, plastic wood, or plaster of Paris. The sodium fluoride powder should be sprinkled along the back of shelfing and drainboards.

If possible blow the powder with a duster, bellows, or an electric power duster into the insects' hiding places. Sodium fluoride is poisonous, however, and should be kept out of food and away from children and pets.

Pyrethrum powder is a non-poisonous roach-killer and may be used in the same way. On exposure to air, however, it loses its effectiveness after a time.

Phosphorous pastes are useful. Spread the paste on a piece of cardboard and roll into a cylinder. Wrap around with a rubber band and then place it wherever the roaches have been destructive.

Sprays of kerosene oil and pyrethrum extract may be used also but it must be remembered that these kill only by contact.

German Cockroach Control

0.5% diazinon. Squirt insecticide into cracks and openings, where roaches run. NOT to be sprayed in open room. The product is POISON and must be properly labelled and registered.

Roach Eradicating Powder

	%/wt.
Sodium Fluoride	60
Wheat Flour	20
Corn Starch	12
Cocoa	8

Proc.: The sodium fluoride should be in a finely powdered form and thoroughly mixed and then sifted to make certain of a homogeneous product. This may be made into a paste with a minimum of water and placed in new or used crown caps, allowed to dry and laid in roach infested places. It may also be dusted as a powder. The filled caps, however, can be used over again and are cleaned up more readily than the powder.

Roach Poison, Young

	pts./wt.
Calcium Sulfate Hemihydrate	50.00
Wheat Floor	50.00
Butter or Vegetable Oil	0.25

Proc.: Mix the ingredients together.

Household Insecticide
(For roaches)
I

	%/wt.
Chlordane	2.5
Pyrethrum Extract	2.5
Deodorized Kerosene	95.0

Proc.: Mix the ingredients together.

II

	pts./wt.
Powdered Sodium Borate	57
Powdered Sugar	10
Flour	25
Casein	8

Proc.: Mix the ingredients by shaking well together or tumbling in barrel. Run through a tablet machine.

Safe Household Spray

	%/wt.
Pyrenone	5
Pyrethrum Extract 1 to 20	2
Deodorized Kerosene	93

Proc.: Mix the ingredients together.

Bed Bugs

The bedbug infests furniture, clothing, baggage, walls, laundry. It is not known to be a disease carrier, but its bite frequently results in inflammation and welts.

The most efficient method of eliminating bedbugs is to employ a professional fumigator working under a license issued by the local health department.

Where this cannot be done bedbugs can be attacked by heating the room or the entire building to a temperature of 120 to 125 degrees. This temperature should be maintained for several hours. If this too is impossible, bedbugs may be attacked by a solution of pyrethrum in kerosene. This should be sprayed so that it comes in contact with the bedbugs. The best way to do this is to use a power sprayer. Filling the room with spray will not work.

An effective but tedious method is the application of kerosene, turpentine, benzene, or gasoline to cracks in bedsteads or to the other hiding places of the bugs. When doing this the windows should be kept open and all fire kept away. The infestation of furniture may be overcome either by fumigation or by liberal application of clear gasoline to the furniture in the open air.

Bed Bug Killing Powder

	%/wt.
Alum, Powdered	80
Boric Acid, Powdered	10
Salicylic Acid	10

Proc.: Powder very finely. Mix ingredients together.

Bed Bug Spray
I

	pts./wt.
Carbolic Acid, Crystallized	1.0
Camphor	1.0
Thyme Oil	1.5
Alcohol	56.5
Carbon Tetrachloride	65.0

Proc.: Mix the ingredients together.

II

	%/wt.
Paradichlorbenzene	1.5
Naphthalene	2.5
Carbolic Acid	.5
Thyme Oil	1.5
Carbon Tetrachloride	45.0
White Spirit	25.0

Alcohol	24.0

Proc.: Mix all the ingredients together.

Silverfish

Silverfish are slender, wingless, scale-covered insects slightly more than 3/8 of an inch long. They are found in damp, warm basements, storerooms, and attics where they feed upon paper, book bindings, wall paper, rayon fabrics, and anything containing starch or sugar.

They may be controlled by the use of a poison bait made of 1½ cups of oatmeal ground to flour, ¼ teaspoon of arsenic, ½ teaspoon of granulated sugar, and ¼ teaspoon of salt. The mixture should be stirred up and then moistened. Dry the bait, pound into small bits, and scatter wherever the silverfish are found. Sodium fluoride may be substituted for arsenic in the formula.

Pyrethrum powder also may be used. This should be dusted or blown into the infested area. Another effective method is spraying with a saturated solution of paradichlorobenzene in carbon tetrachloride. If possible the room sprayed should be closed for 24 hours.

Silverfish Insecticide

	pts./wt.
Oatmeal (Ground to Flour)	100
White Arsenic	8
Granulated Sugar	5
Salt	2½

Proc.: Mix together dry, the oatmeal, white arsenic, sugar, and salt. Moisten the mass and mix thoroughly to bind the substances together. Then thoroughly dry the bait to prevent mold, and work it into small bits so that it can be scattered easily.

One level tablespoon of sodium fluoride powder can be substituted for the white arsenic in the above formula. If the substitution is made, simply mix the materials thoroughly, but do not add moisture.

Rabbit and Rodent Fumigant
(Australian Patent 19,147)

	pts./wt.
Powdered Charcoal	20
Flowers of Sulphur	130
Potassium Nitrate	112
Sawdust	100

Proc.: Mix all ingredients. The above is ignited in holes or burrows which are closed.

Rat and Roach Pastes
I

Phosphorus	1 oz.
Starch	4 oz.
Flour	12 oz.
Glycerin	12 fl. oz.
Water	24 fl. oz.

Proc.: Mix all ingredients together.

II

Red Squill	1 oz.
Fish Meal	10 oz.
Water	6 fl. oz.

Proc.: Rub together to a paste and use at once.

Mouse Control

Use 400 to 500 g. zinc phosphide per 100 kg. cut lucerne; 5 g. per kg. oats and 50 g. per kg. wheat flour.

Rat Repellent
I
(U.S. Patent 2,862,849)

2,3,5,6-Tetrachloronitrophenol butyl ether.

II
(British Patent 790-022)

	%/wt.
Fluoroacetamide	0.1-1
Bran, Oats and Starch	To make 100

Proc.: Mix all ingredients together.

Rat-Repellent for Paper and Cloth

Paper or cloth is treated with a suspension of "Good-rite Z.A.C." to give a concentration of 0.01 g. per sq. in. The suspending medium may be latex, oils, sizes, organosols, plastisols, etc.

For Cordage & Sacks
I
U.S. Patent 2,822,295-6

Dodecyl alcohol or its acetate dissolved in a volatile solvent.

II
(U.S. Patent 2,864,727)

Treat twine with hydrocarbon solution of

	pts./wt.
Quinaldine	2
Naphthenic Acid	2

Proc.: Mix all ingredients together.

Rat Bait

I

	pts./wt.
Ground Dried Bread	.5
Ground Fresh Pork Fat	.5
Ground Fresh Halibut or Haddock or Cod	2.0
Powdered Red Squill	1.0

Proc.: Mix all ingredients together.

II

	pts./wt.
Ground Dried Bread	3.7
Glycerin	.3
Powdered Red Squill	1.0
Fresh Bait	5.0

Proc.: Mix all ingredients together.

Rat Poison for Flour Mills

	%/wt.
Sodium Silicofluoride	70
Diatomaceous Earth	30

Proc.: Dust on floor, keeping away from sacks. Rats lick powder off feet and go out seeking water and thus die outside.

CATTLE AND POULTRY

Cattle Fodder
(U.S. Patent 2,715,067)

	%/wt.
Molasses	25
Shredded Newspapers	50
Corn	25

Proc.: Mix all ingredients together.

Feed Appetizer for Animals

The addition of small amounts of imitation anise oil (250 p.p.m.) or β-ionone (27 p.p.m.) to cattle and hog feeds stimulates them to eat more.

Preserved Fish Silage

Mix minced fish or offal with 50% sulfuric acid to pH 2. Mix with powdered chalk to pH 4 before using as feed.

Feed Additive for Heifers

Add 10 lb. formic acid (90%) to 1 ton unwilted silage. Mix thoroughly. This feed produces more rapid weight gain.

Silage Preservative
(U.S. Patent 2,714,067)
I

	pts./wt.
Anhydrous Sodium Formate	55
Anhydrous Calcium Chloride	45

Proc.: Mix all ingredients together.

II

	pts./wt.
Anhydrous Sodium Formate	51
Anhydrous Calcium Chloride	42
Sodium Nitrite	7

Proc.: Mix all ingredients together.

Cattle Salt Block
(Dutch Patent 62,223)

	pts./wt.
Salt	33.0
Calcium Acid Phosphate	33.0
Calcium Carbonate	33.0
Iron Oxide	0.6
Copper Sulfate	0.3
Cement	12.5
Water	12.5

Proc.: Mix well and mold into blocks.

Salt Block Pressing

Salt-lick and nutritional animal lick-blocks are formulated to contain 1% of "Castorwax." It acts as an internal lubricant during pressure molding as a mold release. In addition it waterproofs the blocks to reduce rain erosion during use.

Chick Starter

	pts./wt.
Finely Ground Shelled Yellow Corn	63.5
Wheat Bran	20.0
Wheat Middlings	10.0
Finely Ground Heavy Oats	30.0
"Menhaden" Fish Meal	30.0
Dried Skim Milk	20.0
Alfalfa Leaf Meal (20-18)	10.0

Finely Ground Quality Barley	10.0
Finely Ground Oyster Shell	2.5
Fish Oil	4.0

Proc.: Mix these thoroughly. The analysis is usually approximately 21% protein, 5% fiber, and 4% fat.

Grower Mash

	pts./wt.
Ground Shelled Yellow Corn	85
Wheat Middlings	30
Wheat Bran	30
Dried Skim Milk	10
Fish Meal	2
Alfalfa Meal (13-33)	20
Ground Limestone	3
Cod-Liver Oil	2

Proc.: Mix all ingredients together.

After spring sunshine comes, the fish oil is usually omitted and the amount of bran increased by 2 lb.

The analysis is approximately 17.5% protein, 6% fiber, and 4% fat.

Laying Mash

	pts./wt.
Finely Ground Heavy Oats	30
Wheat Middlings	25
Ground Shelled Yellow Corn	65
Fish Meal	25
Alfalfa Leaf Meal (20-18)	20
Riboflavin Supplement	10
Wheat Bran	20
Oyster Shell	2
Bone Meal	1
Fish Oil	2

Proc.: Mix all ingredients together.

The analysis is estimated at 18.5% protein, 5.5% fiber, and 4% fat.

Feed Binder

Poultry and animal feedstuff binder—"Lignosol" FG is used to pelletize poultry and animal feeds. Use is approved, up to 4%, in Canada and in the U.S.A. in poultry feeds.

Cow-Teat Ointment

For treatment of infected sphincters of cows which seem to milk hard, daily application of 10% silver nitrate solution, followed by the application of this teat ointment for 3 days, will, as a rule, prove efficient:

	pts./wt.
Acriflavine	2½
Urea	9¼
Sulfathiazole	11¼
White Petrolatum	34

Proc.: Mix the first three ingredients and then incorporate this mixture into the petrolatum by warming until it softens enough for stirring. Cool and fill into jars.

Pink-Eye Powder
(Veterinary)

Tyrothricin	1 g.
Methylrosaniline Chloride	18 g.
Phenocaine Hydrochloride	20 g.
Sulfathiazole	400 g.
Sulfanilamide	248 g.

Proc.: Mix the first three and last two ingredients separately. Then mix them thoroughly with each other and put into plastic sprayer tubes. The pink-eye powder is then simply puffed into the sick animal's eyes.

Animal-Liver Fluke Remedy

Powdered Hexachloroethane	4 lb.
Bentonite	6 oz.
White Flour	1 tsp.
Water	3 qt.

Proc.: Stir thoroughly. Give 6½ fl. oz. by mouth per 100 lb. body weight to mature, not weak cattle. Calves over 3 months of age receive only 3½ oz., sheep and goats 2 oz.

Animal Eye Washes

One of the best eye washes for irrigation and cleansing of the eye and for purulent discharges and conjunctivitis is as follows:

Sodium Bicarbonate	15 gr.
Borax	15 gr.
Sodium Chloride	15 gr.
Glycerin	1 dr.
Distilled Water	8 oz.

Proc.: Mix all ingredients together.

Animal Ear Preparation
I

Gentian Violet	5

Acetone	5
Alcohol	45
Water	45

Proc.: Mix all ingredients together. Take small amount in an eye dropper and place deep into the ear and remove excess so as not to soil the outside.

II

	%/wt.
Phenol	3
Glycerin	97

Proc.: Add boric acid powder until the glycerin will not absorb any more. Let stand over night and strain.

Place one-half eye dropperful in ear and remove the excess.

Bovine Ringworm Ointment
I

	pts/wt.
Compound G-4 (Givaudan)	2
Isopropyl Myristate and Palmitate	10
Cetyl Alcohol	11
Lanolin U.S.P.	5
White Petrolatum	72

Proc.: Mix all ingredients and heat until uniform.

II

	pts./wt.
Compound G-4	1–4.0
Isopropyl Myristate and Palmitate	15.0
Stearic Acid	18.0
Triethanolamine	2.0
Propylene Glycol	10.0
"Tergitol No. 7"	0.5
Water	50.5–53.5

Proc.: Mix and heat all ingredients and stir until uniform.

Screwworm Control

	pts./wt.
Lindane	3
Pine Oil	35
Mineral Oil	42
Emulsifier	10
Silica Aerogel	10

Proc.: Mix and heat all ingredients and stir until uniform.

Cattle-Mange Mite Spray

Wettable Sulfur	10 lb.
Lindane	1½ lb.
Water	100 gal.

Proc.: Mix and heat all ingredients and stir until uniform.

Midge-Control Cream

	pts./wt.
Dimethyl Phthalate	200
Magnesium Stearate	30
Zinc Stearate	70

Proc.: Mix all ingredients together.

Animal Drench
I

	pts./wt.
Phenothiazine	25
Bentonite	1½
Glycerin	1
Water	22½

Proc.: The glycerin is added to approximately 2 gal. water, then the phenothiazine and bentonite are mixed in by rapid stirring. After standing for 12 hours, the balance of water is added, the mixture stirred rapidly, and the stirring continued while it is poured into containers. This smooth, homogeneous suspension contains approximately 12.5 g. of the (or a lamb dose) per fluid ounce; 2 fl. oz. make a sheep dose. The cattle dose is 5 fl. oz.

II

Phenothiazine	40
Molasses	40
Warm Water	To make 100 gal.

Proc.: Mix first two ingredients and add water.

This contains 0.5 oz. phenothiazine per fluid ounce.

Either drench may be administered with a dose syringe or from a bottle.

Easily Disintegrating Phenothiazine Tablets
I

	pts./wt.
Phenothiazine	380.00
Starch	38.00
Tartaric Acid	19.00
Sodium Choleate	9.50
Phenolphthalein	4.50

Sodium Bicarbonate	23.75

Proc.: The sodium bicarbonate is added separately during granulation of the other ingredients. Gelatin solution is used as a binder and starch-talc-stearic acid mixture as a lubricant.

II

Phenothiazine	100.00
Cornstarch	10.00
Citric Acid	2.25
Sodium Bicarbonate	2.50
Lubricant	To suit

Hog Vermifuge
(U.S. Patent 2,422,106)

Santonin	1¼ gr.
Phenothiazine	4 g.

Proc.: Mix the ingredients together.

Cattle & Space Fly Spray

	pts./wt.
"Crag" Fly Repellent	57.8
Synergized Pyrethrins	33.0
Methoxychlor, (90% Technical Oil Conc.)	7.6
Petroleum Distillate	572.0

Proc.: Mix all ingredients together and stir until uniform.

Aerosol Cattle Fly Spray
I

	pts./wt.
Pyrethrins	0.2
Piperonyl Butoxide	2.0
Methoxychlor	3.0
Methylated Naphthalenes	5.0
Deodorized Base Oil (Petroleum)	9.8
Propellent	80.0

Proc.: Mix all ingredients together and fill in aerosol containers and charge with the propellant.

II

	pts./wt.
Pyrethrins	0.2
Piperonyl Butoxide	2.0
Methoxychlor	3.0
"Thanite"	2.0
Methylated Naphthalenes	5.0
Deodorized Oil	17.8

Propellent   70.0

Proc.: Same as for formula I.

Poultry Drink Antiseptic

"Cetrimide"	25 g.
Water-Soluble Bacitracin	25 g.
Dextrose	To make 1 lb.

Proc.: Mix all ingredients.

Combating Blackhead in Turkeys

Blackhead can be controlled to a certain extent by using 2-amino-5-nitrothiazole mixed with the feed.

Fowl Coccidiosis Remedy

Water	10.0 gal.
Caustic Soda	4.0 lb.
Sulfa Quinoxaline	13.6 lb.

Proc.: Use 18 cc. of this solution to 1 gal. drinking water for fowl coccidiosis.

This formula may be modified with borax or with amines to prevent corrosion of metal watering troughs and drug precipitation due to hardness of water.

Chicken Roundworm Remedy

A	Phenothiazine	151 g.
	Bentonite	287 g.
B	Dry Mash	44 lb.

Proc.: Mix A well into B.

WEED KILLERS (HERBICIDES)

Lawn Weed Poisons

Weed	Remedy
Heal All	Iron (ferrous) Sulphate
Orchard Grass, Crab Grass	Ammonium Sulphate
Dandelions	Ammonium Sulphate
Plantain	Ammonium Sulphate
Canada Thistle	Sodium Chlorate
Bindweed	Sodium Chlorate
Quack Grass	Sodium Chlorate
Toadstools	Copper Sulphate, Bordeaux Mixture or Iron Sulphate

Borax Weed Killer
(U.S. Patent 2,711,367)

Aluminum Sulfate	¼-2 lb.
Borax	½-4 lb.
Water	1-5 gal.

Proc.: Mix and heat the ingredients together.

2.4-D Concentrates
I
(Amine Solution)

	%/wt.
2,4-D Free Acid	40
Mixed Ethylamines (*Sharsol 193*) Or	
Triethanolamine	60

Proc.: Mix in a closed mixer by adding the amine slowly, with stirring, to the acid until completely incorporated. Stabilize with *Iminol D* for use in hard water, using 1 to 5%, depending on the degree of hardness toleration desired.

II
(Emulsifiable)

Butyl 2,4-D Ester (2.67 lb. acid equivalent/gal)	
Butyl 2,4-D	42.0
Kerosine	55.0
"Gafac" RM-510	0.9
"Gafac" RM-710	2.1

Proc.: Mix all ingredients until uniform.

Eradicating Poison Ivy

A 20 per cent solution of sodium thiocyanate seems to be the most practical. This may be had by mixing five pounds of the sodium thiocyanate in three gallons of water and this amount of spray should be enough to take care of one application on an area of 150 square feet.

This spray should be applied with a pressure spray pump, completely saturating all the leaves and stems of the ivy plants above ground. The ideal time to apply this treatment is along the latter part of June.

FUNGICIDES

Wettable Phenyl Mercuric Acetate Fungicide

Incorporate 1% "Antarox" A-400 into this fungicide on a laboratory scale by dry mixing to give a free-flowing, non-bleeding powder. However, only 0.1 to 0.5% "Antarox" A-400 is necessary to impart good wettability to the powder.

Copper Naphthenate Fungicide Emulsion

	pts./wt.
Copper Naphthenate	12
"Antaron" S-140	3
Turpentine	3
Lactic Acid	1
Water	81

Proc.: Mix all ingredients until uniform.

Fungi Spray
(U.S. Patent 2,000,843)

	pts./wt.
Soft Soap	33
Cresol Soap (2% Solution)	11
Tobacco Extract (10%)	17
Potassium Permanganate (½ Normal)	22
Vegetable Glue	17
Alcohol	¼-2

Proc.: Mix all ingredients until uniform.

Rope Preservative Fungicide

	I pts./wt.	II pts./wt.
Copper Naphthenate	35	—
80% Hard Asphalt in Mineral Spirits	12	—
Mineral Oil S.A.E. 10	17	19
Scale Wax	33	33
Mineral Spirits	3	3
Zinc Naphtenate	—	45

Proc.: Warm and mix. Apply at 130°F.

Agricultural Fungicide
I

	pts./wt.
Copper Resinate	20.0
Mercury Phenyl Salicylate	0.2
Mineral Diluent	79.8

Proc.: Mix the ingredients until uniform.

II
(British Patent 775,981)

	pts./wt.
"Cetrimide"	2
Beta-Naphthoxyacetic Acid	1

Carboxymethylcellulose Sodium   10
Water   To make 2 gal.

Proc.: Heat and mix all ingredients until uniform.

Mold Control of Tomatoes

40% Peracetic Acid	6 lb.
"Nacconol" NR	13 oz.
Water	100 gal.

Proc.: Spray or dip the tomatoes; drain and wash with water.

Grape-Vine Fungicide
(Italian Patent 459,302)

	pts./wt.
Ammonium Biphosphate	5
Ammonium Sulfate	5
Aluminum Sulfate	26
Copper Sulfate	64

Proc.: Mix all ingredients.

Citrus Fruit Fungicide
(Spanish Patent 235,781)

Immerse fruit in 0.1 to 5.0% chloroacetic acid for 1 to 10 minutes, drain and wash.

Oak Wilt Fungus Spray

	pts./wt.
Tartaric Acid	0.6
Water	99.0
Wetting Agent	0.4

Proc.: Mix and heat all ingredients until uniform.

Wood Preservative (Fungicide) for Pressure Impregnation

	I pts./wt.	II pts./wt.
Copper Naphthenate	12½	12½
No. 2 Fuel Oil	87½	––
Creosote		87½

Proc.: Mix all ingredients together.

III
(For surfaces to be painted)

	pts./wt.
Copper Naphthenate	25
Spar Varnish	40
Mineral Spirits	35

Proc.: Mix all ingredients together.

Wait 48 hours before painting over this coating.

INSECTICIDES

Parathion Aerosol

	%/wt.
Parathion	10
Acetone	10
Methyl Chloride	80

Proc.: This formulation is prepared on a weight basis. It is further suggested that the acetone and parathion are mixed before filling the cylinder.

U.S.D.A. recommendations call for the application of 1 g. parathion per 1,000 cu. ft. greenhouse space. This is comparable to 10 g. 10% parathion aerosol per 1,000 cu. ft. or 1 lb. per 50,000 cu. ft. space. The treating time for 1,000 cu. ft. is approximately 4 seconds based on the rate of delivery of a 5 gal./hour oil-burner nozzle.

Parathion has a residual contact effect and fumigation action for a period of 5 to 7 days against most greenhouse insects when used as directed.

Aerosol application can be made any time when the greenhouse ventilators can be closed for 2 hours following application without hazard of exposing the plants to too high temperatures. Preferably treatment should be made early in the morning, early in the evening or on cloudy days.

Wettable Sulfur for Use with Parathion

	%/wt.
Sulfur (300 Mesh)	91
Binder (Glue)	1
Lime	½
Ground Clay	7½

Proc.: With parathion, use about 0.5% of the following mixture which has been heated and cooked togehter 10 minutes.

Rosin	13
Potassium Hydroxide	6½
Urea	10
Water	50

2,4-D Insecticide

I

	%/wt.
Isopropyl or Butyl Ester of 2,4-D	31
Kerosene (Or Aromatic Solvent when 18% Ester Formulations Are Used)	56

Antarox A-200	5

Proc.: Mix all ingredients.

II

	%/wt.
Triethanolamine Salt of 2,4-D	40
Water	59
Antaron N-185	1

Proc.: Mix all ingredients.

"Dilan" Concentrates

	%/wt.
"Dilan"	25
Pine Oil	70
"Antarox" B-201 or B-290	5

Proc.: Mix all ingredients.

Where high emulsion stability is the prime consideration, "Antaron" B-290 is recommended. Where slightly less stability can be tolerated, the more economical "Antarox" B-201 is suggested.

Toxaphene (Chlorinated Camphene) Products
Crop-Dusting Insecticide
I

Chlorinated Camphene (40% Wettable Powder)	4 lb.
Water	50 gal.

Proc.: Apply as spray. This is particularly good for tomato, celery, cabbage, bean, tobacco, cotton, and peanut insects. As a mole cricket poison, apply to the soil surface with a sprinkling can.

II
(Grasshopper poison)

	%/wt.
Chlorinated Camphene (40% Dust)	25
Talc, Kaolin or Fuller's Earth	75

Apply 10 to 20 lb. per acre as dust.

65% Toxaphene Emulsifiable Concentrate

	%/wt.
Toxaphene	65
Xylol of Kerosene	25
"Arquad" 2C	5
"Ethofat" 142/20	5

Combine all ingredients, heating and stirring until a clear solution is formed. This concentration also disperses easily in water.

Malathon Emulsion

	%/wt.
Malathon	62
Xylol	30
"Emulsifier S-1132"	8

Proc.: Mix all ingredients.

This formula has a viscosity of 4 poises and requires very slight agitation to produce the emulsion of the desired concentration. The emulsion formed by this concentrate is very stable, exhibits very low foaming properties, and has extremely high surface tension. The last property apparently has two advantages. It keeps the insecticide on the foliage without too rapid a run-off and penetrates less into the foliage, thus reducing the possibility of destruction of the plant.

Lindane Insecticide Concentrates

I

	%/wt.
Lindane	20
"Antarox" A-400	10
Xylene	70

II

	%/wt.
Lindane	20
"Antarox" B-290	10
Cyclohexanone	70

Proc.: These concentrates are prepared by dissolving lindane in the solvents and then adding the emulsifier. Storage tests of 1 week's duration indicate that they remain clear and homogeneous at room temperature and at $37°F$.

CPR (Cyclonene, Pyrethrum, and Rotenone) Insecticide Emulsion

	%/wt.
"CPR" Liquid Concentrate	42.7
Pine Oil	10.9
Refined Soybean Oil	16.4
"Ultrasene"	16.2
"Antarox" B-201 or Combination of B-201 with A-401 (7:3)	13.8

Proc.: Mix all ingredients until uniform.

45% Chlordane Emulsifiable Concentrate

	%/wt.
Chlordane	45.0
"Arquad" 2C	2.5
"Ethofat" 142/20	2.5
Kerosene	50.0

Proc.: Simply mix the ingredients to form a clear solution. The concentrate disperses easily in water to form a stable emulsion.

Chlordane Concentrates
I

	%/wt.
Chlordane	50
"Antarox" A-401	15
"Antarox" B-201	35

II

Chlordane	50
"Antarox" B-201	35
Butyl "Cellosolve"	15

Proc.: For all formulas mix all ingredients together.

It is recommended that these concentrates are packed in glass, aluminum, or steel coated with a high-baked phenolic lacquer.

Formula 1 and 2 are solubilized chlordane concentrates which yield dispersions of at least 2 months' stability in hard or soft water. Formula 1 gives the clearest dispersions; formula 3 gives opaque dispersions.

Emulsifiable Chlordane Solution (2%)

	%/wt.
Water	96
Chlordane	2
Chlorsol	2

Proc.: Mix the *Chlordane* and emulsifier thoroughly and add the water slowly with agitation. Filter or settle if necessary. A 50% concentrate may be made from this formula by omitting the water.

Chlordane Wettable Powder

	%/wt.
Chlordane	20
Attaclay	79
"Antarox A-200"	1

Proc.: Dissolve the Antarox A-200 in the Chlordane and spray on the Attaclay.

Dieldrin Insecticide Base

	%/wt.
Technical Dieldrin	50.5
"Attaclay"	22.4

"Velvex"	22.4
Urea	2.2
"Tamol" 731 Dry	2.0
"Duponol" ME Dry	0.5

Proc.: Mix all ingredients until uniform.

Japanese Beetle Insecticide

I

Rosin	10 lb.
Caustic Soda	1 lb.
Fish Oil Soap	1 lb.

Proc.: Boil together and add a water solution of

Corrosive Sublimate	1 lb.
Nicotine Sulphate	1 oz.

II

Lead Arsenate	6 lb.
Menhaden Oil	1 pt.
Water	100 gal.

Proc.: Mix well before use. Apply to ornamental plants about July 1st.

Adhesive for Hydrated Lime in Sprays

A spray of 20 lb. calcium hydroxide and 3 lb. aluminum sulphate in 100 gal. of water will give an adherent white spray residue which is repellent to the Japanese beetle. The mixture may be of value as an adherent for other spray ingredients.

Self-Vaporizing Insecticide
(U.S. Patent 2,633,444)

	%/wt.
Sodium Nitrite	55
Ammonium Chloride	43
Magnesium Oxide	2
Insecticide	82-90

Proc.: Mix all ingredients until uniform.

On ignition, this will continue burning of itself and give off insecticide fumes.

Insecticide Cone (Burning)
(Italian Patent 505,718)

	%/wt.
Lindane	58
Rosin	25
Sugar	12
Mint Oil	5

Proc.: Melt together, mix and pour into forms.

Dormant Spray Oil

	%/wt.
Oil	98-99
"Triton" X-45	1-2

Proc.: Mix the ingredients.

Dormant Oil-Spray Concentrate Insecticide

	%/wt.
"Sun Oil" No. 15, "Sovaspray" No. 2, or "Gulf"	
Superior Type A Agricultural Oil	98
"Antarox" A-401	2

Proc.: Mix the ingredients
This concentrate is clear, and gives satisfactory emulsification in hard or soft water.

Codling Moth Tree Bands
I

Cloth is impregnated with Beta	
Naphthol Crude	1 lb.
Red Engine Oil	1.5 pt.
Apply at 130-132°F.	

Proc.: Mix the ingredients and impregnate cloth.

II

Beta Naphthol Crude	1 lb.
Mineral Oil (200-300 sec.)	1½ pt.
Gasoline	1 pt.

Proc.: Mix the ingredients

Beetle Powders
I

	pts./wt.
Barium Carbonate	10
Borax	20
Sugar	5

Proc.: Mix the ingredients

II

	pts./wt.
Sodium Fluoride	10
Kaolin	10

Proc.: Mix the ingredients.

III

	pts./wt.
Kieselguhr	22
Sodium Fluoride	40
Sodium Chloride	10

Proc.: Mix the ingredients.

Agricultural Air Spray

	%/wt.
Velsicol	62.5
Kerosene	17.5
Glycox 1300	20.0

Proc.: Use 1 pt. of this concentrate to 7 pt. water. The finished emulsion is quite stable and is commonly used in the proportions of 1 gal. to an acre.

Chewing-Insect Spray

Powdered Cryolite	3 lb.
Fish-Oil Soap	2 lb.
Water	50 gal.

Proc.: Mix and heat the ingredients until uniform.

Peach-Borer Control

	pts./wt.
Pine Tar	1
o-Dichlorobenzene	1

Proc.: Paint the areas infested with borers and all places where sap oozes from the tree. Also use it to treat tree injuries.

Non-Poisoning Fruit Spray

I

Diglycol Laurate	5	qt.
Pyrethrum Extract (20 Fold)	¾	pt.
Water	100	gal.

Proc.: Mix and heat the ingredients until uniform.

II

Derris Extract (5%)	1 qt.
Skim Milk Powder	1 lb.
Water	100 gal.

Proc.: Mix and heat the ingredients until uniform.

Peach Tree Spray

A combination of the lead arsenate and zinc-lime sprays is effective not only against chewing insects such as curculio and codling moth, but against bacteriosis. The formula is:

Zinc Sulphate	8 lb.
Hydrated Lime	8 lb.
Water	100 gal.

Add:

Lead Arsenate	3 lb.

Proc.: Mix and heat all ingredients until uniform. The spray should be used as soon as prepared.

Pear Tree Blight Injection
(U.S. Patent 2,017,269)

	oz.
Pine Tar Oil	1
Turpentine	16

Proc.: Mix the ingredients.

Gladiolus Thrip Spray

Manganese Arsenate (26% Arsenic)	4 lb.
Brown Sugar	66 lb.
Water	100 gal.

Proc.: Mix and heat all ingredients until uniform.

Destruction of Insects at Base of Tree

With outbreaks of extreme severity great numbers of the larvae may migrate in late July from the branches down the trunk and congregate either on the bark or about the base of the tree where they transform to pupae. When such a situation occurs the following contact spray may be applied in a belated attempt to reduce the numbers of the insect:

Summer Oil Emulsion	2 gal.
Nicotine Sulphate	1 qt.
Water	100 gal.

While this treatment kills many of the insects, this procedure is not to be considered as a substitute for the treatment with arsenical sprays directed against adults or larvae earlier in the season.

Tree Bands for Caterpillar and Flies
I

		pts./wt.
A	Rosin Oil	9
	Spindle Oil	20
B	Slaked Lime	6-9
	Spindle Oil	65-62

Proc.: Add A to B, stir violently to homogenize. Stir until congealing starts. Allow to set for 24 hours.

II

	pts./wt.
Rosin	30
Linseed Oil* Varnish	20
Beeswax, Yellow	2

* Or Rape Seed Oil, or Wool Fat, when a longer catching period is desired.

Proc.: The melted and well mixed glue is put on the bark of the tree; over it put a ring of cloth, fastened with wire, then put over that again a layer of glue, all around the stock.

Termite Control
I

	%/wt.
Copper Naphthenate	5
Fuel Oil No. 1	95

Proc.: Treat wood and infested soil with liberal amounts.

II

Pentachlorophenol	5
Orthodichlorobenzene	95

Proc.: Dilute this concentrate with 4 gal. fuel oil to 1 gal. concentrate. Use 1 gal. of this to each lineal foot of the soil surface.

HOUSEHOLD PET CARE
Dog Worm Remedy
I

	pts./wt.
Aloes	45
Soap	45
Oleoresin of Male Fern	30

Proc.: Mix and make into 2 pills.

Administer both pills in the morning, the animal to remain fasting for some time.

II

Areca nut, freshly ground, is considered an excellent remedy for worms in dogs. About one dram made into a pill is the dose for an ordinary sized dog. This should be given at night followed by a dose of castor oil in the morning.

Dog Tick Insecticide
(Brown)
I

	%/wt.
Chlordan	2
Inert Powder	98

Proc.: Mix the ingredients together.

II

	%/wt.
γ-Benzene Hexachloride	1
Inert Powder	99

Proc.: Mix the ingredients together.

Dog Mouth Wash

Tincture Iron	1 oz.
Potassium Chlorate	2 oz.
Glycerin	4 oz.
Water	to make 1 gal.

Proc.: Mix and heat the ingredients until uniform.

Liquid Soap for Dogs and Other Animals

	pts./wt.
Palm Kernel Oil	1200
Olein	300
Caustic Potash (50%) about	736
Glycerin	600
Softened or Distilled Water	6800
Carbolic Acid (Phenol), Crude	400
Perfume Oil (e.g., Eucalyptus	50

Proc.: Mix and heat all ingredients until uniform.

Shampoo for Puppies

The following mixture will produce a dry shampoo especially suitable for puppies too young to be bathed:

	%/wt.
Starch	75
Silica Gel	10
Borax or Sodium Bicarbonate	10
Pyrethrum Powder or Paradichlorbenzol	5

Proc.: Add 1% perfume compound. Mix all ingredients together until uniform.

The powder is dusted over the fur of the animal after which a thorough brushing and combing is necessary to remove the powder. The resulting cleansing and disinfecting of the fur is reasonably thorough, but obviously does not extend to the skin itself.

Bird Food Cake

	pts./wt.
Hemp Seed, Whole	300
Hemp Seed, Crushed	400
Poppy Seed	100
Millet, White	50
Elder Berries, Dried	50
Sun Flower Seeds	50
Ant Eggs	50
Beef Tallow	1,400

Proc.: Melt the tallow on the water bath, mix with the seeds and pour the whole into metal forms shortly before solidifying.

Bird Gravel

	%/wt.
Fine River Sand	97.5
Cuttlefish Bone, Powder	2
Pyrethrum Flowers	0.5

Proc.: Mix all ingredients until uniform.

INSECT REPELLENTS (PERSONAL)

I
(Lotion)

	%/wt.
DEET	50.0
Alcohol SDA No. 40, Anhydrous	49.5
Perfume Oil	0.5

Proc.: Mix the ingredients.

II
(Powder)

	pts./wt.
Dimethyl Phthalate	210
Magnesium Stearate	30
Zinc Stearate	70
Diglycol Stearate	10

Proc.: Grind together until uniform.

Suggested Aerosol Formulation:

III
Suntan Insect Repellent Lotion

	%w/w
A Escalol 506	1.0

Cetyl alcohol	1.5
Emulsynt 1060	5.0
Stearic acid XXX	4.0
Cerasynt D	5.0
N, N diethyl toluamide (Meta Delphene-Hercules Powder Co.)	12.0
Ceraphyl 140A	4.0
B Triethanolamine	0.50
Water	66.35
Cellosize QP4400 (Union Carbide)	0.50
C Perfume 8992G	0.15
D Propellant mixture (87% isobutane and 13% propane)	0.15
13% propane)	0.15

Proc.: Pre-dissolve QP4400 gum in water, then add trienthanolamine and heat to 75°C. Add Part B to Part A at 75°C. Cool with propeller agitation to 35°C.

IV
(Insect Repellent and Sun Screen Lotion
(Quick-Breaking Foam)

	%w/w
A Polawax	1.5
Ethyl alcohol, denatured, anhydrous	42.5
Metadelphene	18.0
Dipropylene glycol salicylate	3.0
B Water, deionized	35.0
C Perfume	q.s.

Proc.: Heat A to 100°-110°F. to dissolve the Polawax. Heat B to 100°F. and add to A with constant stirring. Add Part C to the mixture of A and B with stirring. Fill container with concentrate while still warm (100°F.)

Aerosol Formulation	
Concentrate	92.0
Propellant 12/114 (20:80)	8.0
Valve: Standard valve with foam actuator.	
Container: Plastic-coated glass.	

6. Food Products

The preparation of food products is an immense industry, able to encompass a home-operated business as well as factories and supermarkets. The advantages of undertaking this aspect of business lies in its accessibility and familiarity. No doubt you have most of the equipment necessary in your own kitchen, e.g., oven, refrigeration, freezer, large utensils, measuring and weighing devices, thermometer. Anything that you don't already have can be obtained easily and inexpensively. Thus, your initial investment is minimal.

For those of you who are not free to leave the home, this type of business can be the beginning of an independent income. Your spouse's favorite dish can be the key to your fortune.

The marketing of food products is relatively easy beginning with friends, relatives, and neighbors, and extending to small retail establishments and local eateries.

Included in this chapter is a wide range of recipes from wines and beverages through baking products and candies, to fruits, vegetables and main dishes. Choose a recipe listed, or a special of your own. Do all your testing in small batches. Always use good quality ingredients, and of course standards of cleanliness must be the highest. Try it on your friends and family. You will soon develop a group of customers who will be glad to buy your goodies. As your business increases look for new products to add to your line, but always remember that you don't need Mr. Heinz's "57 different varieties" to make a good living or a supplementary income.

Glaze for Smoked Meats

A gelatin dip which is sometimes used on smoked meats to avoid mold and shrinkage is made of the following ingredients:

	lb.
Commercial Gelatin	25
Glucose	35
Water	40

Proc.: Place gelatin and glucose in a double boiler and mix, having temperature of water in bottom of boiler about luke warm. Then add 40 per cent water to gelatin and glucose, mix well and raise temperature gradually to not less than 130°F. and not over 150°. Cook for 1½ to 2 hours.

Wipe each piece of smoked meat carefully to remove surplus grease, salt, etc., then dip into glaze momentarily. If necessary, pieces may be dipped a second time. Then let them hang over dipping vessel so that any drip may be recovered. This glaze is transparent, resilient and amply tough to resist damage in reasonable handling. Meats may be wrapped and shipped in usual manner.

There are also glazes for covering meat loaves and sausage and for baked hams and picnics.

Filling for Ravioli

Beef, ground	5.250 lb.
Parmesan Cheese, grated	11.250 oz.
Eggs, whipped	1.625 lb.
Butter	0.750 lb.
Parsley, minced	0.500 oz.
Lemon Peel, grated	0.250 oz.
Salt and Pepper	to suit

Proc. Blend the ingredients thoroughly into a smooth mixture; package it and freeze.

Pepperoni Sausage

Make a careful choice of meats for use in pepperoni. A high grade product may be used with the following ingredients:

Pork Trimmings, Reasonably Lean	50 lb.
Beef Trimmings, Trimmed	30 lb.
Selected Regular Pork Trimmings	20 lb.

Meats for pepperoni should be coarse cut. Grind lean and regular pork trimmings through ¼-in. plate and beef trimmings through 1/8-in. plate. Then put the ground beef on the rocker and rock for about 5 minutes.

Seasoning.—Add pork trimmings and the following seasoning ingredients:

Salt	3 3/8 lb.
Granulated Cane Sugar	2 oz.
Dextrose (Corn Sugar)	2 oz.
Nitrate of Soda	2 oz.
Cayenne Pepper	8 oz.
Pimiento	8 oz.
Whole Aniseed	4 oz.
Peeled Garlic	1/2 oz.

Proc.: Pork trimmings, beef and seasonings are rocked together
for an additional 10 minutes. This makes a total rocking time of 15
minutes.

Place the chopped meat on shelving pans in the cooler at a
temperature of 38-40°F. The meat should be in layers not over 10
in. thick and should be kneaded well in order to exclude air as far as
possible. Carry the meat on pans in the cooler for 48 hours minimum and 72 hours maximum time.

At close of holding period meat should be mixed for about 3
minutes in the mixer and then taken to the stuffer.

Use selected narrow hog casings or corresponding artificial casings
for this product. Stuff full length of the strand of casing with meat.
Then break the casing off, making allowance for enough casing to
loop over each end where the casing is broken. Then twist in the
center, which gives a 10-in. length on both links. It is necessary to
remove a little meat from the broken end of each casing, so that
there will be about 1/2-in. of casing to fold up against side of the
sausage link.

Puncture casing thoroughly on stuffing bench to prevent air
pockets in product. Then remove to cooler and allow to hang for 24
hours before taking to dry room.

Air conditioning has solved the problem of controlled, year-round
drying in many sausage plants. Such equipment should be installed
wherever it is desired to manufacture a considerable quantity of
constant quality product, independent of weather conditions. However, each air conditioning installation has its individual operating
characteristics and it is impossible to prescribe general rules to fit all
plants. Pepperoni may be prepared in old type room, however, if
care is used.

The pepperoni is placed on the hanging sections in the dry room
and spread carefully so that the sausages do not touch, and so that
there will be free circulation of air. Select outside sections, hanging
in one section and skipping the next. In this way there will be plenty
of ventilation all around the product.

Dry room temperatures should come within a range of 46-53°F., the best temperature being 48°. The dry room should be equipped with steam coils on the floor and side walls. There should also be windows, fans if possible, and overhead ventilation to provide for the venting of old air. If possible, a relative humidity of 55-65% should be maintained.

Operate the floor and side coils occasionally. The heat, combined with proper air circulation, will dry the pepperoni slowly and from center outward. Do not dry this product rapidly as this will result in case hardening. It is then impossible for the air to penetrate through the shell and dry the center of the sausage.

Where the hanging sections are of considerable height it will be found that sausage on the upper part of the sections dries more rapidly than that hanging on the lower rails. It is good practice to transfer the product to another hanging section, placing the sausage which was formerly on the bottom, at the top, and vice versa. In this way all product will dry evenly.

If weather and dry room conditions are such that mold appears on the sausage during the drying process, the mold must be washed off, either by hand with water and a brush, or through a summer sausage washing machine. Then re-hang on sections and carry there till dry. This condition rarely arises when air conditioning is used.

The dry room must be kept under careful supervision of a trained operator. The temperature of the room must be regulated and the windows, doors and ventilators opened and closed according to temperature and humidity conditions. If the dry room is carried at higher temperatures than specified, there is danger of the pork fat rendering out from the heat and turning rancid. All conditions necessary for producing a good sausage are interdependent and must be controlled intelligently.

If it is impossible to market the pepperoni when fully dried, it is best to transfer it to dry, cooler temperatures of 40-42°F.

If pepperoni is stuffed in casings not more than 1 3/8 in. in diameter, it need not be held in the dry room more than 15 days, provided that 20 days have elapsed from the time curing ingredients were added to the meat.

Coating for Sausages

	g.
Paraffin Wax	35
Rosin	62.8
Whiting	2.2

Proc.: Mix all ingredients together using heat. Stir until uniform.

Head Cheese, Country Style

Cured Pork Snouts	40	lb.
Cured Pork Tongues	25	lb.
Cured Hogskins	15	lb.
Cured Hog Ears	20	lb.
Dry Milk Solids	12	lb.
Cooking Water	60	lb.
Fresh Onions	5	lb.
Salt	1½ lb.	
Pepper	10½ oz.	
Marjoram	3½ oz.	

Proc.: Cook all meats in separate nets until tender. Pork tongues will take the longest time, about three hours. Use just enough water to cover meats nicely.

After cooking, grind hogskins and onions through a fine plate and cut all other meat either by hand or with head cheese cutter.

Then place cooking water in mixer and add ground hogskins, snouts, tongues, ears, dry milk solids and spices. Mix well.

After mixing is completed pour in sauce pans. 3 pound and 5 pound pans are mostly used. After filling, place in cooler.

The same formula may be used for stuffing in hog stomachs, beef bungs, or artificial casings. If hog stomachs are used, they should be cooked for 45 minutes at 160°F. after stuffing. If beef bungs are used, 15 minutes at 160°F. will be sufficient; if artificial casings are used it is only necessary to rinse them in hot water, to clean outside of casings.

Then place them in cooler. When they are set they are ready to market.

Pastrami

Beef of high quality is the base for manufacture of good pastrami. The shoulder clod of kosher cattle may be used for the Jewish trade; otherwise, the rib or tender parts from the hindquarters are used, particularly the round. If plate beef is used it will not require so long to cure and smoke.

The meat is boned, cut into 6- to 7-lb. pieces and trimmed so that fat covering is not too heavy. It is held in the cooler for 24 hours before curing. The very best sanitary conditions must prevail in making pastrami. The curing vat must be absolutely sweet and clean before any meat is put in it.

A dry curing mixture is up of in the proportion of 3½ lb. of sodium nitrate for each 100 lb of salt. About 5 lb. of this mixture is used for each 100 lb. of meat. Bottom of the curing vat is sprinkled with the cure and a layer of meat is then packed on top. Cover this layer with curing mixture and sprinkle some ground black pepper and ground garlic on top of it. Then put in another layer of meat, treating it in same manner as first layer, and continue this building-up process until the vat is full. The pieces should be packed compactly. After 24 hours weighted wooden covers are put on the meat.

The meat now begins to form its own brine, which should cover the top layer. Any shortage in brine should be repaired with 65° plain pickle. The meat should be left in cure for 60 days at 38 to 40° F., without overhauling. If overhauled it is difficult to get it repacked as compactly as is desirable. This product must not be over-salted as it is not soaked after curing.

Remove meat from vat after curing and rub each piece with a spice mixture consisting of 65 per cent black pepper, not too fine, and 35 per cent allspice. Another spice mixture used for this purpose consists of:

	%/wt.
Ground Coriander (not too fine)	60
Allspice	25
White or Black Pepper Garlic Flour	15

Hold the meat overnight after rubbing with the spice mixture. Next morning put a hanging string in each piece and hang about four pieces to the smokestick.

Place in smokehouse where product is roasted through at a temperature of 320°F. About 6 to 7 hours may be needed to complete the treatment. Because of the high temperature used, specially designed smokehouses, or smoke-roasting houses, are desirable. These are about 5 ft. deep, 3 ft. wide and 12 ft. high. The heat is furnished by a gas heating system on the floor. There should be at least 6 ft. between the fire and the meat.

Chop Suey
(To Freeze)

A	Onions, sliced	28.12 lb.
	Lard	4.50 lb.
B	Celery, 1/4 inch pieces	18.00 lb.
	Water	3.38 gal.
C	Meat concentrate (Armour's Vitalox)	3.38 lb.
D	Mushrooms, sliced	4.50 lb.
	Green pepper, sliced	4.50 lb.
	Bean sprouts	13.50 lb.

Place A in a kettle and lightly brown it. Introduce B into A and simmer for 1/2 hour. Now add C, dissolved in a little water. Next, add D and bring to a boil. Finally, thicken the whole batch with the following paste:

Cornstarch	1.50 lb.
Cold water	2.25 pt.

Proc.: Mix thoroughly. Fill into suitable containers, and freeze.

Chow Mein Noodles

Flour	23.67 lb.
Eggs	4.78 lb.
Water	4.84 lb.
Salt	8.00 oz.

Proc.: Mix thoroughly to form a smooth dough. Roll out the dough very thin, dry partially and cut into thin strips. Dry the resultant noodles, and then deep-fat fry them at 375°F.

DAIRY PRODUCTS

Process Cheese Spreads
I
(Pimento Cream Spread)

	lb.
Cream Curd from 6% Fat Milk	252
Pimento Through 5/16-in Grinder Plate	62½
Vinegar-Sugar Solution	40
Gum Paste	31
Salt	5
Cheese Color	6
Cayenne Pepper	1-1/3
Sugar (Granulated)	4

Mix thoroughly before putting into cooking kettle. Heat to 170°F. Run through filling machine into sterilized glass containers of desired capacity. Use of emulsifier optional. If used, emulsifiers are: Sodium citrate, 1%, or di-sodium phosphate, 2%, calculated on weight of batch mix.

II
(Relish Spread)

	lb.
Cream Curd from 6% Fat Milk	250
Sweet Pickle Relish Vinegar-Sugar Solution	40
Pimento (Through an 1/8-in. Grinder Plate)	27
Condensed Whey or Whey Powder Plus Water	30
Gum Paste	40

An emulsifier may be used; some operators prefer its use, some do not.

Mix constituents thoroughly, heat to 165°. or slightly higher, depending upon flavor desired, and run through filling machine into sterilized glass containers of desired capacity.

III
(Limburger Cheese in Glass)

	lb.
Limburger Cheese	180

Cream Curd from 6% Fat Milk	60
Water	10
Disodium Phosphate	10
Salt	2

Mix thoroughly and heat to 174°F. Run through filling machine into sterilized glass containers of desired capacity.

IV
(Special "Sharp" Spread)

	lb.
American Cheddar Cheese (Smooth)	34.0
Salt	.25
Disodium Phosphate	.6
Condensed Whey	8.4
Water	4.0
Cheese Color, To suit	

If needed to firm up the batch, 1 to 2 lb. of sodium citrate or of sodium metaphosphate may be added.

The American cheddar cheese should be partly broken down. One half should be of clean flavor and aged.

Add cheese to heated kettle slowly as in making the limburger product. Add 15 lb. of the water at start. Do not add the condensed whey until after the batch has been heated to about 135°F.

Finish by heating to 160°F. and packaging in containers other than glass. Make into packages similar to those of processed cheeses.

(Pecorino Romano)

This is the only unique type of sheep milk cheese which belongs distinctly to Italy. It gets its name from the Roman fields where it is produced on a large scale and from where it is exported. Today, its manufacture is extended through southern Italy and into Sardinia.

It is manufactured in a sort of cottage or barn built with poles and straw, using the following utensils:

1. A copper kettle or boiler.
2. "La chiova"—a long ruler-like piece of wood used for stirring.
3. "Cascine"—woven forms for soft cheese—made by weaving willows or twigs.
4. Cheese table or platform for working cheese.

The milk, as soon as it is drawn, is taken into the barn or hut and strained through a linen cloth stretched across the kettle. Then it is heated to about 38°C. (100°F.) and a paste of rennet is added, which is made from dried baby lamb stomachs cut up and kneaded in brine.

Coagulation: The coagulation should take place in 15 to 20 minutes.

It is left to harden further for 10 to 15 minutes, after which it is broken with the "chiova."

It is cooked for 15 or 20 minutes, constantly stirring the particles (using the same device) and raising the temperature to 50°C. (122°F.). The kettle is removed from the fire and is left to allow the curd to settle for 10 to 15 minutes until it is gathered at the bottom in a mass.

In the meantime, some pails of fresh water are poured into the kettle to lower the temperature of the whey.

Then the cheesemaker very carefully compresses the cheese at the bottom of the kettle, turns it over, and continues the pressing on all sides. In doing this, always working under the whey, with a string he cuts the mass in one or more pieces and turns them over to his assistants, to be kneaded in the forms on the table.

Molding the cheese: The cheese is pressed in the forms, to overflowing, in such a fashion as to make a great part of it rise over the rim of woven mold. This becomes manipulated, by hand, into a sort of cone shape. This is worked until the vertex of the cone is flush with the rim of the form. The whey, meanwhile, is being squeezed out of the cheese and trickles out through the openings on the side of the form. From these same holes, punctures are made in various directions into the mass to let out as much whey as possible. In the meantime, the cheese is still under hand pressure. This operation is called "friction" and lasts around one-quarter hour. The pressing is continued for 20 minutes with much care and skill in order to eliminate all perforations and make the mass entirely solid.

Salting: After two days, the salting is begun, in the salting rooms or chambers, by applying dry salt abundantly. During the salting operation, the cheese is removed from the form and deposited on a sort of stone trough and then, after a few days, they are transported to a wooden table, situated near the same place. Every other day, they are turned over and rubbed with fresh salt and placed in piles of twos or threes. The salting period lasts around two months.

From the salting chambers, the cheeses pass into the ripening rooms where they are immediately scraped and cleaned.

The only cure which they then receive is turning over and a rubbing by hand (comparable to massaging).

The cheeses are properly ripened in one year.

Yield: 100 l. (200 pounds) of milk yield 20 kg. (44 pounds) of fresh cheese which during ripening, become reduced by 18 to 30% or to about 16 or 17 kg. (35.2 to 37.4 pounds). Each form is from 12 to 14 cm. (4.8 to 5.6 inches) high with a diameter of 24 to 26 cm. (9.6 to 10.4 inches), and weighs on the average 7 kg. (15.2 pounds).

The whey, remaining after the process, is of milky appearance and is highly esteemed in the Roman country for its high residual of albumin and fat content which makes possible a 10 to 12 kg. (22 to 26.4 pounds) yield of ricotta of a fine and delicate quality.

(American Style Process Cheese)

Green (Young) Cheddars	5½ Cheddars
Sharp (Aged) Cheddars	1 Cheddar
Skim Milk Cheddars	¼ Cheddar
Salt	3 lb.
Sugar	2 lb.
Diosodium Phosphate (Anhy.)	13 lb.
Water	4 gal.

Dissolve the salt, sugar, and disodium phosphate in the water. Slowly add the cheese, ground medium fine, to the above solution at 170°F. with constant stirring until uniform and smooth. Pour into box molds.

Mold Control in Cheese Making

Simple means of mold control are now in use in the cheese industry. Washing of walls, ceilings and shelves with a 10% copper sulfate solution, or a 5% borax solution, has been of great aid. Likewise, cheddar cheeses can be dipped in a 5% borax solution before being paraffined. Or, if the paraffin is agitated during dipping, the borax can be added directly to the paraffin. This not only controls molds, but also prevents fungus or rind rot from developing.

It has also proved possible, in mixing water paints or cement coatings for cheese rooms, to substitute 10% copper sulphate solution for the water. Whether such a paint is used or not, walls and ceilings should be well made and kept well painted. Every possible effort should be made to keep foreign matter and mold from walls and ceilings.

If condensation on the ceiling is hard to control, then the ceiling should be washed more frequently, with the 5% borax solution. In any case, walls and ceilings should be washed at regular, frequent intervals with borax solution, after being first scrubbed with soap and water.

Borax should be spread on floors that are washed and kept dry.

Yogurt
I

(Yogurt is the purest, most efficacious and easily digestible way of supplying dairy milk protein. It stimulates the digestion.)

Ready-prepared yogurt can be used if a good quality can be obtained from a dairy or a health food store.

There is some kitchen equipment available which makes it possible to produce yogurt at home. The makers' directions must be closely followed.

Milk	5 cups (2½ pt.)
Yogurt	½ cup

Proc.: Bring milk to the boil. Cool to 100°F. Whisk yogurt together with milk. Pour into a glass receptacle. Place the glass receptacle in the top of a double boiler. Cover up the pan and keep the water at a temperature of 100°F. As soon as the milk has solidified put it in the refrigerator. When cold it is ready to eat.

Cultured Buttermilk

Solution:

(a) The buttermilk must contain 10 per cent of total solids, therefore, 500 pounds of buttermilk must contain (500 x 0.10) 50 pounds of total solids.

(b) Since dry milk solids contain approximately 96 per cent total solids it will be necessary to use (50 ÷ 0.96) 52.08 pounds of dry milk solids and 447.92 pounds of water in the preparation of 500 pounds of cultured buttermilk.

1. Place water at about 70°F. in vat and add dry milk solids in two or three installments, mixing thoroughly to avoid lumping and to secure complete solution.

2. Heat to 180°F. for 30 minutes; cool to 68° to 70°F., and culture with 5 per cent starter.

3. Allow the cultured buttermilk to stand until an acidity of 0.95 to 1.00 per cent has developed.

4. Where possible, cool the curd to 50°F. prior to breaking. When this is impractical, satisfactory results can be secured by cooling the curd with slow, intermittent agitation in the vat. Satisfactory results have also been secured by breaking the curd in the vat, drawing it off into cans, and placing the cans in ice water or in a cold room with frequent agitation to reduce it to the desired temperature of 50°F. or lower.

5. In every case the curd of fermented reconstructed milk should be broken to a smooth consistency with an agitator operating at a speed which results in a minimum amount of air incorporation.

6. In so far as possible, avoid pumping.

Ice Cream Mix
I
4% Fat

Cream, 40%	7½ lb.
Concentrated Skim Milk, 30%	31½ lb.
Cane Sugar	9 lb.
Corn Syrup Solids	4½ lb.
Stabilizer	6 oz.
Water	22-1/8 lb.

II
4% Fat

	pts./wt.
Butter	3.53
Dry Skim Milk	8.51
Sugar	12.00
Gelatin, U.S.P.	0.38
Egg Yolks (fresh yolks)	0.75
Water	49.85

III
6% Fat

	pts./wt.
Butter	5.25
Dry Skim Milk	9.19
Sugar	12.00
Gelatin, U.S.P.	0.34
Glyceryl Monostearate (Aldo 28)	0.11
Water	48.07

IV
10% Fat

	lb.
Butter	8.81
Dry Skim Milk	8.50
Sugar	12.00
Gelatin, U.S.P.	0.34
Glyceryl Monostearate (emulsifier)	0.11
Water	45.24

V
10½% Fat
(Soft)

Cream, 40%	21.0 lb.
Concentrated Skim Milk, 30%	29.0 lb.
Liquid Sugar	17.0 lb.
Water	20.0 lb.
Egg Yolk, powdered	8.0 oz.
Stabilizer	4.4 oz.

Proc.: Mix the skim milk with the water and stir in the liquid sugar, powdered egg yolk, and stabilizer. Now stir in the 40% cream, and mix well. Pasteurize this mixture and then homogenize it. Allow it to cool, then place the mixture in a freezing tank where it is beaten until smooth. Finally, stir in any desired flavor.

Pasteurize at 165°F. for about 30 minutes. Homogenize at 3,500 pounds pressure.

JELLIES & PRESERVES

Grapefruit Jelly

Peeled Grapefruit	1 lb.
Water	2 pts..
Sugar	¾ lb.

Proc.: Cut up grapefruit and cook with water until thoroughly soft. Drain juice through bag, bring to a boil, add the sugar, and boil to 220°F. Seal in glasses covered with melted paraffin wax.

Wine Jelly

Pectin Solution:	
100 Grade Exchange Citrus Pectin	
Slow Set No. 451	11 oz.
Granulated Cane or Beet Sugar	1.5 lb.
Water	9 lb. 6 oz.
Sweet Wine	4.75 gal.
Granulated Cane or Beet Sugar	59 lb.
*Standard Citric Acid Solution	4.5 fl. oz.

Proc.: First prepare the pectin solution by thoroughly mixing the pectin with the sugar and stirring it into the boiling or nearly boiling water. Shut off heat and stir until the pectin has all dissolved. Draw off and set aside until needed.

Add the wine and the sugar (larger amount shown in the table) to the kettle. Warm and stir until the sugar has dissolved. Add the pectin solution, stir well and heat to boiling. Shut off heat and stir in the citric acid solution. Allow to stand until the bubbles have risen to the surface (not over 15 minutes), skim, and package.

Fig Jam for Fig Bars

	pts./wt.
Sugar	15.00
Water	4.88
Dextrose, U.S.P.	1.50
Figs, ground	16.13

Proc.: Place the sugar, dextrose, and water in a clean kettle and boil to 222°F., then stir in the ground figs and continue to boil to 222°F.

Peach Preserves

Peeled Sliced Cling Stone Peaches	10 lb.
Sugar	7 lb.
Water	3 pints
Peach Kernels	10

Proc.: Bring sugar and water to a boil, add the peaches and

kernels. Cook until the fruit is clear when lifted from the syrup. Pack in sterilized containers and seal.

Orange Marmalade
I

Oranges	3 lb.
Lemons	3
Water	1½ pt.
Sugar	3 lb.

Proc.: Wash, remove the peel and seeds, cutting one half of the peel into very thin strips, and add it to the pulp and balance of the peel, which has first had the yellow portion grated off and has been passed through a food chopper with the pulp. Cover with water and let stand overnight. Boil for 10 minutes the next morning, allow to stand for 12 hours, add the sugar and again stand overnight. Cook it rapidly next morning until the jelly test can be obtained (about 222°F.). Cool to 176°F. pour into sterilized glasses, and seal with paraffine.

II

Peeled Sweet Oranges	1 lb.
Sugar	7/8 cup
Water	1 qt.
1/4 peel removed from each orange.	

Proc.: Wash fruit, remove 1/4 of the peel. Cut this peel into thin slices and then boil for about 10 minutes until tender. Now cut fruit into small pieces, add to it the white of the Orange Peel, and boil about 20 minutes. Now strain the juice, then place a kettle and bring to a boil. For each pound of fruit taken add 7/8 cup of sugar. When this comes to a boil, add, the peel which has been cooked tender, and cook to the jellying point.

Spice Tea

	oz.
Allspice	4.25
Celery Seed	25.00
Parsley Seed	22.50
Sweet Basil	3.50
Thyme	5.25

Proc.: Place spices in 5 gallons of cold water for 5 minutes and agitate, while simmering for 15 minutes.

Strain and add enough water to make 5 gal.

BEVERAGES

Orange Nectar

Orange Oil	.85 lb.
Acacia, Powdered	.40 lb.
Water	.70 pt.
Emulsifying and add to the following:	
Tartaric Acid	15.0 lb.
Sugar	20.0 lb.
Orangeade Color	.2 gal.
Water	To Make 10 gal.

Frozen Punches

Grape

Concord Grape Juice	1 qt.
Basic Lemon Water Ice	1 gal.
Roman Punch Flavor	2 oz.
Orange Extract	2 oz.

Tutti-Frutti

Basic Lemon Water Ice	1 gal.
Assorted Crushed Fruit	1 qt.
Cinnamon Extract	1 oz.
Cordial Flavor	To suit
Red Food Color	To suit

Wild Cherry

Basic Lemon Water Ice	1 gal.
Clarified Apple Cider	1 pt.
Wild Cherry Essence	½ oz.
Cherry Color	To suit

Ginger

Basic Lemon Water Ice	1 gal.
Chopped Candied Ginger	1 lb.
Ginger Extract	½ oz.
Orange-Peel Extract	½ oz.

Claret

Basic Lemon Water Ice	1 gal.
Clarified Apple Cider	1 pt.
Cordial Claret Flavor	1 oz.
Cinnamon Extract	½ oz.

Orange

Basic Lemon Water Ice	1 gal.
Orange Juice	1 pt.

Orange-Peel Extract	1 oz.
Maraschino Flavor	1 oz.
Orange Color	To suit

Cheap Apple Cider

Boiled Cider	2 gal.
Granulated Sugar	25 lb.
Tartaric Acid	¾ gal.
Water	30 gal.

Proc.: Color to suit with sugar color. Thoroughly mix; let stand three days, then draw off and add one ounce of benzoate of soda to each ten gallons of cider. Keep in a cool place.

Creme de Menthe

I

Place 2 oz. green mint leaves in a jar and pour over them 1 qt. 95% alcohol. Let steep 8 to 10 days, then press out leaves. Next, add 3 gills sugar syrup (30%), green food coloring to suit, and 10 drops peppermint oil. Filter and bottle.

II

Peppermint Oil	10 cc.
Lemon Oil	1 cc.
Clove Oil	2 drops
Cinnamon Oil	2 drops
Sugar	4,000 cc.
Ethyl Alcohol (95%)	10,000 cc.
Distilled Water	8,000 cc.
Green Food Coloring	To Suit

Proc.: Mix all together, let stand 3 or 4 days, and filter.

Prune Cordial

Boil ½ lb. dried prunes 5 minutes, drain off water, and cool the prunes. Next, fill a jar with them and cover with a good grade of whiskey (80 to 100 proof). Add 2 teaspoonfuls sugar, and let set a week or so. Drain off the cordial, crush the prunes, filter, and use.

CONDIMENTS

Mushroom Sauce

1 heaping tablesp. vegetable fat
1 onion, chopped
¾ lb. mushrooms
2-3 tablesp. flour
1-1½ cups vegetable stock
salt

1 teasp. lemon juice
4½ tablesp. cream
1 egg yolk

Proc.: Melt fat and saute onion in it. Add mushrooms, thinly sliced. Saute together, cover and cook for ¼ hr. Sprinkle flour over. Add stock, salt and lemon juice and cook for 10 min. Add cream and egg yolk to improve the flavor of the sauce.

Worcestershire Sauce

Vinegar	1 qt.
Powdered Pimento	2 dr.
Powdered Cloves	1 dr.
Powdered Black Pepper	1 dr.
Powdered Mustard	2 oz.
Powdered Jamaica Ginger	1 dr.
Common Salt	2 oz.
Shallots	2 oz.
Tamarinds	4 oz.
Sherry Wine	1 pt.
Curry Powder	1 oz.
Capsicum	1 dr.

Proc.: Mix all together, simmer for 1 hour, and strain. Let the whole stand for a week, strain it, and fill in bottles. Worcestershire sauce is never quite clear; straining to remove the coarser particles is all that is necessary.

Mayonnaise
I

10 Egg Yolks
15 tablesp. lemon juice
10 cups oil
salt
Marmite or other yeast extract

Proc.: Whisk egg yolk with a few drops of lemon juice. Add oil drop by drop whisking constantly. Add a little more lemon juice if the mayonnaise becomes too thick. Season last of all.

Note: The oil should be at room temperature before it is used, not allowed to become too hot or too cold.

II

	pts./wt.
Egg Yolks	14
Vinegar	10
Cotton Seet Oil (Prime Summer Yellow)	70
Salt	1½
Sugar	3½

Mustard	¾
Pepper	¼

Proc.: Mix thoroughly in mixing bowls and run through a colloid mill* with a clearance of .005".

Tomato Ketchup

Take 8 lb. tomatoes and stew until tender, together with one or two shallots or onions cut up, then put through a sieve. Return to the saucepan, and add a tablespoon salt, 1 tablespoon ground cloves, 1 tablespoon ground ginger, ½ tablespoon allspice, and 1 pint vinegar. Boil until reduced to one-third.

Chili Sauce

Ripe Tomatoes (Finely Chopped)	5 lb.
Vinegar	2 pt.
Garlic	1 oz.
Red Pepper	1 dr.
Salt	2 oz.
Lemon Juice	5 oz.

Proc.: Boil for 1 hour. Then force through a colander and bottle while warm. Cork tightly.

Tartar Sauce

	pts./wt.
"Kelcoloid" HVF	3.26
Water	180.00
Sweet pickle relish	132.74
Vegetable oil	117.94
Spirit vinegar (100 gr.)	76.28
Sugar	31.40
Salt	24.74
Fresh whole eggs	23.00
Chopped capers	6.60
Dry mustard	3.84
Chopped parsley, dry	0.20

Proc.: Pre-mix "Kelcoloid" HVF with 12 grams sugar until uniform. (If "Kelcoloid" DH is used, there is no need for a sugar pre-mix.) Dissolve pre-mix in 75-80°F. water with strong agitation, 8-10 min. While still mixing, add eggs and mix 1 min. While mixing add vinegar, sugar, mustard and salt and mix 1 min. While mixing add vegetable oil slowly and mix 1 min. Homogenize at 1,000 psi (if desired). Combine homogenized mixture with parsley, capers, and sweet relish and mix 2 min slowly. Fill at room temperature with mechanical vacuum at 10-14 in.

Nonfat Gravy Powder

	pts./wt.
Milk, Powdered (low heat)	12.50
"Starbake"-100 Powdered, (edible wheat starch)	52.75
"HVP-A" Seasoning Powder	17.20
Salt	3.20
"Huron" MSG (monosodium glutamate)	2.10
Onion Powder	4.40
Caramel Color, Powdered	3.20
Pepper, Soluble	0.50
Paprika	0.50
Celery, Soluble	0.10
Sugar, Granulated	3.50
Bay, Soluble	0.05

Proc.: Mix all ingredients together. Use 20 grams to package.

USES:

STEWS (beef or lamb): Brown meat in hot fat. Add about ½ of package of powder, cover meat with water, and cook until done. Add vegetables and cook until they are soft. The other half of the powder should be added just before serving the stew.

CHICKEN FRICASSEE OR BRAISED CHOPS (pork, veal, or shoulder lamb chops): Place 1 package of powder in a paper bag. Add chops or chicken parts and shake bag so the meat is well coated. Brown in hot fat. After all pieces are browned, add ¼ cup cold water, cover and cook for about 15 minutes. Sprinkle powder remaining in bag over meat and continue cooking with cover on until done; baste occasionally. Serve rich brown sauce over meat, rice, or potatoes.

CHOWDERS: Cover fish with water, add ½ package of powder and cook until fish is done. Add potatoes and desired vegetables and cook until they are soft. Add milk, butter, and serve.

GRAVY: Add 1 package of powder to 1 cup cold water. Bring to boil, stirring constantly. Continued boiling thickens gravy. A tablespoon of meat juices and fat or vegetable fat ("Crisco" "Spry", etc.) can be added if desired. A cup of vegetable broth or meat stock may be substituted for the water.

VEGETABLE BEEF SOUP: Cover meat and bones with water and add 1 package of powder. Cook until meat starts to fall off bones. Remove meat from bones and use meat and broth as "stock" called for in your favorite recipe.

VEGETABLE SOUP—NO MEAT: Cover vegetables with water. Add 1 package of powder and cook until vegetables are soft.

MEAT LOAF: Blend 1 package of powder into 1 pound of meat. Form into a loaf and bake until done.

SAUCE FOR HAM: Add 1 cup water to 1/2 cup seedless raisins.

Bring to boil. Add 1 package of powder and 1/4 cup brown sugar and blend. Serve over baked ham. If desired, 1/3 cup orange juice and 2/3 cup water may be substituted for 1 cup water.

Pepper-Onion Relish

Onions, finely chopped	4 qt.
Sweet Red Peppers, finely chopped	8 cups
Green Peppers, finely chopped	8 cups
Salt	2½ oz.
Sugar	32 oz.
Vinegar (5-6% acetic acid)	1 gal.

Proc.: Mix all the ingredients, bring slowly to a boil, and cook until slightly thickened. Finally, pour into clean, hot, sterile containers; fill containers to top, and seal tightly.

Corn Relish

Corn (52-56 ears)	8 qt.
Sweet Red Peppers, diced	4 pt.
Green Peppers, diced	4 pt.
Celery, chopped	4 qt.
Onion, sliced	4 cups
Sugar	32 oz.
Vinegar (5-6% acetic acid)	1 gal.
Salt	3 oz.
Celery Seed	2 oz.
Mustard, dry	3 oz.
Flour	4 oz.
Water	1 pt.

Proc.: Remove the husks and silks from the corn, and place it in boiling water. Allow to simmer for about 10 minutes, then dip the ears into cold water. Drain and remove corn from the cob, then measure out 8 quarts of it. Mix the peppers with the celery, celery seed, onion, salt, sugar, and vinegar, and boil for about 15 minutes. Mix the mustard with the flour, stir in the water, and add this mustard-flour combination and the corn to the pepper mixture. Stir and boil for about 5 minutes. Pack the mixture into clean, hot, suitable containers, filling to ½ inch from top. Attach the lids and boil containers for about 10 minutes in boiling-water bath. Remove containers, and seal tightly. If more yellow color is desired, mix about 1 oz of curcuma powder with the mustard-flour mixture.

Powdered French Salad
Seasoning

Salt, finely ground	9.00 lb.
Sugar, finely ground	4.50 lb.
Black Pepper, finely ground	1.50 lb.
Dehydrated Onion Powder	1.20 lb.

Monosodium Glutamate, finely ground	1.20 lb.
Soluble Celery (sugar base)	9.50 oz.
Dehydrated Celery Stalk, finely ground	9.50 oz.
Mustard Flour	8.00 oz.
Red Pepper, finely ground	2.25 oz.
Curcuma, finely ground	2.25 oz.

Proc.: Mix all thoroughly. Add 3/5 oz. of this powdered season-ing to the following liquid mixture:

Vinegar, 10% acetic acid (100 gr)	2 1/6 oz.
Water	1 oz.
Corn oil, refined	5 oz.

Proc.: Mix and shake well. The resultant product is a good French dressing.

FLAVORS, EXTRACTS, AND SYRUPS

Honey Flavor (Synthetic)

	pts./wt.
Ethyl Phenylacetate	5.250
Propyl Phenylacetate	4.500
Phenyl Ethyl Salicylate	1.500
Phenyl Ethyl Phenyl Acetate	1.050
Allyl Phenylacetate	0.750
Ethyl Pelargonate	0.750
Vanillin	0.300
Protovanol "C"	0.150

Proc.: Mix all ingredients together until smooth.

Honey (Miel) (Synthetic)

	pts./wt.
Citronellol, Extra	3.00
Hydroxycitronellal dimethylacetal	3.30
Phenylacetic Acid, Pure	2.70
Phenylethyl Alcohol, Extra fine	1.80
Dimethyl Octanyl Acetate	0.90
Benzophenone, C.P.	0.75
Ethyl Phenylacetate	0.75
Methyl Ionone	0.75
Rhodinyl Phenylacetate	0.75
Terpineol, Extra	0.75
Amyl Cinnamic Aldehyde	0.60
Benzyl Propionate	0.60
Methyl Acetophenone	0.60
Isoeugenol, Extra	0.45

Musk Ambrette	0.45
Alpha Ionone	0.30
Benzyl Acetate Coeur	0.30
Isobutyl Phenylacetate	0.30
Methyl Phenylacetate	0.30
Veronol Aldehyde, 10% in Dimethyl Phthalate	0.30
Benzyl Salicylate	2.25

Proc.: Mix well and allow to stand a week or more with frequent shaking.

Strawberry Flavor, Imitation
I
(Concentrated)

Orris Oil	2	fl. oz.
Neroline	1	fl. oz.
Vanillin	2	fl. oz.
Ionone	¼	fl. oz.
Amyl Acetate	4	fl. oz.
Palergonic Ether	8	fl. oz.
Butyric Ether	24	fl. oz.
Alcohol	3	gal.
Rhatany Extract	2	fl. oz.
Strawberry Juice	6	pt.

Proc.: Mix solution together until uniform.

II

Yara Yara (10% Solution in Benzyl Benzoate)	30 g.
Amyl Acetate	100 g.
Aldehyde C_{14}	50 g.
Iris Concrete (10% Solution)	25 g.
Bitter Almond Essence	10 g.
Rose Essence	3 g.
Jasmine Essence	1 g.
Ionone	1 g.
Bourbonal	10 g.
Alcohol (95%)	375 g.
Glycerin	200 g.

Use straight for in 1:1000 dilution; for a 1:250 essence use as follows:

Strawberry Flavor (Above)	250 gr.
Alcohol (95%)	550 gr.
Glycerin (28° Bé.)	100 gr.
Water	100 gr.

Proc.: Mix ingredients together until uniform.

Cherry Flavor, Imitation
I

	pts./wt.
Eugenol	5
Cognac Oil	10
Cinnamic Aldehyde	10
Benzoic Aldehyde	95
Coumarin	20
Ethyl Caprylate	20
Amyl Acetate	80
Ethyl Acetate	80
Bourbonal	50
Caramel Color	50
Black Current Ester	5
Rose Essence	5
Sweet Orange Essence	50
Lemon Essence	400
Iris Essence (1/70)	50

II

	pts./wt.
Eugenol	1.75
Cinnamic Aldehyde	4.50
Anisyl Acetate	6.25
Anisic Aldehyde	9.25
Ethyl Oenanthate	12.50
Benzyl Acetate	15.50
Vanillin	25.00
Aldehyde C_{16}	25.00
Ethyl Butyrate	37.25
Amyl Butyrate	50.00
Tolyl Aldehyde	125.00
Benzaldehyde	558.00
Alcohol 95%	130.00

Proc.: Mix all ingredients together until uniform.

Imitation Raspberry Flavor

	pts./wt.
Dimethyl Anthranilate	1.25
Corps Praline	1.25
Dimethyl Sulfide	1.50
Citral	12.00
Diethyl Succinate	13.50
Aldehyde C_{14}	15.50
Oil of Celery	16.00
Anethol	21.50

Ethyl Valerate	21.50
Vanillin	40.00
Ethyl Acetate	58.00
Ionone Beta	593.50
Imitation Jasmin	180.00

Proc.: Mix all ingredients together until uniform.

Imitation Apple Flavor

	oz.
Amyl Valerianate	6
Ether Acetic	3
Spirits of Nitrous Ether	3
Amyl Butyrate, Absolute	1
Aldehyde	½
Essence of Peach Blossom	½
Alcohol 95 percent, enough to make	1 qt.

Proc.: Mix solution until uniform.

Household Lemon Extract
I

Lemon Oil (66% Citral)	3 dr.
Alcohol	½ gal.
Distilled Water	½ gal.

II

Lemol Oil	10 cc.
Alcohol	90 cc.

Color with 2 drops of 0.5% alcoholic solution of Dimethyl-amidoazobenzol (Butter Color).

Proc.: Mix solution until uniform.

Synthetic Apricot Flavor

	pts./wt.
Anisic Aldehyde	10
Aldehyde C_{14}	100
Amyl Acetate	150
Iris Concrete (10% Solution)	50
Bitter Almond Essence	50
Ionone (10% Solution)	10
Jasmine Absolute Solution	1
Vanillin	50
Ethyl Caprorate	50
Ethyl Caprylate	65
Ethyl Acetate	100
Amyl Butyrate	50
Amyl Valerianate	50

| Eugenol | 5 |
| Alcohol | 285 |

Proc.: Mix solution until uniform.

Imitation Pineapple Flavor

	pts./wt.
Ethyl Isovalerate	180.0
Ethyl Butyrate	180.0
Heptilate	140.0
n-Butyl Acetate	100.0
Allyl Caproate	100.0
Propionate	80.0
Allyl Cyclohexane Propionate	60.0
Ethyl Oenanthate	60.0
Vanillin	20.0
Amyl Acetate	20.0
Oil of Orange	30.0
Oil of Lemon	10.0
Pineapple Ether	10.0

Proc.: Mix all ingredients together until uniform.

Grenadine

Mash 2 kg. of black sweet cherries and 1 kg. of dark sour cherries. Heat in porcelain to thick consistency. Add 1 kg. mashed raspberries and 3 kilos 90% ethanol. Let the mixture stand for four weeks, then add 4 kg. of sugar dissolved in 2.5 to 3 kg. of water.

Cane and Maple Syrup

Maple Syrup	15 gal.
Water	28 gal.
Corn Syrup	3 gal.
Sugar	435 lb.
Maple Flavor (Imitation)	2 oz.

Proc.: Mix and heat to 219°F. The maple syrup used is usually one of the dark grades. This product may also be made with maple sugar, using 8 lb. of sugar to each gallon of maple syrup, and making up the difference in volume with water.

Malted Milk Powders

	pts./wt.
Powdered Malt Extract	50
Powdered Skimmed Milk	20
Cane Sugar	30

Proc.: Mix well. One teaspoonful when added to 8 ounces of a mixture of chocolate syrup, milk and ice cream and then mixed with the malted milk machine will make a delicious malted milk drink.

Chocolate Syrup
I

Sugar	3 lb.
Corn Syrup	1 1/2 lb.
Water	3 lb.
Cocoa Powder	1 1/2 lb.
Vanilla Extract	3/4 oz.

Proc.: Bring the water to a strong boil and dissolve the sugars in it. Add the cocoa powder, rubbing it into a smooth paste. When cool, stir in the vanilla extract and continue stirring until the product is homogeneous.

II

Sugar	2 1/2 lb.
Corn Syrup	12 oz.
Water	5 1/4 lb.
Chocolate Liquor, bitter	5 oz.
Vanilla Extract	1/2 oz.

Proc.: Place all the ingredients, except the vanilla extract, in a clean kettle and cook to a density of 30° Baumé. Stir in the vanilla extract when cool.

Walnuts in Imitation
Maple-Flavored Syrup
(Prepared Sundae Dressing)
I

"Sweetose"	48 lb.
Granulated Sugar	17 lb. 4 oz.
Water	7 lb. 12 oz.
Maple Flavor	To suit
Caramel Color	To suit

II

"Sweetose"	60 lb.
Granulated Sugar	7 lb. 4 oz.
Water	5 lb. 12 oz.
Maple Flavor	To suit
Caramel Color	To suit

Proc.: Bring the "Sweetose," sugar, and water to a boil and then add the flavor and color.

Place the desired quantity of walnut pieces in each container; then fill with hot syrup and seal at once. A half-gallon container will require about 2 lb. of nut meats and a No. 3 can, about 1 pound.

Orange Juice Syrup

Concentrated Orange Juice	12	fl. oz.
Concentrated Lemon Juice	3	fl. oz.

Orange Juice-Pulpy Type	19	fl. oz.
Water	42 1/2	fl. oz.
Granulated Sugar	5 1/4	lb.
15% Orange Oil-Pectin Emulsion	1 1/4	fl. oz.
Certified Orange Color		
(F.D. & C. Yellow No. 6)	As desired	
Benzoate of Soda U.S.P.	3	g.

Proc.: If necessary, add water to make a total of 128 fluid ounces of the finished syrup.

Dissolve the benzoate of soda in the water and then add the orange juice-pulpy type. Stir well, add the sugar, and continue the stirring until the sugar is dissolved. (Do not heat to hasten solution of the sugar.) Add the concentrated orange juice, concentrated lemon juice, orange oil-pectin emulsion, and certified orange color. Mix thoroughly.

1 fluid ounce of the finished orange juice syrup is diluted with 5 fluid ounces of ice water to produce the finished drink that will contain not less than 15% fruit juice (orange and lemon) by volume.

Imitation Grape-Flavored Syrup

Granulated Sucrose	685	lb.
"Flex-Sol" Imitation Grape Concentrate		
F-1229	40-50	oz.
"Givconental" Blend Grape Fruit		
Extract (Optional)	200	oz.
50% Tartaric Acid Solution	144	oz.
Sodium Benzoate Solution	100	oz.
Amaranth Color Solution	50	oz.
Brilliant Blue Color Solution	6 1/4	oz.
Water	To make 100	gal.

Proc.: Mix ingredients until uniform.

Imitation Strawberry Flavored Syrup

Granulated Sucrose	685 lb.
"Flex-Sol" Imitation Strawberry	
Concentrate F-1228	40-50 oz.
"Givconental" Blend Strawberry	
Fruit Extract (Optional)	200 oz.
50% Citric Acid Solution	150 oz.
Sodium Benzoate Solution	100 oz.
Amaranth Color Solution	20 oz.
Ponceau 3R Solution	5 oz.
Water	To make 100 gal.

Proc.: Mix ingredients until uniform.

Imitation Raspberry Flavored Syrup

Granulated Sucrose	685 lb.
"Flex-Sol" Imitation Raspberry	
Concentrate F-1232	40-50 oz.
"Givconental" Blend Raspberry Fruit	
Extract (Optional)	200 oz.
50% Citric Acid Solution	175 oz.
Fivefold Concentrated Orange Juice	
(Reduce the Citric Acid	
Acid Solution to 150 oz.) (Optional)	1 gal.
50% Citric Acid Solution	200 oz.
Sodium Benzoate Solution	100 oz.
Sunset Yellow Color Solution	64 oz.
Water	To make 100 gal.

Proc.: Mix ingredients until uniform.

Lemon-Flavored Syrup

Granulated Sucrose	685 lb.
"Flex-Sol" Lemon Concentrate F-1226	50 oz.
Fivefold Concentrated Lemon Juice (Reduce the Citric Acid Solution to 150 oz.) (Optional)	1 gal.
50% Citric Acid Solution	200 oz.
Sodium Benozate Solution	100 oz.
Tartrazine Color Solution	20 oz.
Water	To make 100 gal.

Proc.: Mix ingredients until uniform.

Lemon-Lime (or Lime) Flavored Syrup

Granulated Sucrose	685 lb.
"Flex-Sol" Lemon-Lime	
Concentrate F-1227	50 oz.
Fivefold Concentrated Lemon Juice	
(Reduce the Citric	
Acid Solution to 150 oz.) (Optional)	1 gal.
50% Citric Acid Solution	200 oz.
Sodium Benzoate Solution	100 oz.
Tartrazine Color Solution	25 oz.

Root Beer Flavored Syrup

Granulated Sucrose	685 lb.
"Flex-Sol" Root Beer	
Concentrate F-1233	20-25 oz.
Foamy Caramel Color	2 gal. 96 oz.

| Sodium Benzoate Solution[1] | 100 oz. |
| Water | To make 100 gal. |

Proc.: Mix ingredients until uniform.

The amount of sugar indicated in the formulae will give approximately a 34 Bé.[†] syrup which is recommended for household use. For fountain use, a 32° Bé. syrup, made by using 630 lb. granulated sucrose, is recommended and for vending machines, a 29 to 30° Bé. syrup, made by using 565 lb. sucrose, is desirable. Citric acid is the usual acidulant, while tartaric acid is generally used for grape drinks. The market price, however, is very often the deciding factor. One ounce of 50% tartaric acid solution will replace 1.75 oz. of 50% citric acid solution.

The better grades of refiner's syrups and small amounts of corn or malt syrups may be used for sweetening, where price is a prime consideration. Additional cloud may be obtained in the citrus drinks by the addition of "Flex-Sol" Cloud TR-58-1.

The sodium benzoate solution may be eliminated if the syrups are packed sterile.

NDIES

Fudge
Vanilla

Sugar	6 lb.
Corn Syrup	2 lb.
Sweet Cream, 18%	4 pt.
Butter	4 oz.
Evaporated Milk	2 pt.
Vanilla Flavor	to suit

[1] Sodium Benzoate Solution 20 oz. sodium benzoate dissolved in sufficient water to make 1 gal.

1 oz. sodium benzoate solution will preserve 1 gal. of syrup, emulsion, non-alcoholic extract, etc.

[2] Brilliant Blue Color Solution

Brilliant Blue Powder	4 oz.
Sodium Benzoate Solution	1 oz.
Water	To make 1 gal.

[3] Orange I Color Solution

Orange I Powder	2 oz.
Sodium Benzoate Solution	1 oz.
Water	To make 1 gal.

[4] 50% Citric Acid Solution

5 lb. citric acid crystals dissolved in sufficient water to make 1 gal.

A 50% tartaric acid solution is made in exactly the same manner. [5] Amaranth, Ponceau 3R, Tartrazine and Sunset Yellow Color Solution

Color Powder	8 oz.
Sodium Benzoate Solution	1 oz.
Water	To make 1 gal.

[†] °Bé - °Baumé which is the scale on a Baumeé hydrometer which measures specific gravity.

Proc.: Mix all the ingredients and cook to 242°F. Remove from the fire and keep stirring while adding the vanilla flavor. When the mass is thick, pour it onto oiled wax paper spread over a wooden tabletop. When cold, mark with cutter and cut into squares.

This recipe may be used for making chocolate fudge by adding 16 oz. of sweet chocolate and 8 oz. of chocolate liquor at the end of the cooking.

Chocolate

A	Granulated Sugar	45.00 lb.
	Water	12.00 lb.
B	Corn Syrup	22.50 lb.
	Invert Sugar	7.50 lb.
	Skim Milk Solids (in solution)	0.75 lb.
	Salt	1.50 oz.
C	Extra-high-strength Gelatin in 2¼ lb. of hot water	0.60 lb.
D	Chocolate Liquor and Flavor	9.00 lb.

Proc.: Bring A to a boil and add B. Continue to boil to 236°F. Cool the batch to 165°F. and add C. Stir the mass thoroughly and add D. Whip the whole at 130°F. until the weight per gallon is 9 lb. and deposit it in trays.

Maple

Sugar	18.0 lb.
Sugar, dark brown	7.5 lb.
Corn Syrup	4.5 lb.
Sorbitol Solution	3.0 lb.
Vegetable Butter, m.p. 92° (refined coconut oil)	18.0 oz.
Cream (20%)	9.0 pt.
Evaporated Milk	3.0 pt.
Maple Flavor	q.s.
Butter	1.5 lb.
*Mazetta	18.0 oz.
Salt	0.75 oz.
Walnuts or Pecans	6.00 lb.
Egg Whites, fresh	3.00 oz.

Proc.: Mix the sugars, syrup, sorbitol, vegetable butter, and cream, and heat to boiling point. Add the evaporated milk and flavor and cook until the temperature reaches 244°F. Remove from fire and pour into a clean cream beater (preferably ball-type). Allow it to cool to 150°F., then add the butter, mazetta, and salt and start the beater. As soon as signs of graining appear, add the walnuts. When it appears that the batch would hold its shape if the beating were stopped, add the egg whites, and allow about four more turns. Ladle the mass out immediately in "globs" on paper-lined trays, and slice these "globs" for packaging.

Marzipan Candies

Basic Almond Paste	15 lb.
Short or Standard Casting Fondant	10 lb.
Finely Powdered Sugar	15 lb.
"Convertit"	1/2 lb.
Salt	1 oz.
Color and Flavor	As desired

Proc.: Place all ingredients into a mixing machine, or mix and knead by hand, to obtain a paste or dough sufficiently firm to retain its shape when the marzipan is later formed.

The basic almond paste, fondant, sugar and invertase may be mixed together to form a master batch, then later divided into sections to which color and flavor may be added to each portion or section.

Honey Paste

		pts./wt.
A	Sugar, Granulated	12 1/2
	Sugar, Brown	12 1/2
	Baking Soda	1/8
	Salt	1/8
	Corn Syrup	5
	Honey	5
	Water, hot (variable)	3
	Butter	4
B	Margarine	4 1/2

Proc.: Mix A together, then stir in B and continue to mix until a smooth paste is obtained.

Almond Fig Paste

Corn Syrup	32 lb.
Sugar	16 lb.
Water	2 gal.

Proc.: The above is cooked to 240°F., and then add 80 pounds ground figs. Cook to 242°F. Now add 16 pounds chopped roasted almonds, cook for one minute, and then remove from kettle.

Turkish Candy

Granulated Sugar	5.5 lb.
Inverted Sugar Syrup	4.5 lb.
Water	11.0 lb.
Thin Boiling Starch	1.1 lb.
Cream of Tartar	.3 oz.

Proc.: Put 5 lb of the water into a candy kettle. Add the granulated sugar and invert syrup, boil. Mix the starch and cream of tartar with the remaining 6 lbs. of water. Add this to the boiling sugar and cook all to a heavy sheet. Cast to the desired thickness. This batch should finish to a weight of 12.8 pounds.

Hard Candy (Coffee)

Granulated Sugar	7.00 lb.
Water	1.68 lb.
Invert Sugar	11.20 oz.
Dry, Soluble Coffee (this can be varied to suit one's taste)	6.00 oz.

Proc.: Place the sugars and water in a gas-fired copper vessel and gradually bring to a boil. Wash down the sugar crystals adhering to the sides while bringing it to a boil. Cover the boiling syrup for 3 minutes. Remove the cover, and quickly heat to 320°F. Pour the batch onto a greased cooling slab. Wair 3 minutes, then turn in the cool edges. When partly cooled, fold the batch, add the dry coffee, and then work it into the hot portion of the batch. Finally, form into suitable shapes and cut. Allow it to cool, and then wrap.

Stabilized Caramel

Corn Syrup	15.00 lb.
Sugar	7.25 lb.
Sweetened Condensed Whole Milk	12.50 lb.
Standard Vegetable Shortening	2.50 lb.
Butter	1.25 oz.
Salt	1.00 oz.
Lecithin	0.63 oz.
Ammonium Carbonate Solution	1.25 oz.
3 oz. to 1 pt. water.	

Proc.: Place all the ingredients in a kettle equipped with scrapers, and cook to 190°F. Now prepare a dry blend of 5 oz. of sugar and 224 gr. of Sea Kem Irish moss extractive, and dissolve this blend in 1½ lb. of water previously heated to 180°F. Finally, stir the solution of Irish moss and sugar into the main batch, and cook to about 240°F. Allow it to cool, and stir in 7 dr. of vanilla flavor. Pour the mass onto a greased slab and cool. While cooling, cut into small blocks of any desired size.

Chewing Taffy

Granulated Sugar	5 lb.
Corn Syrup	5 lb.
Invert Syrup	2 lb.
Water	1 lb.
Vegetable Fat, 86 to 96	0.5 lb.
Salt	0.5 oz.
Color and Flavor	To suit

Proc.: Place the syrups, sugar, and water in a kettle and heat to the boiling point. Wash down the sides of the kettle. Add the fat and cook to medium firm ball (254-262°F.). Pour the batch onto an oiled cooling slab, add the salt, and fold in the edges when partly cool. When sufficiently cool, pull well on the hook. Add the flavor and color while pulling. Spin, cut, and wrap in cellophane or waxed paper.

Taffy Apples
(Clear or Colored)

Granulated Sugar	6 lb.
Corn Syrup	½ lb.
Water	1 qt.

Proc.: Put the water, corn syrup, and granulated sugar into a kettle and heat to 240°F. After it has boiled to the proper consistency, this syrup may be divided into portions and colored red, yellow, or any color desired. Pure food colors, of course, must be used for this purpose. Various flavors may also be added if desired. In dipping a highly colored red or green apple, some producers prefer to use a clear syrup as it permits the natural beauty of the fruit to show to advantage.

Any sort of good eating apple can be used. Skewer the apples on sticks, dip into the hot taffy, and cover the apple completely. Let the surplus drip off, then allow to cool.

Weather conditions play an important part in the success of the coating. In humid or rainy weather, it may be necessary to cook at a slightly higher temperature.

Coconut Mounds

	pts./wt.
Sugar	15.00
Water (variable—the finished mix should be firm)	6.00
Corn syrup	11.25
Invert Syrup	15.00
Macaroon coconut, desiccated	30.00

Proc.: Mix the sugar and water and heat to about 180°F., but do not boil. Add the syrups and mix thoroughly. Finally, incorporate thoroughly the desiccated macaroon coconut. Use a wire cut machine to deposit the cookies, or shape into mounds by hand, on paper-lined pans. Bake at 465°F. only long enough to brown the outside surface.

Do not allow the cookies to be baked through.

"Arctic Ice" Dried Fruit Candy

High Melting Fat (90°F. or Above)	7	lb.
Powdered Sugar	10	lb.
Powdered Milk	3 1/2	lb.
Chopped or Ground Dried Fruit	20	lb.
(Figs, Apricots, Prunes or Raisins		
or a Mixture of Any Two or More)		
Vanilla Flavoring, To Suit		
Chopped Walnuts or Almonds	3-5	lb.

Proc.: Melt the Fat. Stir in the powdered milk and sugar. Allow to cool until it begins to thicken. Add the fruits and nuts. Mix well. Spread on slab or paper to harden. Cut to desired size.

Florentines
(A Continental Favorite)

	pts./wt.
Sugar, granulated	21.00
Cream	15.00
Almonds, finely chopped	12.00
Orange Peel, candied and finely chopped	9.00
Butter	1.75
Flour	3.00

Proc.: Mix and cook to 220°F. Bag the mixture out, as pats 1-1/8 to 1-1/4 inch in diameter, onto slightly greased, flour-dusted pans.

Spiced Gum Drops

	pts./wt.
Gum Arabic	40
Glucose	7
Sugar	28
Ground Mixed Spice[†]	4
Caramel Color	To suit

Proc.: The gum is dissolved in 2¼ gal. water in a steam kettle and

[†] Coriander, ginger, nutmeg and cinnamon, 25% each.

passed through a hair sieve. Simultaneously, the spices are tied in a muslin bag and placed in 2 gal. water in a steam kettle. The water is brought to boiling, and, after simmering for 1 hour to extract the spice oils, the bag is removed and the solution passed through a hair sieve.

The sugar, glucose, and spiced water are then cooked at 254 to 255°F. in a steam kettle, the gum solution added, and the whole set aside in a warm room (110°F.) to allow the impurities to rise in a scum. After this has been removed, the batch is deposited in small pastille-shape impressions in warm dry starch.

Marshmallow

I

Water	7 1/2 lb.
Gelatin, medium strength	12 oz.
Sugar	18 lb.
Plastic Invert Sugar	7 1/2 lb.
Corn Syrup	7 1/2 lb.

Proc.: Place the water in a kettle and, while mixing, sift the gelatin into it. Apply a little heat and continue to stir until the gelatin is in solution. Add at once the sugar, and continue the heating and stirring until the sugar is dissolved. Remove the heat, stir in the invert sugar and corn syrup, and mix thoroughly. Transfer the batch to a beater and whip until light. Add color and flavor if so desired. Use this marshmallow for bagging on top of biscuits, cakes, crackers, sugar wafers, etc.

II

Water	28 1/2 oz.
Gelatin	6 oz.
Granulated Sugar	9 3/8 lb.
Water	43 1/2 oz.
Invert Syrup	2 1/4 lb.
Corn Syrup	12 oz.
Vanilla	q.s.

Proc.: Place the gelatin in the 28 1/2 oz. water, and allow it to soak for about 20 minutes. Dissolve the sugar in the 43 1/2 oz. water, and cook to 225°F. Stir in the syrups and bring the temperature back to 225°F. Remove from the fire, stir in the gelatin mixture, and continue stirring until the gelatin is in complete solution. Pour into a sterile container and allow it to cool. Use any amount required at a time for beating and continue beating until the desired consistency is reached. Add the vanilla flavor just before the beating is completed.

Butterscotch Wafers, Chocolate

Sugar	6½ lb.
Corn Syrup	2½ lb.
Salt	1 teaspoon
Water	1 qt.
Butter	12 oz.
Melted Bitter Chocolate	10 oz.
Vanilla Flavor	1 teaspoon

Proc.: The sugar, corn syrup, water, and salt are placed in a copper pan on the stove and stirred until dissolved. When the batch begins to boil, it is covered and allowed to steam for a few minutes. The sides are washed down and a thermometer is placed in the batch, which is cooked to 290°F. The butter, chocolate, and flavor are stirred in rapidly and mixed through well. The batch is then run into wafers through a funnel. The wafers are placed on a greased slab and when cool, they are loosened and packed in tins or jars.

This formula should yield approximately 10 pounds.

BAKERY PRODUCTS

English Muffins

Mix.

Water	4 lb.
Salt	2 oz.
Sugar	1 oz.

Add and mix well

Yeast	4 oz.

Add and mix well

Flour	6 lb. 12 oz.

Proc.: Beat 5 minutes. Cover—let rise 1 hour. Divide into 3 oz. pieces, round, place on trays dusted with white corn meal. Bake in muffin rings on soapstone. Turn once.

Brown 'N Serve Rolls

Flour	100 lb.
Water	68 lb.
Nonfat Milk Solids	4 lb.
Salt	2 lb.
Sugar	12 lb.
Shortening	12 lb.
Eggs	6 lb.
Yeast	2 lb.
Yeast Food	4 oz.
Calcium Propionate (Mold Inhibitor)	4 oz.

Proc.: Dissolve the yeast in some of the water. Place the milk solids, salt, sugar, shortening, eggs, yeast food, mold inhibitor, and the balance of the water into the machine and stir together. Add the flour, then the yeast solution, and mix until a uniform dough is obtained.

This dough should have a temperature of 90°F. when mixed. Ferment at 80°F.

To alter consistency of dough, water may be reduced to 66 or 64 lb.

Bagels
I

Water	1 gal.
Flour (Variable)	15 lb.
Salt	4 oz.'
Sugar	8-12 oz.
Nonfat Milk Solids	12-16 oz.
Shortening	12 oz.
Yeast	4- 6 oz.

Straight Dough Method

Proc.: The amount of sugar to use depends upon the sweetness desired in the finished product. Dissolve the yeast in a small portion of the water. Put the remainder of the water into the mixer, add the salt, sugar, and dried milk, and stir to dissolve these ingredients. Then introduce the flour, begin mixing, and add the yeast. When the dough is partially mixed, add the shortening and continue mixing until the dough is smooth.

Assuming it takes 3 hours to mature this dough, divide the fermentation period as follows:

First rising	1¾ hours
Second rising	1 hour

Allow the dough to rest 15 minutes, after which it should be divided.

II

High Gluten Flour (a wheat flour from which a large part of the starch has been removed)

	7 1/2 lb.
Water (variable)	3 3/8 pt.
Sugar	9 , ,oz.
Yeast	1 1/2 oz.
Salt	1 1/2 oz.
Corn Oil, refined	3 3/4 oz.
Eggs	6 oz.

Proc.: Mix all the ingredients until the dough is smooth and pliable. Remove the dough from the mixer at approximately 80°F.

Allow it to ferment for about 30 minutes. Shape into bagel form, and let them rest for 15 minutes, then place them in boiling water. As soon as the bagels appear on the surface, remove them from the water, and bake at about 450°F.

Premium-Quality Bread

This loaf commands a premium price. It is made with lavish quantities of the finest ingredients and makes delicious toast.

Flour	12 lb., 8 oz.
Water	7 lb., 8 oz.
"Parlac" Dry Whole Milk	8 oz.
Milk	8 oz.
Salt	5½ oz.
Yeast	3½ oz.
Honey	10 oz.
Unsalted Butter	1 lb.
Invert Syrup†	10 oz.

Proc.: Mix in an upright mixer 2 minutes at low speed, 6 minutes at medium speed. The temperature is 78°F. The first rise is 2½ hours and the second rise, 1 hour. Use 5½ cu. in. pan volume for each ounce of dough. The loaf need not be rounded. Flatten the dough as scaled. When scaling is complete, mold the loaves by hand, expelling as much gas as possible and molding tight with a smooth skin.

Proof only until the rounded center extends over the tops of the pans. Bake at 400°F. to golden brown. Cool at room temperature at least 2 hours or until the loaf reaches 90°F. This safeguards against mold and holds the delicious crispness of the crust.

There is a class in every community to which this bread is very appealing and which will have no other. An unbleached short patent spring wheat flour should be used.

† Invert Syrup

Granulated Sugar	20 lb.
Water	9 lb.
Tartaric Acid Crystals	1/3 oz.

Bring the sugar and water to a boil and then add the tartaric acid crystals.

Protein Bread

	I	II
	Dough	Sponge
	(g or lb)	(%)†
Patent Flour	40.0	60.0
Water	29.4	34.6

†The percentage figures are based on the total flour weight in the finished mix.

Yeast	—	3.5
Malt	—	1.25
Shortening	—	3.5
Salt	1.75	—
Sugar	3.5	—
Non-fat Milk Solids	3.5	—
"Promine"-R of "Soyalose" 105	5.0	—

Proc.: Yeast food may be added to the sponge at a level of 0.25 to 0.50, based on the total flour in the finished mix, depending on the nature of the flour used.

To prepare sponge: Sift flour and add to mixing bowl. While mixing using dough hook at low speed, slowly add water, followed by the addition of yeast, malt (yeast food, if used) and shortening. Mix at medium speed until well mixed (dough will pull from bowl). Temperature of sponge should be about 75-78°F when finally mixed. Add sponge to trough and proof at 75-80°F for 3-4 hours and punch. The sponge may be placed in a retarding box at 40°F. overnight for use the following day or immediately mixed with the dough for finishing.

To prepare dough: Mix milk solids, salt, sugar and soy flour (or "Promine"-R) with remaining water (warm), flour, and sponge. Mix about 10 minutes at low speed and 5 minutes at medium speed. The temperature after mixing should be about 78°F. Let dough rest 30 minutes. Divide, mold, pan, and proof for 30-45 minutes. Bake at 410-425°F. for about 25 minutes.

Cracked-Wheat Hearth Bread

A	Water	3 lb.
	Honey	6 oz.
	Cracked Wheat	2 lb.
	Whole Wheat	2 lb.
B	Clear Flour	9 lb.
	Water	5 lb.
	Yeast	4 oz.
	Salt	4 oz.
	Shortening	6 oz.
	"Breadlac"	3 oz.

Proc.: Mix A to a dough at 78°F. and let it stand for 3 hours. (This is not a sponge, but a soaking dough.)

Add B to the first mix and make a dough at 78°F. First rise is 2 hours; second rise, 50 minutes. Divide, round, and let stand 15 minutes.

Mold as rye bread, place on mealed boards, and give medium proof. Cut on the peel as rye bread and bake on the hearth with steam at 425°F. The yield should be 22 lb. 9 oz.

Resembling rye bread in appearance, though somewhat darker, this loaf has a delicious flavor that is derived from the blend of flours with honey.

Sesame Egg Twist Bread

Flour (Variable)	13	lb.
Water (Variable)	1	gal.
Yeast	5	oz.
Salt	4	oz.
Egg Yolks	10	oz.
Powdered Skim Milk (or Equiv.)	10	oz.
Yeast Food	¾ oz.	
Malt	2	oz.
Sugar	12	oz.
Shortening	14	oz.

Proc.: Cream shortening, sugar and egg yolks, then add other ingredients and mix as for any straight dough. Dough temperature, 80°F. Time: first punch, 1 hour 30 minutes; second punch, 45 minutes; take in 15 minutes. May be run as sponge if desired. Scaling weight, two 9-oz. pieces twisted for 1-lb. loaf. After twisting dip in milk solution and roll in sesame seed before panning. Oven temperature, 450°F.; baking time, 35 minutes. Use small amount of steam at finish of baking. Suggested pans, 1 lb., bottom 10 x 4 in.; outside, top 10½ x 4½ in.; inside depth, 3 in. Milk solution should be 1 part of unsweetened evaporated milk to 2 parts of water.

DeLuxe Raisin Bread
Sponge

Bread Flour	8 lb. 4	oz.
Water	5 lb. 3	oz.
Yeast	4½ oz.	
Malt Extract	1	oz.

The temperature should be 76 to 78°F.; the time, 4½ hours.

Dough

Bread Flour	5 lb. 5	oz.
Yeast	2½ oz.	
Water	3 lb.	
Sugar	12 oz.	
Salt	4 oz.	
Shortening	7 oz.	
Egg Yolks	1 lb., 8 oz.	
"Breadlac" Nonfat Dry Milk Solids	2 oz.	

Proc.: Beat the sugar, milk, shortening, and yolks together. Add the sponge and other ingredients. Mix to a smooth dough at 80°F.

Before the dough is completely mixed, incorporate the raisins at low speed.

Prepare the raisins in advance by placing 10 lb. in a cloth sack and soaking in cold water for 5 hours and then draining.

Let the dough stand 15 minutes out of the mixer. Scale and round. Proof 15 minutes. Mold into loaves and give medium pan proof.

The oven temperature should be 425°F., the time, about 30 minutes, and the yield 25 lb. 4 oz.

This is a premium-quality loaf, having 80% raisins and an abundance of top-notch ingredients.

Old-Fashioned Vienna Bread
Sponge

Hard Patent Flour	65 lb.
Water	40 lb.
Yeast	2½ lb.
Yeast Food	6 oz.
Malt	12 oz.
Shortening	2 lb.
Salt	4 oz.

Dough

Hard Patent Flour	35 lb.
Water	16 lb.
Sugar	3 lb.
Salt	2 lb.

Proc.: Temperature of sponge, 72°F.; temperature of dough, 80°F.; sponge fermentation, 4 hours; dough fermentation, 15 minutes. Mix dough 8 minutes and sponge 5 minutes at 65 r.p.m. This formula is made for hand method of make-up.

Old Southern Pecan Pie

Granulated Sugar	1	lb. 2	oz.
Cornstarch		6	oz.
Salt		½	oz.
Emulsified Shortening		8	oz.
"Breadlac" Nonfat Dry Milk Solids		2	oz.
Cream together in a mixer.			
Scalding Water	1	lb.	,
Corn Syrup	4	lb. 8	oz.
Vanilla		½	oz.

Proc.: Have the solution hot (180 to 200°F.) and incorporate.
Whole Eggs 1½ lb.
Fluff with a hand whip and incorporate.
This makes eight 8-in. pies.

Place 2 to 4 oz. pecans, whole or pieces, in each shell and fill at the oven. Use unbaked or partially baked shells. Bake at 375 to 400°F. with strong bottom heat. Remove from the oven when the pies start to puff.

Home-Type Orange Pie

Water	7 lb. 8	oz.
Sugar	3 lb. 8	oz.
Shortening	4	oz.
Bring to a brisk boil.		
Cornstarch	14	oz.
Powdered Orange Juice	1 lb.	
Powdered Lemon Juice	4	oz.
Salt	½ oz.	
Mix together dry.		
Water	1 lb. 10	oz.
Egg Yolk	6	oz.

Proc.: Add the dry mix to this, stirring to a smooth cream.

Pour the cream mixture into the briskly boiling syrup. Cook until the mixture thickens, stirring constantly. Fill into baked shells while the filler is warm. When the pies are partially cooled, spread with meringue. Sift powdered sugar lightly over the top and brown in a medium oven. This makes ten 9-in. pies.

Starchless Rich Custard Pie

Water	8 lb.	
"Parlac" Dry Whole Milk	1 lb. 8	oz.
Dissolve in advance.		
Granulated Sugar	1 lb. 8	oz.
Salt	1/4 oz.	
Nutmeg	1/16 oz.	
Whole Eggs	2 lb.	

Proc.: Rub until the sugar is dissolved, then stir in the milk solution gently to avoid excess foam.

Bake at 425°F. Pour into unbaked shells at the oven. Remove before the filling has entirely set.

This makes ten 9-in. pies.

Pie Crust

Pastry Flour	12½ lb.
Emulsifier-Type Shortening	9½ lb.
Salt	5 oz.
Ice Water	3 lb.

Flaky-Type Pie Crust

Proc.: Mix all the shortening with the flour to obtain an irregular mixture, with lumps of fat varying from the size of a pecan to the size of a pea. Add the cold water and mix just enough to absorb the water. You may find that this mix will absorb a little more water than the other two mixes.

Short, Mealy-Type Crust

Proc.: Mix all the shortening thoroughly with half of the flour. Add the balance of the flour and use it to break up the creamed mass that was mixed first. Add the cold water and mix just enough to incorporate.

Short, Flaky-Type Pie Crust

Proc.: Mix half of the shortening with all of the flour, until you obtain a complete distribution of the fat. Add the balance of the shortening, mix in just enough to get it incorporated. This added shortening must be in little chunks throughout the dough. Add the cold water and mix just enough to get a dough.

Mincemeat
I

Granulated Sugar	125 lb.
Brown or "C" Sugar	25 lb.
Seedless Raisins	50 lb.
Currants	25 lb.
Cider Jelly	25 lb.
Boiled and Comminuted Meat	25 lb.
Suet	12 lb.
Apples	6 bu.
No. 1866 "Spiceolate" Mincemeat No. 1	5 oz.
Citron Pulp	20 lb.

Proc.: Boil the first six ingredients for about 20 minutes, chop the suet and citron with the apples, mix the "Spiceolate" with a few pounds of the sugar, spread evenly over the batch, and mix thoroughly.

If desired, 100 lb. of glucose can be substituted for 75 lb. of granulated sugar, but the product must be labeled "containing glucose."

If a preservative is to be added, use 1 lb. of benzoate of soda for the batch and label accordingly.

II

A Raisins	18.00 lb.
Currants	2.50 lb.
Orange Peel	2.40 lb.

	Evaporated Apples	10.00 lb.
	Citron Peel	1.00 lb.
B	Water	3.60 gal.
	Sugar	20.00 lb.
	Corn Syrup	10.00 lb.
	Molasses	3.20 pt.
	Vinegar	1.60 pt.
	Salt	8.80 oz.
	Benzoate of Soda	1.40 oz.
C	Cinnamon	5.60 oz.
	Dry Mustard	1.80 oz.
	Allspice	0.70 oz.
D	Nutmeg	0.35 oz.
	Cloves	0.20 oz.
	Ginger	0.15 oz.

Proc.: Grind A and soak overnight. Place B in a steam-jacketed kettle and mix thoroughly. Add A to the kettle and bring it to a boil. Turn off the steam and blend into the batch. While the batch is cooling, stir in 6.40 oz. of rum and 6.40 oz. of artificial rum flavor.

Allow this mincemeat to age in a wooden barrel.

Puff Pastry

A	Bread Flour	7 1/2 lb.
	Butter	1 7/8 lb.
	Salt	1 oz.
B	Water, cold (variable)	5 1/4 lb.
	Color	q.s.
	Pastry Butter	5 5/8 lb.

Proc.: Mix A at low speed for about 2 minutes. Add the cold water and color and mix until the dough is mature. Remove dough from the agitator with a scraper, then add the pastry butter, and stir only enough to incorporate it. Roll and fold the dough a few times. Now roll out the finished dough to a thickness of about 1/7 inch, then cut from it disks five inches or more in diameter. Brush the outside edge with egg wash, then fill center with the following mixture:

	Currants or raisins	3.75 lb.
	Cake crumbs	15.00 oz.
	Juice and rind of	5 lemons
	Cinnamon	0.75 oz.

Raspberry or other jam, enough to make a paste.

Proc.: When the disks of puff paste are filled, fold the edges into the center, with minimal overlapping. Flatten them by pressing down, then turn them over and place on wet pans. Allow them to rest for one hour, then bake at 375°F. When baked and cool, cover the tops with powdered sugar.

Linzer Torte

Powdered Sugar	2 2/3 lb.
Butter	3 lb.
Eggs	1 1/2 lb.
Almonds, finely ground	2 lb.
Cake Flour	5 1/3 lb.
Baking Powder	1 1/3 oz.

Proc.: Mix the sugar with the butter, and cream thoroughly. While creaming, stir in the eggs. Finally add the ground almonds, cake flour, and baking powder. Continue to mix until a smooth product is obtained. Chill, then scale as required. Shape into balls, then roll these fairly thin as for cookies. Line the bottom and sides of round pans with this rolled paste. Fill with preserves or any desired fruit fillings. Cover with latice strips made from this paste. Now brush with egg wash, and bake at 375°F. Cool, and besprinkle with powdered sugar.

Jelly Rolls

Eggs	4 lb.
Sugar	4 lb.
Salt	1/2 oz.
Cake Flour, high moisture content	4 3/4 lb.
Sugar	10 1/2 oz.
Baking Powder	2 oz.
Butter	1 1/3 lb.
Milk	29 1/3 oz.
Flavor	to suit

Proc.: Mix the eggs, sugar, and salt, and heat over hot water to 140°F., stirring constantly to prevent coagulation of the eggs. Then whip until the beaten mass will hold a crease for a few seconds when drawing a spatula through the top of the batter. The crease should not close completely for 25 seconds. Mix and sift the cake flour, baking powder, and sugar, and fold this carefully into the whipped mass. Finally, mix the butter, milk, and flavor, heating slowly until the butter is melted, and then fold this in, carefully but thoroughly. Scale into standard sheet pans, 18 by 26 inches and bake at 380°F. Roll as usual. This makes a jelly roll of excellent keeping qualities.

Yeast-Raised Doughnuts

Milk	2 qt.
Water	2 qt.
Yeast	½ lb.

Sugar	2	lb.
Shortening	2	lb.
Eggs	1	pt.
Mace	½	oz.
Vanilla	½	oz.
Bread Flour	15	lb.
Salt	2	oz.

Proc.: Dissolve the yeast in 1 pt. of the water at a temperature of 80°F. Cream together the sugar, shortening, and salt and then stir in the eggs. Add the flour, the balance of the water, and when partly mixed in, add the yeast dissolved in the water. Mix until smooth, getting it out at a temperature of 80 to 82°F., and put in a bowl. Cover and let rise until the dough breaks, then fold over, give another full rising and fold again. When it again shows life, take to the bench. Roll the dough out on the bench to 1/4 in. thickness and cut with a regular doughnut cutter. Set on screens or racks to proof to double size and fry as usual.

Yeast Pastry
(for fruit, cheese and onion tarts)

1 1/3 cups flour
2 tablespoons butter
1/4 teaspoon salt
1/3 oz. yeast
6 tablespoons milk

Proc.: Sift flour and salt into a bowl, make a well in the center and put in a warm place. Cut the butter into small pieces and dot on the flour. Mix the yeast with 2 tablespoons of milk, pour into the well and mix with a little flour to make a thick paste. Cover bowl with a cloth and put in a warm place till the dough doubles its size. Add the rest of the milk and knead into the dough. Beat dough until smooth. Roll out and line pie pan.

NOTE: For a cheese tart, leave pastry to rise before adding the filling. For fruit and onion tarts, add filling and then leave to rise.

Low-Calorie Oatmeal Cookies

Sifted Enriched Flour	5 lb.			
Salt (optional)			3/4 oz.	
Baking Powder	3	oz.		
"Sucaryl" Powder			1-1/2 oz.	
Shortening, melted and cooled	8 lb.	1	, oz.	
Oatmeal, uncooked	3 lb.	10-1/4 oz.		
Skim Milk	3 lb.	1/2 oz.		
Chocolate Pieces			3-1/2 cups	
Vanilla			2-1/2 oz.	

Proc.: Mix and sift flour, salt and baking powder. Add Sucaryl sweetener to melted shortening with eggs and oatmeal, stir thoroughly; add half the flour mixture. Add milk, chocolate pieces and vanilla, then remaining flour mixture.

Drop by teaspoons on greased baking sheet. Flatten slightly with a fork, if desired, or allow to stand up with peaks. Bake in hot oven (400°F.) 15 min. or until golden brown.

Each cookie contains approximately 0.5 g protein; 2 g fat; 3 g carbohydrate; 32 cal. If made with sugar, each cookie would supply approximately 43 cal.

Peanut-Butter Brownies

Peanut Butter	2 lb.
Shortening	1 lb.
Vanilla Extract	4 teaspoons
Brown Sugar	4 pt.
Whole Eggs	16
All-Purpose Flour	2 lb.
Salt	2 teaspoons
Chopped Walnuts, Peanuts, Pecans and Almonds	4 pt.

Proc.: Blend the peanut butter, shortening, and vanilla extract in a bowl. Then add brown sugar gradually, creaming in well. Add the eggs, two at a time, beating in well. Finally, blend in the flour, salt, and nuts. Bake 25 minutes, or until the center is firm, in lightly greased pans (8 X 16 X 2) in an oven at 350°F. The formula yields enough for four pans. On removing from the oven, cut into 2 in. squares while hot. Remove from the pans and cool on wire racks.

Gingerbread

A	Sugar	45	oz.
	Shortening	15	oz.
	Eggs, mixed	5	oz.
B	Baking Soda	1 3/4	oz.
	Salt	1/2	oz.
	Ginger, ground	1 1/4	oz.
	Cinnamon, ground	3/4	oz.
	Eggs, mixed	22 1/2	oz.
	Light Molasses	88	oz.
	Cake Flour	5	lb.
	Bread Flour	22 1/2	oz.
	Cream of Tartar	5/8	oz.
	Water (variable)	60	oz.

Proc.: Mix and cream A thoroughly. While creaming, add gradually, first the eggs and then the light molasses, and incorporate them thoroughly. Sift together the cake flour and bread flour, and add it to the mixture, alternating with the cream of tartar dissolved in the water. Continue the mixing until the batter is smooth. Scale as required, and bake at about 375°F.

Low Calorie Orange Dream Cake

"Sucaryl" Powder	1.0 oz.
Salt (optional)	.5 oz.
Butter or Margarine, melted	2 lb. 8.0 oz.
Eggs, separated	4 lb.
Sifted Cake Flour	5 lb.
Double-Action Flour	5 lb.
Orange Juice	3 lb.
Lemon Juice, fresh	1 lb. 1.5 oz.

Proc.: *Topping:* 1/4 oz. Sucaryl in juice and pulp of 16 oranges. Add Sucaryl and salt to melted butter blend. Add egg yolks, beat well. Sift flour and baking powder together; add alternately with the combined fruit juices to the egg yolk mixture.

Beat egg whites stiff; fold in carefully. Bake in moderate oven.

Apple Crunch Cake

		oz.
A	Sugar, fine granulated	27
	Shortening	9
	Baking Soda	3/4
	Salt	3/8
	Cinnamon	1/2
	Raisins, ground	6
B	Eggs	10 1/2

Proc.: Cream thoroughly A, and add gradually, while creaming, B. Finally add alternately the following:

	oz.
Cake flour sifted together with the	
Baking powder	3/4
Applesauce	33
Buttermilk	6

Continue the mixing until a smooth product is obtained. Weigh into heavily greased suitable molds lined with a liberal amount of the following crunch:

A	Coconut, long shred	13 1/3	oz.
	Sugar, powdered	6	oz.
	Salt	40	gr.
	Nutmeats, chopped fine	9	oz.
	Eggs	3 3/4	oz.
B	Lemon gratings	1 1/2	oz.

Proc.: Mix together A, then spread the mass evenly over a pan, and toast at 325°F. until dry—not too dark. When dry and not too hot, run it through a coarse food chopper and mix it together thoroughly with B.

Apple Sauce Cake

A	Cake flour, high moisture	3 3/4	lb.
	Baking Powder	2 1/4	oz.
	Baking Soda	1 1/8	oz.
	Salt	1/8	oz.
	Non-fat dry milk	3	oz.
	Sugar, fine granulation	4 7/8	lb.
	Shortening, emulsifying type	30	oz.
	Cinnamon	1/4	oz.
	Allspice	40	gr.
	Cloves	40	gr.
	Nutmeg	40	gr.
B	Applesauce	2 1/4	lb.
C	Eggs	39	oz.
	Applesauce	3 3/4	lb.

Proc.: Sift the flour, baking powder, and baking soda together, and mix A very thoroughly. Add at slow speed B and mix until smooth. Now mix together the eggs and applesauce, and add this in three separate portions, mixing until smooth after each addition.

Scrape the bottom and sides of the mixing bowl several times during the mixing operation. Finally, scale into loaf tins and bake at 365°F.

Old-Fashioned Apple Cake

Medium Brown Sugar	4 lb.	4	oz.
High-Grade Shortening	1 lb.	12	oz.
Soda		1	oz.
Cinnamon		1/2	oz.
"Parlac" Dry Whole Milk		4	oz.
Cream.			
Whole Eggs	2 lb.		
Add, scraping the bowl.			
Water	1 lb.	12	oz.
Incorporate.			
High-Grade Cake Flour	4 lb.	8	oz.
Cream of Tartar		1/2	oz.

Proc.: Sift together twice. Incorporate and cream to finish the mix.

Peel and chop fresh apples to 1-in. pieces. Batch weight 20 lb. 10 oz. When ready to scale, fold in the apples. Scale 12 oz. to a 7-in.

high-walled layer tin. If the tin is less than 2 in. deep, cut the weight to 10 oz. Sprinkle the top with brown sugar and bake at 325°F.

Coffee Crumb Cake

A	Sugar, granulated	4 lb.
	Honey	10 2/3 oz.
	Brown Sugar	10 2/3 oz.
	Shortening, emulsifying type	2 1/4 lb.
	Non-Fat Dry Milk	6 2/3 oz.
	Eggs	8 oz.
	Salt	1 1/2 oz.
	Flavor	to suit
B	Eggs (part yolks)	30 2/3 oz.
	Water (variable)	66 2/3 oz.
	Cake Flour	94 2/3 oz.
	Baking Powder	4 1/3 oz.

Proc.: Mix the cake flour with the baking powder and sift it. Cream A thoroughly. Add while creaming A the eggs (part yolks), water, and the sifted flour mixture. Weigh out about 14 oz. into tin pans, measuring 5-1/2 x 10-5/8 x 1-5/8 inches.

Low Caloric Cinnamon Coffee Cake

Sifted Cake Flour	5 lb.	
Baking Powder		3-1/4 oz.
Salt		1/4 oz.
Soft Butter		15 oz.
Skim Milk	3 lb.	12 oz.
"Sucaryl" Powder		1 oz.
Yellow Food Coloring, liquid		0.1 oz.
Eggs		1 lb. 3-1/2 oz.
TOPPING		
Melted Butter		5 oz.
Cinnamon		1/2 oz.
"Sucaryl" Powder		1/4 oz.
Toasted Dry Bread Crumbs		9 oz.

Proc.: Preheat oven to 375°F. Combine topping ingredients, stirring until blended; Set aside. Sift flour, baking powder and salt into mixing bowl. Cut in butter on low, then medium speed, 3-5 min., until mixture is completely blended and looks like fine cornmeal. Add 40 fl. oz. of the milk mixed with Sucaryl and coloring. Beat 1/2 min. on medium speed. Add remaining milk and beat 1 min. longer. Add unbeaten eggs and beat 1 min. more. Pour into greased pan; sprinkle with cinnamon crumbs. Bake in moderate oven (375°F) 20 min. This formula represents a reduction of approximately 39% in cal.

Rum Chiffon Cake

		pts./wt.
A	Cake Flour	40 1/2
	Sugar	15
	Sugar, light brown	12
	Baking powder	2 1/16
	Salt	3/4
B	Corn oil	21
	Egg yolk	21
	Water (variable)	1 1/2
C	Rum	7 1/2
D	Egg whites	42
	Sugar	28 1/2
	Cream of Tartar	3/8

Proc.: Mix A thoroughly by sifting. Add B and stir until the mix is smooth. Add the rum C and stir thoroughly, but do not over-aerate. Now mix C beat this to a firm peak, and fold the batter portion into this whipped mass, thoroughly but carefully. Scale as required, and bake at 365°F. Remove the cake from oven, allow it to cool, then dip it in the following rum glaze:

Baba au Rum

This is one of the most delicious of yeast-raised products. Very often it is baked in a miniature size and packed in air-tight jars for wide distribution at fancy prices.

Granulated Sugar	13 lb.
Water	10 lb.
Cinnamon	1 small stick
Orange Peel	1 oz.

First prepare this rum syrup and boil at 220°F. Gage to 28 on the sugar scale.

Proc.: Cool the syrup to lukewarm. Add 1 qt. dark rum and incorporate thoroughly. If desired the rum may be replaced with high-grade rum flavor. When the time comes for dipping, have the syrup hot and the babas cold. Before using the syrup, strain to remove the orange peel and cinnamon.

"Parlac" Dry Whole Milk	11 oz.
Lukewarm Water	4 lb. 13 oz.

Prepare in advance. Place the "Parlac" on top of the water and dissolve with a hand whip.

Sponge

Milk (from the Above)	4 lb.
Yeast	1 lb. 4 oz.
Bread Flour	4 lb.

Bring the sponge to 80°F. and ferment about 1 hour.

Dough

Whole Eggs	2 lb.
Yolks	2 lb.
Milk (Balance of the Above)	1 lb. 8 oz.
Salt	3 oz.
Sugar	1 lb. 2 oz.
Melted Butter and Shortening	4 lb. 8 oz.
Bread Flour	7 lb.
Chopped Glace Cherries	2 lb.
Currants	2 lb.

Proc.: Beat the eggs with the sugar and salt. Add the milk. Incorporate the flour alternately with the melted butter. Add the sponge and fruits. Mix until the dough is dry and not sticky. Scale at once. The dough should be at 77°F. Without shaping, place the dough in Turk-head pans. The dough should fill the pans about half. Proof about 3/4 volume and bake at 350°F.

When cool, dip into the hot rum syrup and drain.

Babka
(Polish Cake)

Yeast	2 oz.
Lukewarm Water	1 pt.
Scalded and Cooled Milk	1 pt.
Sugar (Sucrose or Dextrose)	12 oz.
Egg Yolks	1 1/2 pt.
Melted and Cooled Butter	1 lb.
Salt	2/3 oz.
Flour	3 lb. 6 oz.
Almond Extract	1/2 oz.
Ground Walnuts	1 lb.
Cinnamon	1/3 oz.
Apple Jelly	12 oz.

Proc.: Dissolve the yeast in the water. Combine the milk, one half of the sugar, egg yolks, salt, and butter. Add two thirds of the flour and the yeast solution, and mix until smooth.

Let this stand about 2 1/2 hours then add the remaining sugar, flour and the almond extract and mix until smooth. Then combine the cinnamon and walnuts and work these into the dough just enough to distribute them.

Place in greased round 9-inch cake pans, 4 inches deep. Brush the batter with beaten egg white and allow it to rise until double in bulk. Bake in slow oven at 325°F., for about 1 hour. Remove, cool, melt the apple jelly, and brush over the cake.

English Fruit Cake

Sugar	1 1/2 lb.
Butter	1 1/2 lb.
Eggs	1 1/2 lb.
Flour	1 7/8 lb.
Currants, thoroughly cleaned	18 oz.
Sultanas, thoroughly cleaned	15 oz.
Mixed Peel, finely cut	6 oz.

Proc.: Cream the sugar and butter and place it in a beating kettle. Add the eggs gradually, stirring and beating well after each addition. Finally, stir in the flour, currants, sultanas, and mixed peel. Blend thoroughly and weigh the mixture into a heavy paper-lined cake mould. Bake at a slow heat for about 3 hours.

Old-Fashioned Poundcake

120% Granulated Sugar	12 lb.
Butter	3 lb. 8 oz.
High-Grade Shortening	3 lb. 8 oz.
Mace	1/4 oz.

Cream light and fluffy.

High-Grade Cake Flour	10 lb.
"Parlac" Dry Whole Milk	1 lb. 4 oz.
Salt	5 oz.

Sift together and add, incorporating thoroughly.

Water	1 lb. 14 oz.

French Cheese Cake

A Bakers' Pot Cheese	1 lb. 10 oz.
Cake Flour	5 oz.
Salt	1/8 oz.
Granulated Sugar	5 oz.
Dried Milk	3 oz.
B Egg Yolks	4 oz.
C Melted Butter or Shortening	4 oz.
D Water	8 oz.
Lemon Juice	1 1/2 oz.

	Grated Lemon Rind	1/8 oz.
E	Egg Whites	1 lb.
F	Granulated Sugar	5 oz.
	Purity C Bakers' Starch	1 1/2 oz.
	Water	1 lb.
G	Granulated Sugar	5 oz.

Proc.: Blend A on low speed, add B, and mix 5 minutes. Incorporate C, then D. Prepare a cooked meringue by cooking F until clear and thick, dissolving G, and then beating this hot syrup into the whites, E, already whipped to a soft peak. Fold the meringue into the cheese batter, gently but evenly.

Prepare the pan with fat and crumbs. Fill the pan, set it in a second pan, and bake at 350°F.

Basic Cream Icing

Icing Sugar	12 lb.
Butter	3 lb.
Shortening, emulsifying type	2 2/5 lb.
Salt	33 gr.
Condensed Milk	33 1/2 oz.
Flavor	to suit

Proc.: Cream thoroughly at medium speed the sugar, butter, shortening, and salt. Then add the condensed milk very slowly, at slow speed on a three-speed mixing machine. Finally stir in the flavor. Chocolate or cocoa, candied fruits, coconut, nutmeats, etc., may be added to this icing.

Baker's Topping Cream

	pts./wt.
Butter	100
Shortening	140
Water	340
Whole-Milk Powder	40

Proc.: The whole-milk powder and water are mixed together and heated to 160°F. The butter and shortening are then added, with stirring, and the mixture homogenized at 2,000 lb. pressure. The topping cream is cooled and bottled. When cool, this cream may be whipped in the same manner as the usual 40% heavy cream.

Chocolate Fudge Icing

A	Sugar	18 lb.
	Salt	1/2 oz.
	Water	5 5/8 pt.
B	Bitter Chocolate	9 lb.

	Invert Sugar	4 1/2 lb.
C	Powdered XXXX Sugar	6 3/4 lb.
D	Shortening	3 3/8 lb.
	Margarine	1 1/8 lb.
	Vanilla Flavor	1 1/8 oz.

Proc.: Boil A to 230°F. Remove from fire and add B, stir until smooth and then transfer the whole to a mixing bowl. While mixing the batch at low speed, add C. When the batch stiffens, add D.

Malted-Chocolate Frosting

Malt (Syrup or Powdered)	1 lb.
Boiling Water	10 oz.
Salt	1/8 oz.

Proc.: Dissolve the malt and the salt in the water, then sift and add:

Icing Sugar	4 lb.
Milk Powder	4 oz.

Pour in:

Vanilla Flavor	1/2 oz.

Beat in machine until smooth, then add:

Margarine	1/2 lb.

Last, add:

Melted Bitter Chocolate	1 lb.

To use, heat the frosting to approximately 100°F.

Coffee Cake Icing

Icing Sugar	9 lb.
Pectin Syrup	1 lb.
Egg Whites	3 oz.
Water	1 1/4 lb.
Flavor	As desired

Proc.: Place pectin syrup, egg whites, and 1/2 pound of water in mixing bowl and whip until light. Add half the sugar and water, and when that is thoroughly mixed in, add the balance of sugar, water, and flavor. Cream until light. This icing will remain lustrous and soft for a long time. It will blend beautifully with chocolate, cocoa, fruits, and nuts for other delicious varities.

Edible Cake Decorations

Gelatin	2 oz.
Water	4-6 oz.
Glucose	6½ oz.
Granulated Sugar	2 lb.
Water	12 oz.

Icing Sugar	2½ lb.
Cornstarch	12 oz.

Proc.: Soak the gelatin in water. While the gelatin is soaking, boil the sugar, glucose, and water to 250°F. Cut off the heat and add the soaked gelatin, mixing it in well. The mass is then strained and allowed to cool to approximately 100°F., at which time approximately one-third of the icing sugar is worked into the syrupy mass. It is to be worked in until the mass turns creamy white in color. The starch is then added and mixed in thoroughly. The balance of the icing sugar is then added and the mass mixed to a smooth and rather firm paste. If the paste is too stiff for handling or shaping, expose it to indirect heat, reworking it to the desired consistency.

The paste may be colored, as desired, with food colors.

Winemaking

This must start as a hobby, but if your are so inclined and are willing to spend the time and money in experimentation you may find you have a good hobby-industry. Before the industry part can start you must contact your local I.R.S. office for local regulations and licensing. For small home consumption none of this is necessary—so experiment and have fun.

Basic Equipment:

2 gal. capacity of boiling water
2 gal. capacity plastic pail
1 gal. capacity glass fermentation jars
1 air lock and bored bungs for each fermentation jar.
Empty wine bottles
Supply of bottle corks
Plain bungs to fit fermentation jars (for storing while the wine is clearing)
Nylon sieve at least 6 inches diameter
Funnel at least 6 inches diameter
Four feet of rubber or plastic siphon tube
Corking tool (You could bang the corks in with a flogger; but you'd be sorry!)
Hydrometer and trial jar (Not essential, but you must have one if you want consistent results)

Basic Ingredients:

A pinch of yeast
Yeast nutrient
Campden tablets for sterilizing and inhibiting mould growth
¼ lb. citric acid
A small tube of Tannin
A small packet of Depectinizer
3 lb. sugar for each gallon of wine made.

When you have the ingredients, try out the procedure. But read through the recipe several times.

Marigold Wine (dry)

Marigold Flowers	3 qt.
Sugar	2½ lb.
Citric Acid	½ oz.
Campden Tablet	1
Water	1 gal.
Yeast: Hock	
Nutrient	

Proc.: Bring the water to the boil, adding the sugar. Allow to cool. Add the main ingredients along with Campden or depectinizer if required. Add all other ingredients except yeast. Introduce yeast or starter. When the must is fermenting leave in a container covered with a cloth in "aerobic" fermentation for the number of days indicated in the recipe. Strain off into fermentation jar and fit air lock. Aerobic fermentation four days.

Plum Wine (dry)

Stoned, Mashed Plums	4 lb.
Sugar	2 ½ lb.
Depectinizer	½ oz.
Water	1 gal.
Yeast: Bordeaux, Burgundy or Port	
Nutrient	

Proc.: Place the main ingredient and the sugar in a container. Pour boiling water over and stir. Allow to cool. Continue as in procedure for marigold wine. Aerobic fermentation five days.

Grape Wine

Measure out 1 quart clean Concord grapes in a cup. Place these in a 1 gallon jug and add a sugar solution composed of 3 to 3½ pounds of sugar, depending on what sweetness is wanted. For a medium wine, use 1 quart of water and 3¼ pounds of sugar dissolved therein. Now pour this over the grapes in the jug, add a small piece of a yeast cake dissolved in 1 cup of water, and fill the jug to within 4-5 inches of the top. Next shake the jug to mix the grapes, water, sugar water, and yeast and then tie a piece of cloth over the top of the jug to keep out gnats, and let it ferment. DO NOT STOPPER THE JUG DURING FERMENTATION. After about 3-4 weeks the fermentation has about ceased, if the jug and contents have been kept in a warm place. Next, strain off the grapes, press out the remaining juice and pomace and let work or ferment for another 2 weeks. At the end of this time filter or strain or through a double folded cloth, bottle, and set away to age. At the end of 5-6 months nearly a full gallon of ruby red grape wine will be ready.

HOME CANNING OF FRUITS AND VEGETABLES

General Procedure
Selecting Fruits and Vegetables for Canning

Choose fresh, firm fruits and young, tender vegetables. Can them before they lose their freshness. If you must hold them, keep them in a cool, airy place. If you buy fruits and vegetables to can, try to get them from a nearby garden or orchard.

For best quality in the canned product, use only perfect fruits and vegetables. Sort them for size and ripeness; they cook more evenly that way.

Washing

Wash all fruits and vegetables thoroughly, whether or not they are to be pared. Dirt contains some of the bacteria hardest to kill. Wash small lots at a time, under running water or through several changes of water. Lift the food out of the water each time so dirt that has been washed off won't go back on the food. Rinse pan thoroughly between washings. Don't let fruits or vegetables soak; they may lose flavor and food value. Handle them gently to avoid bruising.

Filling Containers

Raw pack or hot pack.—Fruits and vegetables may be packed raw into glass jars or tin cans or preheated and packed hot. In this publication directions for both raw and hot packs are given for most of the foods.

Most raw fruits and vegetables should be packed tightly into the container because they shrink during processing; a few—like corn, lima beans, and peas—should be packed loosely because they expand.

Hot food should be packed fairly loosely. It should be at or near boiling temperature when it is packed.

There should be enough sirup, water, or juice to fill in around the solid food in the container and to cover the food. Food at the top of the container tends to darken if not covered with liquid. It takes from ½ to 1½ cups of liquid for a quart glass jar or No. 2½ tin can.

Head space.—With only a few exceptions, some space should be left between the packed food and the closure. The amount of space to allow at the top of the jar or can is given in the detailed directions for canning each food.

Closing Glass Jars

Closures for glass jars are of two main types:

Metal screwband and flat metal lid with sealing compound. To use this type, wipe jar rim clean after produce is packed. Put lid on, with sealing compound next to glass. Screw metal band down tight by hand. When band is tight, this lid has enough give to let air escape during processing. Do not tighten screw band further after taking jar from canner.

Screw bands that are in good condition may be reused. You may remove bands as soon as jars are cool. Metal lids with sealing compound may be used only once.

Porcelain-lined zinc cap with shoulder rubber ring. Fit wet rubber ring down on jar shoulder, but don't stretch unnecessarily. Fill jar; wipe rubber ring and jar rim clean. Then screw cap down firmly and turn it back ¼ inch. As soon as you take jar from canner, screw cap down tight, to complete seal.

Porcelain-lined zinc caps may be reused as long as they are in good condition. Rubber rings should not be reused.

Exhausting and Sealing Tin Cans

Tin cans are sealed before processing. The temperature of the food in the cans must be 170°F. or higher when the cans are sealed. Food is heated to this temperature to drive out air so that there will be a good vacuum in the can after processing and cooling. Removal of air also helps prevent discoloring of canned food and change in flavor.

Food packed raw must be heated in the cans (exhausted) before the cans are sealed. Food packed hot may be sealed without further heating if you are sure the temperature of the food has not dropped below 170°F. To make sure, test with a thermometer, placing the bulb at the center of the can. If the thermometer registers lower than 170°F.—or for the length of time given in the directions for the fruit or vegetable you are canning.

Remove cans from the water one at a time, and add boiling packing liquid or water if necessary to bring head space back to the level specified for each product. Place clean lid on filled can. Seal at once.

Cooling Canned Food

Glass jars. —As you take jars from the canner, complete seals at once if necessary. If liquid boiled out in processing, do not open jar to add more. Seal the jar as it is.

Cool jars top side up. Give each jar enough room to let air get at all sides. Never set a hot jar on a cold surface; instead set the jars on a rack or on a folded cloth. Keep hot jars away from drafts, but don't slow cooling by covering them.

Tin cans. —Put tin cans in cold, clean water to cool them; change water as needed to cool cans quickly. Take cans out of the water while they are still warm so they will dry in the air. If you stack cans, stagger them so that air can get around them.

Day-After-Canning Jobs

Test the seal on glass jars with porcelain-lined caps by turning each jar partly over in your hands. To test a jar that has a flat metal lid, press center of lid; if lid is down and will not move, jar is sealed. Or tap the center of the lid with a spoon. A clear, ringing sound means a good seal. A dull note does not always mean a poor seal; store jars without leaks and check for spoilage before use.

If you find a leaky jar, use unspoiled food right away. Or can it again; empty the jar, and pack and process food as if it were fresh. Before using jar or lid again check for defects.

When jars are thoroughly cool, take off the screw bands carefully. If a band sticks, covering for a moment with a hot, damp cloth may help loosen it.

Before storing canned food, wipe containers clean. Label to show contents, date, and lot number—if you canned more than one lot in a day.

Wash bands; store them in a dry place.

Storing Canned Food

Properly canned food stored in a cool, dry place will retain good eating quality for a year. Canned food stored in a warm place near hot pipes, a range, or a furnace, or in direct sunlight may lose some of its eating quality in a few weeks or months, depending on the temperature.

Dampness may corrode cans or metal lids and cause leakage so the food will spoil.

Freezing does not cause food spoilage unless the seal is damaged or the jar is broken. However, frozen canned food may be less palatable than properly stored canned food. In an unheated storage place it is well to protect canned food by wrapping the jars in paper or covering them with a blanket.

On Guard Against Spoilage

Don't use canned food that shows any sign of spoilage. Look closely at each container before opening it. Bulging can ends, jar lids, or rings, or a leak—these may mean the seal has broken and the food has spoiled. When you open a container look for other signs—spurting liquid, an off odor, or mold.

It's possible for canned vegetables to contain the poison causing botulism—a serious food poisoning—without showing signs of spoilage. To avoid any risk of botulism, it is essential that the pressure canner be in perfect order and that every canning recommendation be followed exactly. Unless you're absolutely sure of your gage and canning methods, boil home-canned vegetables before tasting. Heating usually makes any odor of spoilage more evident.

Bring vegetables to a rolling boil; then cover and boil for a least

10 minutes. Boil spinach and corn 20 minutes. If the food looks spoiled, foams, or has an off odor during heating, destroy it.

Burn spoiled vegetables, or dispose of the food so that it will not be eaten by humans or animals.

HOW TO CAN FRUITS, TOMATOES, PICKLED VEGETABLES

Fruits, tomatoes, and pickled vegetables are canned according to the general directions on pages 5 to 8, the detailed directions for each food on pages 11 to 16, and the special directions given below that apply only to acid foods.

Points on Packing

Raw pack.—Put cold, raw fruits into container and cover with boiling-hot sirup, juice, or water. Press tomatoes down in the containers so they are covered with their own juice; add no liquid.

Hot pack.—Heat fruits in sirup, in water or steam, or in extracted juice before packing. Juicy fruits and tomatoes may be preheated without added liquid and packed in the juice that cooks out.

Sweetening Fruit

Sugar helps canned fruit hold its shape, color, and flavor. Directions for canning most fruits call for sweetening to be added in the form of sugar sirup. For very juicy fruit packed hot, use sugar without added liquid.

To make sugar sirup.—Mix sugar with water or with juice extracted from some of the fruit. Use a thin, medium, or heavy sirup to suit the sweetness of the fruit and your taste. To make sirup, combine—

4 cups of water or juice

2 cups sugar	For 5 cups THIN sirup.
3 cups sugar	For 5½ cups MEDIUM sirup.
4¾ cups sugar	For 6½ cups HEAVY sirup.

Heat sugar and water or juice together until sugar is dissolved. Skim if necessary.

To extract juice.—Crush thoroughly ripe, sound juicy fruit. Heat to simmering (185° to 210°F.) over low heat. Strain through jelly bag or other cloth.

To add sugar direct to fruit.—For juicy fruit to be packed hot, add about ½ cup sugar to each quart of raw, prepared fruit. Heat to simmering (185° to 210°F.) over low heat. Pack fruit in the juice that cooks out.

To add sweetening other than sugar.—You can use light corn sirup or mild-flavored honey to replace as much as half the sugar called for in canning fruit. Do not use brown sugar, or molasses, sorghum, or other strong-flavored sirups; their flavor overpowers the fruit flavor and they may darken the fruit.

Canning Unsweetened Fruit

You may can fruit without sweetening—in its own juice, in extracted juice, or in water. Sugar is not needed to prevent spoilage; processing is the same for unsweetened fruit as for sweetened.

Processing in Boiling-Water Bath

Directions.—Put filled glass jars or tin cans into canner containing hot or boiling water. For raw pack in glass jars have water in canner hot but not boiling; for all other packs have water boiling.

Add boiling water if needed to bring water an inch or two over tops of containers; don't pour boiling water directly on glass jars. Put cover on canner.

When water in canner comes to a rolling boil, start to count processing time. Boil gently and steadily for time recommended for the food you are canning. Add boiling water during processing if needed to keep containers covered.

Remove containers from the canner immediately when processing time is up.

Processing times.—Follow times carefully. The times given apply only when a specific food is prepared according to detailed directions.

If you live at an altitude of 1,000 feet or more, you have to add to these processing times in canning directions, as follows:

	Increase in processing time if the time called for is—	
Altitude	20 minutes or less	More than 20 minutes
1,000 feet	1 min.	2 min.
2,000 feet	2 min.	4 min.
3,000 feet	3 min.	6 min.
4,000 feet	4 min.	8 min.
5,000 feet	5 min.	10 min.
6,000 feet	6 min.	12 min.
7,000 feet	7 min.	14 min.
8,000 feet	8 min.	16 min.
9,000 feet	9 min.	18 min.
10,000 feet	10 min.	20 min.

To Figure Yield of Canned Fruit From Fresh

The number of quarts of canned food you can get from a given quantity of fresh fruit depends upon the quality, variety, maturity, and size of the fruit, whether it is whole, in halves, or in slices, and whether it is packed raw or hot.

Generally, the following amounts of fresh fruit or tomatoes (as purchased or picked) make 1 quart of canned food:

Apples	2½ to 3
Berries, except strawberries	1½ to 3
	(1 to 2 quart boxes)

Cherries (canned unpitted)	2 to 2½
Peaches	2 to 3
Pears	2 to 3
Plums	1½ to 2½
Tomatoes	2½ to 3½

In 1 pound there are about 3 medium apples and pears; 4 medium peaches or tomatoes; 8 medium plums.

Applesauce

Make applesauce, sweetened or unsweetened. Heat to simmering (185°–210°F.); stir to keep it from sticking.

In glass jars.—Pack hot applesauce to ¼ inch of top. Adjust lids. Process in boiling-water bath (212°F.)—

| Pint jars | 10 minutes |
| Quart jars | 10 minutes |

As soon as you remove jars from canner, complete seals if necessary.

Beets, Pickled

Cut off beet tops, leaving 1 inch of stem. Also leave root. Wash beets, cover with boiling water, and cook until tender. Remove skins and slice beets. For pickling sirup, use 2 cups vinegar (or 1½ cups vinegar and ½ cup water) to 2 cups sugar. Heat to boiling.

Pack beets in glass jars to ½ inch of top. Add ½ teaspoon salt to pints, 1 teaspoon to quarts. Cover with boiling sirup, leaving ½-inch space at top of jar. Adjust jar lids. Process in boiling-water bath (212°F.)—

| Pint jars | 30 minutes |
| Quart jars | 30 minutes |

As soon as you remove jars from canner, complete seals if necessary.

Berries, Except Strawberries

Raw Pack.—Wash berries; drain.

In glass jars.—Fill jars to ½ inch of top. For a full pack, shake berries down while filling jars. Cover with boiling sirup, leaving ½-inch space at top. Adjust lids. Process in boiling water bath (212°F.)—

| Pint jars | 10 minutes |
| Quart jars | 15 minutes |

As soon as you remove jars from canner, complete seals if necessary.

Hot Pack.—(For firm berries)—Wash berries and drain well. Add ½ cup sugar to each quart fruit. Cover pan and bring to boil; shake pan to keep berries from sticking.

In glass jars. —Pack hot berries to ½ inch of top. Adjust jar lids Process in boiling-water bath (212°F.)

Pint jars	10 minutes
Quart jars	15 minutes

As soon as you remove jars from canner, complete seals if necessary.

Fruit Juices

Wash; remove pits, if desired, and crush fruit. Heat to simmering (185°–210°F.). Strain through cloth bag. Add sugar, if desired—about 1 cup to 1 gallon juice. Reheat to simmering.

In glass jars. —Fill jars to ½ inch of top with hot juice. Adjust lids. Process in boiling-water bath (212°F.)

Pint jars	5 minutes
Quart jars	5 minutes

As soon as you remove jars from canner, complete seals if necessary.

Peaches

Wash peaches and remove skins. Dipping the fruit in boiling water, then quickly in cold water makes peeling easier. Cut peaches in halves; remove pits. Slice if desired. To prevent fruit from darkening during preparation, drop it into water containing 2 tablespoons each of salt and vinegar per gallon. Drain just before heating or packing raw.

Raw Pack. —Prepare peaches as directed above.

In glass jars. —Pack raw fruit to ½ inch of top. Cover with boiling sirup, leaving ½-inch space at top of jar. Adjust jar lids. Process in boiling-water bath (212°F.)—

Pint jars	25 minutes
Quart jars	30 minutes

As soon as you remove jars from canner, complete seals if necessary.

Hot pack. —Prepare peaches as directed above. Heat peaches through in hot sirup. If fruit is very juicy you may heat it with sugar, adding no liquid.

In glass jars. —Pack hot fruit to ½ inch of top. Cover with boiling liquid, leaving ½-inch space at top of jar. Adjust jar lids. Process in boiling-water bath (212°F.)—

Pint jars	20 minutes
Quart jars	25 minutes

As soon as you remove jars from canner, complete seals if necessary.

Tomatoes

Use only firm, ripe tomatoes. To loosen skins, dip into boiling water for about ½ minute; then dip quickly into cold water. Cut out stem ends and peel tomatoes.

Raw Pack. —Leave tomatoes whole or cut in halves or quarters.

In glass jars. —Pack tomatoes to ½ inch of top, pressing gently to fill spaces. Add no water. Add ½ teaspoon salt to pints; 1 teaspoon to quarts. Adjust lids. Process in boiling-water bath (212°F)—

Pint jars	35 minutes
Quart jars	45 minutes

As soon as you remove jars from canner, complete seals if necessary.

Hot Pack. —Quarter peeled tomatoes. Bring to boil; stir to keep tomatoes from sticking.

In glass jars. —Pack boiling-hot tomatoes to ½ inch of top. Add ½ teaspoon salt to pints; 1 teaspoon to quarts. Adjust jar lids. Process in boiling-water bath (212°F.)—

Pint jars	10 minutes
Quart jars	10 minutes

As soon as you remove jars from canner, complete seals if necessary.

Tomato Juice

Use ripe, juicy tomatoes. Wash, remove stem ends, cut into pieces. Simmer until softened, stirring often. Put through strainer. Add 1 teaspoon salt to each quart juice. Reheat at once just to boiling.

In glass jars. —Fill jars with boiling-hot juice to ½ inch of top. Adjust jar lids. Process in boiling-water bath (212°F.)—

Pint jars	10 minutes
Quart jars	10 minutes

As soon as you remove jars from canner, complete seals if necessary.

Beans, Snap

Raw Pack. —Wash beans. Trim ends; cut into 1-inch pieces.

In glass jars. —Pack raw beans tightly to ½ inch of top. Add ½ teaspoon salt to pints; 1 teaspoon to quarts. Cover with boiling water, leaving ½-inch space at top of jar. Adjust jar lids. Process in pressure canner at 10 pounds pressure (240°F.)—

Pint jars	20 minutes
Quart jars	25 minutes

As soon as you remove jars from canner, complete seals if necessary.

Hot Pack.—Wash beans. Trim ends; cut into 1-inch pieces. Cover with boiling water; boil 5 minutes.

In glass jars.—Pack hot beans loosely to ½ inch of top. Add ½ teaspoon salt to pints; 1 teaspoon to quarts. Cover with boiling hot cooking liquid, leaving ½-inch space at top of jar. Adjust jar lids. Process in pressure canner at 10 pounds pressure (240°F.)—

Pint jars	20 minutes
Quart jars	25 minutes

As soon as you remove jars from canner, complete seals if necessary.

Corn, Whole-Kernel

Raw Pack.—Husk corn and remove silk. Wash. Cut from cob at about two-thirds the depth of kernel.

In glass jars.—Pack corn to 1 inch of top; do not shake or press down. Add ½ teaspoon salt to pints; 1 teaspoon to quarts. Fill to ½ inch of top with boiling water. Adjust jar lids. Process in pressure canner at 10 pounds pressure (240°F.)—

Pint jars	55 minutes
Quart jars	85 minutes

As soon as you remove jars from canner, complete seals if necessary.

Hot Pack.—Husk corn and remove silk. Wash. Cut from cob at about two-thirds the depth of kernel. To each quart of corn add 1 pint boiling water. Heat to boiling.

In glass jars.——Pack hot corn to 1 inch of top and cover with boiling-hot cooking liquid, leaving 1-inch space at top of jar. Or fill to 1 inch of top with mixture of corn and liquid. Add ½ teaspoon salt to pints; 1 teaspoon to quarts. Adjust jar lids. Process in pressure canner at 10 pounds pressure (240°F.)—

Pint jars	55 minutes
Quart jars	85 minutes

As soon as you remove jars from canner, complete seals if necessary.

Peas, Fresh Green

Raw Pack.—Shell and wash peas.

In glass jars.—Pack peas to 1 inch of top; do not shake or press down. Add ½ teaspoon salt to pints; 1 teaspoon to quarts. Cover with boiling water, leaving 1½ inches of space at top of jar. Adjust jar lids. Process in pressure canner at 10 pounds pressure (240°F.)—

Pint jars	40 minutes
Quart jars	40 minutes

As soon as you remove jars from canner, complete seals if necessary.

Hot Pack.—Shell and wash peas. Cover with boiling water. Bring to boil.

In glass jars.—Pack hot peas loosely to 1 inch of top. Add ½ teaspoon salt to pints; 1 teaspoon to quarts. Cover with boiling water, leaving 1-inch space at top of jar. Adjust jar lids. Process in pressure canner at 10 pounds pressure (240°F.)—

Pint jars	40 minutes
Quart jars	40 minutes.

As soon as you remove jars from canner, complete seals if necessary.

HOME CANNING OF MEAT

Cut-Up Meat

Follow directions for cutting up meat (see illust. 1, 2)

Cut tender meat into jar- or can-length strips. Strips should slide into jars or cans easily, with the grain of the meat running the length of the container. Strips may be any convenient thickness, from 1 or 2 inches to jar or can width.

Cut less tender meat into chunks or small pieces suitable for stew meat.

Small, tender pieces may be packed by themselves, with meat strips, or with stew meat.

Hot Pack

Put meat in large shallow pan; add just enough water to keep from sticking. Cover pan. Precook meat slowly until medium done. Stir occasionally, so meat heats evenly.

Glass jars.—Pack hot meat loosely, Leave 1 inch of space at top of jars. Add salt if desired: ½ teaspoon to pints or 1 teaspoon to quarts. Cover meat with boiling meat juice, adding boiling water if needed. Leave 1 inch of space at top of jars. Adjust lids. Process in a pressure canner at 10 pounds pressure (240°F.)—

Pint jars	75 minutes
Quart jars	90 minutes

Raw Pack

Cut up meat (see illust. 1,2). Pack containers loosely with raw, lean meat.

Glass jars.—Leave 1 inch of space above meat. To exhaust air, cook raw meat in jars at slow boil to 170°F., or until medium done (about 75 minutes). Add salt if desired: ½ teaspoon per pint or 1 teaspoon per quart. Adjust lids. Process in a pressure canner at 10 pounds pressure (240°F.)—

Sausage

Hot pack.—Use any tested sausage recipe.

Use seasonings sparingly because sausage changes flavor in canning and storage. Measure spices, onion, and garlic carefully. Omit sage—it makes canned sausage bitter.

Shape sausage meat into patties. Precook, pack, and process as directed for hot-packed ground meat.

Meat-Vegetable Stew

Raw Pack.

Beef, lamb, or veal, cut in 1½ inch cubes	2 quarts
Potatoes, pared or scraped, cut in ½-inch cubes	2 quarts
Carrots, pared or scraped, cut in ½ inch cubes	2 quarts
Celery, ¼-inch pieces	3 cups
Onions, small whole, peeled	7 cups

Combine ingredients. Yield is 7 quarts or 16 pints.

Glass jars.—Fill jars to top with raw meat-vegetable mixture. Add salt if desired: ½ teaspoon per pint or 1 teaspoon per quart. Adjust lids. Process in a pressure canner at 10 pounds pressure (240°F.)—

Pint jars	60 minutes
Quart jars	75 minutes

How To Can MEAT—raw pack

Cut meat carefully from bone. Trim away most of fat without unduly slashing the lean part of meat.

Cut meat in jar-length pieces, so grain of meat runs length of jar. Fill jars to 1 inch of top with one or more pieces of meat.

Set open, filled jars on rack in pan of boiling water. Keep water level 2 inches below jar tops. Insert thermometer in center of a jar (above), cover pan, and heat meat slowly to 170°F. Without thermometer, cover pan; heat slowly for 75 minutes.

Remove jars from pan. Add salt if desired. Wipe jar rim clean. Place lid so that sealing compound is next to glass (above). Screw the metal band down tight by hand. When band is screwed tight, this lid has enough "give" to let air escape during processing.

Have 2 or 3 inches of boiling water in pressure canner—enough to keep it from boiling dry during processing. Put jars in canner (above); fasten cover. Let steam pour from open petcock or weighted-gage opening 10 minutes. Shut petcock or put on gage.

When pressure reaches 10 pounds, note time. Adjust heat under canner to keep pressure steady. Process pint jars packed with large pieces of meat 75 minutes; process quart jars 90 minutes. When processing them is up, slide canner away from heat.

Let pressure fall to zero (30 minutes). Wait a minute or two, then slowly open petcock. Unfasten cover, tilting far side up to keep steam away from your face.

Set jars on rack to cool overnight. Keep them away from drafts, but do not cover. When jars are thoroughly cool, remove metal bands and wipe jars clean. Label and store.

Home Freezing

Quick freezing at $0°$F. is one of the most healthful methods of preserving foods whether for your own use or for sale. However the importance of proper packaging of foods to be frozen cannot be over-emphasized. Don't skimp when buying packaging materials. Never buy plain wax paper or butcher paper. We recommend the following: laminated freezer paper; cellophane; polyethylene, vinyl or pliofilm bags; bag-and-box combinations, paper tubs or plastic containers; heavy aluminum foil; gummed freezer tape; stockinette.

Freezing vegetables - You will never again have to rely on seasonal crops or pay premium prices for out-of-season vegetables. Use the produce from your own garden, or buy the fresh vegetables in season at their lowest prices.

In order to preserve flavor and nutritive value of the food, freeze, it within two to three hours after picking it from the garden.

Water Scalding: Allow one gallon of water for every 1 to 1½ pounds of vegetables. When the water is boiling, lower the vegetables into the water using a colander. Begin timing it immediately for the required length of time. Keep the heat high. It is necessary to change the water every third or fourth batch.

Steam Scald: After placing 2 to 3 inches of water into a large sauce pan, place rack above the water level. Let the water come to a boil. Place the vegetables on the rack and cover. Begin timing when the steam begins to come out around the cover. Remove vegetables from rack at the end of the required time.

Cooling: It is necessary to chill vegetables well before packaging them, otherwise they will continue to cook in their own heat. Cooling may be done by plunging the vegetables immediately into a large quantity of very cold water—at least $50°$F. Change the water frequently to maintain this temperature. If your tap water runs this cold, you may hold the vegetables under a spray until they are cooled.

What follows is a few examples on how to freeze vegetables:

After cooling the vegetable should be packaged securely and frozen immediately.

Asparagus: Wash and cut tips into lengths to fit containers or into 1" pieces.

Small Stalks —   Water Scald — 3 min.
                                Steam Scald — 4 min.

| Larger Stalks ≈ | Water Scald — 4 min. |
| | Steam Scald — 5 min. |

Beans (green and waxed): Wash and cut into lengths 1" long, or slice lengthwise.

Water Scald — 3 min.
Steam Scald — 4 min.

Broccoli: Wash and trim off outer leaves and imperfect stalks. Immerse ½ hour in brine (¼ cup salt to 1 qt. water); rinse. Split lengthwise so that heads are no more than 1" in diameter.

Water Scald — 4 min.
Steam Scald — 5 min.

Corn on Cob: Chill for 15 min. after scalding to insure proper cooling. Select small or medium size ears; husk and silk. Scald in large kettle using about 4 medium size ears per gallon of boiling water.

Water Scald —	Midget — 7 min.
	Small — 8 min.
	Medium — 10 min.

MISCELLANEOUS

Dietary Salt Substitute

	%/wt.
Calcium Glutamate	10
Dextrose	20
Tartaric Acid	5
Calcium Chloride	9
Potassium Chloride	50
Sucrose	5
Calcium Stearate	1

Proc.: Mix all of the ingredients together. Use it like table salt.

Fruit and Vegetable Peeling Liquid

	%/wt.
"Miranol" C2M Conc.	1
"Carbitol"	1
Caustic Soda	20
Water	78

Proc.: Mix everything together until uniform. This mixture loosens up the peels from various vegetables. It can be washed off with water.

7. Miscellaneous

There are many chemical specialties which did not fit into the chapter specifically set aside for them. We have thus compiled a potpourri of formulas that were just too good to omit from a book of this kind. Unlike previous chapters, because of the great diversity of material here, there is no developmental order in this section. So, just thumb through the following pages until something catches your interest.

Water & Oil Repellent Finish
Scotch Guard Type)

	%/wt.
Perfluorobutylacrylate	45
Perfluoro-octanic Acid	55

Proc.: The above are mixed together. 2 to 3 Parts of metallic chromite dissolved in chromic acid should be added to the above mixture.

Water Repellent

	%/wt.
Silicone	39.5
Water	59.3
Ethomid C/15, HT/15, or O/15	1.2

Proc.: The Ethomid is dispersed in water heated to 40-50°C. The silicone is added to the aqueous system with vigorous agitation. To obtain optimum particle size and increased stability, the emulsion is run through a colloid mill* or homogenizer. Laboratory work indicates that with the use of a homogenizer, approximately one-half of the quantity of Ethomid produces an emulsion of equal quality.

Nonstiffening Waterproofing for Cloth

	pts./wt.
Linseed Oil	1
China-Wood Oil	1
Xylol	94
Aluminum Stearate	2
Stearic Acid	2

Proc.: Mix the two oils with the xylol, and heat them on a hot plate, away from open flames,

Waterproofing for Shoes

	pts./wt.
Wool Grease	8
Dark Petrolatum	4
Paraffin Wax	4

Proc.: Melt together in any container.

Removal of Mildew

Proc.: Fresh mildew stains may be removed by washing the articles in a solution of soap and water, then rinse them well and dry them in sunlight. If the mildew stain persists, prepare the following solution:

Sodium perborate	1/2 oz.
Warm, lukewarm	1 pt.

Sponge the stain with this solution and allow it to stand a minute or two; then rinse well. Use extreme care on colored goods.

Modelling Clay
I

	oz.
Glycery Oleate	10
Red Oil	50
Beeswax, Crude	20
Castor Oil	15

Pipe-Clay, Powdered sufficient to suit
Proc.: Heat all ingredients to 100°F. and stir until homogeneous.

II

	oz.
Tallow	19
Gum Mastic	30

Beeswax, Crude	3
Ozokerite	2
Paraffin Wax	4
Gypsum	12
Pipe Clay	60

Proc.: Heat all ingredients to 100°F. and stir until homogeneous.

Grafting Wax

	pts./wt.
Wool Grease	11
Rosin	22
Paraffin Wax	6
Beeswax	4
Japan Wax	1
Rosin Oil	9
Pine Oil	1

Proc.: Melt together until clear and pour into tins. This composition can be made thinner by increasing the amount of rosin oil and thicker by decreasing it.

Modelling Wax

	g.
Beeswax	10
Mastic	10
Ceresin	7.5
Paraffin Wax, Hard	15
Tallow	65
Sulphur (Flowers)	90
Calcium Sulphide	42.5
Kaolin	120

Proc.: Grind the powders into the melted waxes. A mineral pigment e.g. Armenian bolus, is added. (4%). Stir thoroughly and cool.

Contact-Lens Fluid

Sodium Bicarbonate	0.813 g.
Sodium Chloride	0.813 g.
Triple Distilled Water	100.000 cc.

Proc.: The osmotic pressure of the diluted fluid is equivalent to that of a 1.4% solution of sodium chloride.

"Soap Type Eraser"

Corn Oil	4 lb. 6 oz.
Light Magnesium Oxide	2½ oz.
Sulphur Chloride	16 oz.
Ground Pumice (Milled to 200 mesh)	1 lb. 8 oz.

Proc.: Add the magnesium oxide and the pumice to the corn oil, and stir until the solids are uniformly distributed throughout the mass. Now add slowly, and with vigorous agitation, 8 ounces of sulphur chloride at such a rate that the temperature does not rise above 90°F. The mixture is now allowed to stand for 12 hours, whereupon the balance of the sulphur chloride may be added somewhat more rapidly than the initial quantity, but with equally vigorous agitation.

Upon the addition of the second quantity of sulphur chloride the mass thickens somewhat and is poured into wooden channels, open at the top, and having a cross section approximately one inch square. The mixture is allowed to stand for several hours and is removed from the molds when solidified, cut into three inch lengths, imprinted and packed.

Spectacle Lens Cleaner and Mist Preventer
(U.S. Patent 2,333,794)

Lens or tissue paper is impregnated with ½% sodium stearate. The lens is wetted slightly before cleaning.

	%/wt.
Cleaning Solution for Lenses	
Tincture of Green Soap	8
Ammonia	12
Distilled Water	80

Proc.: Mix all ingredients together until uniform.

PYROTECHNICS

Waterproofing Matches

One of the simplest ways to waterproof matches requires only that they be coated with a well-mixed solution of:

	pts./wt.
Glycerin	1
Collodion	50

Proc.: The coating should be applied only to the heads of the matches. In working with the liquid collodion care should be taken to keep open flames away.

Strike-Anywhere Match Tip and Base Compositions

	Tip pts./wt.	Striking Base pts./wt.
Animal Glue	11.0	12.0
Starch	4.0	5.0
Paraffin Wax	—	2.0

Potassium Chlorate	32.0	37.0
Phosphorus Sesquisulfide	10.0	3.0
Sulfur	–	6.0
Rosin	4.0	6.0
Dammar Gum	–	3.0
Infusorial Earth	–	3.0
Powdered Glass and Other Filler	33.0	21.5
Potassium Dichromate	–	0.5
Zinc Oxide	6.0	1.0

Proc.: Heat and mix all ingredients together until uniform.

Underwater Flare

	pts./wt.
Linseed Oil	8
Manganese Dioxide	1
Magnesium Powdered	16
Aluminum, Powdered	12
Barium Sulfate	40
Barium Nitrate	32

Proc.: Mix all ingredients together.

Smokeless Signal Flare
(U.S. Patent 3,262,824)

A consumable pyrotechnic mixture consisting of about 74.2%, by weight, of ammonium perchlorate, about 11.1%, by weight, of stearic acid, about 3.6%, by weight, of paraffin, and about 11.1%, by weight, of copper dust.

Sparklers

	I pts/wt.	II pts./wt.
Potassium Perchlorate	60.0	–
Barium Nitrate	–	50.0
Potassium Chlorate	–	–
Aluminum, Dark Pyro	30.0	8.0
Dextrin	10.0	10.0
Steel Filings	–	30.0
Charcoal Fine	–	0.5
Neutralizer	–	1.5

Proc.: The outside of the composition is coated with magnesium-aluminum alloy grit.

Candles

	pts./wt.
Paraffin Wax	30.0

Stearic Acid	17.5
Beeswax	2.5

Proc.: Melt together and stir until clear. If colored candles are desired, add a very small amount of any oil-soluble dye. Pour into vertical molds in which wicks are hung.

Oxygen--Generating Candle
(Incense)

	pts./wt.
Sodium Chlorite	360
Pyrolusite Powder	20
Magnesium Dioxide	
Mix in a kneader while adding:	
Sodium Dichromate	1
Water	12
Add:	
Magnesium Powder	24

Proc.: Mix until uniform. Compress into cylindrical molds. Dry at 95°C. A candle weighing 750 g. continues to burn for 40 minutes and produces 210 l. of oxygen.

Incense

	oz.
Sandalwood, Powdered	6.0
Cascarilla, Powdered	3.0
Cardamom, Powdered	0.8
Cubebs, Powdered	0.7
Myrrh, Powdered	0.2
Willow Charcoal, Powdered	2.4
Potassium Nitrate, Powdered	0.6
Benzoin Siam	3.0
Balsam Peru	0.6
Oleo-resin Orris	0.2
Bergamot Oil	0.6
Neroli Oil	0.4
Cassia Oil	0.3
Patchouli Oil	0.2
Clove Oil	0.1
Iso-butyl Cinnamate	0.6
Methyl Ionone	0.2
Musk Grains	0.1

Proc.: Mix the aromatic powders with the charcoal and potassium nitrate. Mix the oils with the gum resins, aromatic chemicals, and musk grains. Shake well and allow to stand until solution is complete.

Add the oils-gum resins mixture to the powders in such proportions that the finished powder is damp and adherent. Finally, press the finished powder into shapes of cones. Place each cone on an ash tray and light it.

Incense Powder
I
(Hindu)

	pts./wt.
Sandalwood, Powdered	80.00
Orris Root, Powdered	4.00
Licorice Root, Powdered	12.00
Patchouly Leaves, Powdered	2.50
Potassium Nitrate, Pulverized	0.50
Oriental Rose Compound	1.00

Proc.: Place all of the ingredients in a roller mixer and mix at low rate of speed for 30 minutes.

II
(Pine)

	pts./wt.
Cedarwood, Powdered	25.00
Pine Needles, Powdered	25.00
Orris Root, Powdered	25.00
Sandalwood, Powdered	20.00
Myrrh, Pulverized	1.00
Olibanum, Pulverized	2.50
Potassium Nitrate, Pulverized	0.50
Pine Incense Bouquet	1.00

Proc.: Place all of the ingredients in a ribbon mixer and mix at low speed for one-half hour.

Silver Mirrors on Glass

Clean the glass thoroughly with soap and water until the glass does not show "water break" (irregularities in the running off of rinsing water). They lay the glass on a flat surface and cover with dilute stannous (tin) chloride while preparing the following solutions.

A	Distilled Water	25 cc.
	Granulated Sugar	2 g.
	Nitric Acid (Conc.)	10 drops

Boil for five minutes and cool before using.

B	Distilled Water	40 cc.
	Silver Nitrate	2 g.
	Sodium Hydroxide	1 g.

C Distilled Water	15 cc.
Silver Nitrate	1 g.

Proc.: Add just enough ammonium hydroxide to solution B to redissolve the brown precipitate. Then add solution C to B until a slight darkening is produced.

Pour off the stannous chloride and rinse the glass several times with distilled water. Do not dry. Lay on a flat surface and pour on a freshly made mixture of one part of A to four parts of B. Just enough of the solution should be used to cover the surface without any running off over the edges. The silvering should begin at once and should be complete in 15-20 minutes.

Caution: Solution B or any mixture containing it may explode after standing a few hours. Discard such solutions at the end of a work period.

Staining Glass
(U.S. Patent 2,622,035)
I
(Green)

	pts./wt.
Copper Sulfide	13.0
Silver Sulfide	1.0
Zinc Sulfide	9.0
Ochre	80.0
Hydrochloric Acid	1.0
Water	100.0
Temperature	1000°F.

II
(Reddish-Amber)

	pts./wt.
Silver Nitrate	1.2
Cuprous Chloride	1.0
Basic Copper Acetate	26.0
Basic Zinc Acetate	18.0
Ochre	86.0
45% Ammonium Sulfide	40.0
Water	120.0
Temperature 1000°F.	

III
(Green)

	pts./wt.
Silver Nitrate	1.2
Copper Carbonate	13.0
Zinc Sulfide	9.0

Ochre	80.0
45% Ammonium Sulfide	29.0
Hydrochloric Acid	2.0
Water	To make a smooth slip
Temperature	1050°F.

IV
(Red-Green Mix)

	pts./wt.
Silver Oxide	0.9
Cuprous Chloride	13.0
Zinc Sulfide	9.0
Ochre	80.0
45% Ammonium Sulfide	29.8
Water	To make a smooth slip
Temperature	1050°F.

Glass Etching Fluid

Hot Water	12fl. oz.
†Ammonium Bifluoride	15 oz.
Oxalic Acid	8 oz.
Ammonium Sulfate	10 oz.
Glycerin	40 oz.
Barium Sulfate	15 oz.

Proc.: Dissolve all ingredients in hot water.
†Corrosive.

Deep Etch for Aluminum

Copper Chloride	81.1 g.
Ammonium Bromide	16.0 g.
Hydrochloric Acid, Conc. C.P.	14.2 cc.
Calcium Chloride Solution (40° Bé.) to make 1 qt.	

Proc.: Mix all ingredients together until uniform. This is a controlled etch for aluminum since its action produces no undercut. The depression produced has almost straight walls, closing in slightly at bottom.

Silver Etching

Decorative designs can be etched on silver with chlorine free nitric acid. For matt effects, use the acid at 24° Bé acid. Another mordant for silver is:

Nitric Acid	1 oz.
Hydrochloric Acid	1 oz.
Water	12 oz.

Proc.: Mix all ingredients together. Other suggested etching agents for silver are chromic acid or the analogous mixture of potassium bichromate with sulphuric or nitric acid, or a mixture of dilute sulphuric acid with potassium permanganate.

Odorless Firelighter

	pts./wt.
Sawdust	15.0
Cork dust	15.0
Paraffin	24.0
Sugar	3.0
Potassium chlorate	3.0

Proc.: Dissolve the sugar and potassium chlorate in the smallest amount of water, and mix thoroughly with the sawdust and cork dust. Place this mixture in a heated dough mixer, and pour in the melted paraffin. Keep stirring until a uniform mixture is obtained, and cast in blocks.

Cigarette-Lighter Fluid
I

Dioxane is unexcelled for spark lighters. The addition of 5% of an aromatic solvent such as xylene or toluene makes the flame luminous.

II

Cyclohexane has been found superior to the commercial hydrocarbons commonly used.

Cold Fire

	%/wt.
Carbon Disulfide	50
Carbon Tetrachloride	50

Proc.: Place 25 to 30 cc. of this mixture in an evaporating dish. It is inflammable when ignited with a match, but will not set fire to clothing, hair, etc. It can be put on hands, in a pocket, on the hair, etc., with impunity. In a darkened room, it can be used for a very striking demonstration.

Dry Fire Extinguisher

	oz.
Ammonium Sulphate	15
Sodium Bicarbonate	9
Ammonium Phosphate	1
Red Ochre	2
Silex	23

Proc.: Use powdered materials only; mix well and pass through a fine sieve. Pack in tight containers to prevent "lumping."

Fireproofing Solution for Wood

The solution is made by dissolving separately 1¼ oz. of sodium bichromate and 7 oz. zinc chloride and then combining in a total of 1 gal. of water.

Color may be added for decorative, identification or other purposes. However, the solution exerts a powerfully destructive action on most coal-tar colors. Of many checked only the following three are unattacked:

Amacid Blue V
Kiton Red S
Amacid Yellow T Ex.

These may be combined to produce a number of shades.

The wood is brushed generously or allowed to soak in the solution and then dried.

Wood-Dough Plastic

	pts./wt.
Collodion	86
Powdered Ester Gum	9
Wood Flour	30

Proc.: Allow the first two ingredients to stand until dissolved, stirring from time to time. Then, while stirring, add the wood flour, a little at a time, until uniform. This product can be made softer by adding more collodion.

Wood Hardener

Wood is impregnated by vacuum-pressure method with the following mixture:

	pts./wt.
"Aroclor" 4465	70
Microcrystalline Wax	20
Sulfur	10

Proc.: Heat all ingredients and stir until uniform.

Wood Preservative

Decay (rot) in wood is caused by low forms of plant life called fungi. This decay can be stopped by poisoning the food supply in this case, the wood itself. The wood can be preserved by putting into it poisonous chemicals that will either kill the decay fungi outright, or keep them from growing. The same preservative is effective in controlling attack by termites. These insects are most abundant in moist, warm soil where there is a plentiful supply of food in the form of wood.

Zinc chloride, commercial	54 oz.
Sodium dichromate, commercial	11½ oz.
Water	10 gal.

Proc.: Dissolve the chemicals in the water and shake well. Soak either green or seasoned wood in this chromated zinc chloride solution for 1 or 2 weeks. If time is very limited, shorten the soaking period to 4 or 5 days. This steeping treatment has been found to be definitely beneficial to posts made of ash, jack pine, lodgepole pine, ponderosa pine, red pine, and Scotch pine.

Chromated zinc chloride is more resistant than simple zinc chloride to leaching (losing soluble parts when water passes through). It gives equal protection against termites, and greater protection against the decay fungi. Wood treated with water-soluble preservatives can be painted satisfactorily after it has been seasoned in order to remove the water added during the treatment.

Determining the Polarity of Batteries

Moisten a piece of soft paper with a 4% aqueous potassium iodide solution and touch the paper with the leads from the battery or cell, keeping the leads about ½ in. apart. A brown spot will form at the positive terminal, provided the battery is active.

Cut-Flower Preservative
I
(Australian Patent 165,666)

$2,2^1$-bis (4,6-Dichlorophenol) Sulfide water solution. Use up to 5 p.p.m.

II
(German Patent 952,753)

	pts./wt.
Hydrazine Sulfate	12
Sodium Chloride	6
"Pril" (Detergent)	1

Proc.: Add 1 g. of above to 1 liter of water.

Gas Warning Odorant

(U.S. Patent 2,068,614)

Mercaptan	0.5 lb.
Alkylnitrosoamine	0.1 lb.
Sudan III Dye	0.75 lb.
Gasoline	1 gal.

Proc.: The above solution will odorize 500,000 cu. ft. of gas.

Dyeing of Nacre or Mother-of-Pearl

Quickly and easily a great variety of shades can be laid on nacre.

Besides natural shades most beautiful artificial shades and lusters may be produced.

First the objects are treated with a weak pickling solution in order to remove grease. For this purpose they are put in a solution of 1 part of pure white potash to 10 water; keep at a steady temperature for about ¼ of an hour, drain and repeatedly rinse and dry. Treated this way, nacre will eagerly absorb dyes, particularly those dissolved in alcohol; in a short time it will be colored through-out. Because of the density of the material, colorations are rather dark. Weak solutions are therefore advisable. The longer the material is allowed to stay in the bath the deeper the dye will penetrate. It is therefore up to the dyer to dye the whole body equally and uniformly. After dying the objects are rinsed in pure water and dried in saw dust.

Often light or irregularly colored gray or gray-black mother-of-pearl is to be blackened. Many shells are colored only around the edge or a few cm. toward the center, the center itself being white.

Nacre is treated in such a way that silver chloride is formed and then exposed to sunlight. A saturated solution of silver chloride in ammonia is used. Put the nacre in a well sealed container for several days in a dark room, shaking the same from time to time. Then put the pieces on blotting paper, expose them to sunshine and after 3 days dark colored nacre is the result.

Now the finished and polished articles of nacre are put into a glass container and the above solution is poured over them. From time to time the container is shaken in order to change the position of the pieces so that the same spots are not always covered. After 24-60 hours the pieces are taken out, put on blotting paper and exposed to direct sunlight.

The blackening effect goes rather deep. Disks of about 10 mm. thickness which have been in the solution for 48-60 hours, when broken apart, appear, inside, a uniform dark gray. They may be ground or polished and no white spots will appear. The outer appearance of nacre, dyed this way, is exactly like natural black nacre. The longer the solution and the sunlight are allowed to act on the material, the darker it will turn. Iridescence will be the higher, the more brilliant the original piece of nacre.

It has recently been asserted that admission of air to the solution will produce an even more attractive coloring than exclusion of air.

Waxless Carbon Paper Ink
(U.S. Patent 2,820,717)

	pts./wt.
A "Vinylite" VYHH	10.0
Mineral Oil	27.5
Pigment	7.5

B Ethyl Acetate   45.0
  Toluol   15.0

Proc.: Grind A in warm ball mill* until uniform. Then add B.

Typewriter Ribbon Ink
(U.S. Patent 2,160,511)

	pts./wt.
Carbon Black	6
Tricresyl Phosphate	15
Nigrosine Base	9
Diglycol Laurate	15

Proc.: Grind all items in a ball mill until uniform.

White Ink for Photographic Prints

1	Water	2 oz.
2	Copper Sulfate	20 g.
3	Salt	10 g.
4	Concentrated Ammonia	q.s.
5	HYPO	10 g.

Proc.: Dissolve 2 in 1. Add 3. When dissolved add 4 with stirring until mixture turns a deep clear blue color. Add 5 and dissolve. Apply ink to photo with a gold or glass pen.

Photographic Film Marking Ink
(U.S. Patent 2,879,168)

	pts./wt.
2-Furaldehyde	80-2
Cellulose Acetate Butyrate	1-2
Yellow Azo Dye	10
Blue & Orange Azo Dye	2-6

Proc.: Grind all ingredients together.

Fluorescent Printing Ink
(U.S. Patent 2,845,023)
I
(Hot Melt)

	pts./wt.
Fluorescent Pigment	12
Ethylcellulose	1
Beeswax, White	9
Boiled Linseed Oil	2

II
(Plastisol)

Fluorescent Pigment	45

Polyvinyl Chloride, Powdered	23
Dioctyl Phthalate	27
Naphtha	5

Proc.: Grind in a warm ball mill* all ingredients until uniform.

Purple Copy Ink

	pts./wt.
Glycerin	2.00
Yellow Dextrin	1.00
White Dextrin	1.00
Methyl Violet	0.33
Water	1.00

Proc.: Heat not over 200°F. until dissolved.

Hectograph Composition
I
(U.S. Patent 2,412,200)

Polyvinyl Alcohol	15.0
Glycerin	50.7
Antimony Trifluoride	0.2-2.0
Titanium Dioxide	4.0
Ethylene Glycol	13.0
Calcium Chloride	4.0

Proc.: Grind in a warm ball mill* all ingredients until uniform.

II

Ground Dried Glue	½ oz.
Dried Gelatin	2 oz.
Glycerin	18 fl. oz.

Proc.: Dissolve the glue in hot water, the gelatin in warm water, and mix the two together. Then add the glycerin and mix thoroughly. Pour into a pan or tray and let cool. This makes a hectograph of 12 x 9 x ½ in.

Duplicating Ink

	%/wt.
Nigrosine Dye (water soluble)	5.0
Wetting Agent (nacconol N.R.S.F.)	1.0
Glycoxal (30% Sol.)	1.5
Ethylene Glycol	35.0
Ammonium Polyacrylate 25% Water Sol.	56.0
Zinc Oxide	1.5

Proc.: Make a 30% solution of glycoxal. Instead of glycoxal pyrulic aldehyde can be used. Make ammonium polyacrylate 25%

solution in water. Ammonium polyacrylate is produced by Goodrich Chemical Company, Cleveland, Ohio. The name is K-705. In this ammonium polyacrylate solution disperse zinc oxide. Add ethylene glycol and naccanol N.R.S.F. or other wetting agent. Heat until all ingredients are dissolved. Add water soluble dye. Instead of nigrosine lamp black/aquablack can be used. Stir to room temperature.

Flexographic Ink

	pts./wt.
"Argo" Brand Zein G200	100
Maleic or Fumaric Modified Rosin Ester	100
Proprietary Alcohol	320
Water	20
Iron Blue Pigment	115
Sodium Polyacrylate (dry basis)	5

Proc.: After the Zein and resin are dissolved, the pigment is ground into the vehicle. This is done in a ball mill* or with soft pigments in a colloid mill.*

Letterpress Black Ink

	pts./wt.
Carbon Black	18
Milori Blue	5
"Bentone" 34	3-4
Resin (Maleic Rosin Modified)	28.8
Triethanolamine Resinate	14.4
Diethylene Glycol	33.8

Proc.: Grind in a ball mill* all ingredients until uniform.

Glass-Marking Ink
(Canadian Patent 424,925)

	pts./wt.
Lampblack	50
Titanium Dioxide	10
Silver Oxide	3
Glycerin	150

Proc.: Mix the solid ingredients intimately, stir them into the heated glycerin, and pass through a paint mill.

Ceramic Ink for Marking Glass and Porcelain

A very good ceramic ink can be made by using the following common reagents:

	pts./wt.
Potassium Carbonate	1
Sodium Tetraborate	1

Litharge	2
Cobalt Nitrate	2

Proc.: Grind these together in a mortar to a fine powder, dry if necessary, mix intimately with copaiba balsam mixture composed of four parts copaiba balsam, one part clove oil and one part lavendar oil. If such a mixture is not available use raw linseed oil or any other drying oil. Mix the powder glass color with just enough oil so it will run slowly from a pen. The desired markings are made on the clean surface with a pen or a small brush. The marked article is then warmed evenly over a flame to dry the oil and also prevent cracking in the final heating. The place where the mark has been made has been heated by holding against the side of a flame of a burner; the flame touching the marking on a tangent, the article being rotated part of a circle. The marking will first turn black, then it will begin to glow a dull red. At this point the article is removed, allowed to cool a little, and reheated till the characters, not the glass, begin to glow. Markings thus obtained present a smooth and shiny surface which cannot be removed by mechanical or usual chemical means.

Ink for Marking Metal

I

	pts./wt.
Gum Copal	10
Turpentine	12
Carbon Black	2

Proc.: Mix all ingredients togehter.

II
(For tinplate)

	pts./wt.
Alkyd Varnish Solvent	10
Phenolic Varnish (15 gal.)	1
Channel Black	2
Methyl Violet or Blue Toner	½

Proc.: Grind all ingredients together until uniform. For coloring, use 5 to 10 lb. pigment. Reduce with toluol.

Meat Stamping Inks

	pts./wt.
A. Red	
Carmine	16
Ammonium Hydroxide	120
Glycerin	45

Proc.: Sir until dissolved then stir in

Dextrin	20

Proc.: Mix all ingredients. Use heat if necessary, and stir until uniform.

B. Blue

Pure Food Blue Dye	30
Dextrin	20
Glycerin	82
Water	70

Disappearing Ink

	pts./wt.
Phenolphthalein	0.2
Denatured Alcohol	18.0
Water	6.0
Ammonium Hydroxide	2.0

Proc.: Mix all ingredients together until uniform. This writes red and disappears in a few minutes.

Ball Point Pen Ink
I
(U.S. Patent 2,882,172)

	pts./wt.
Carbon Black	12
Mineral Oil	50
Oleic Acid	6
Hydroabietyl Alcohol	32
Rosin	3.5

Proc.: Mix all ingredients together until uniform.

II
(Indelible)
(German Patent 1,064,663)

	pts./wt.
Zapon Echt Orange	4.0
Hydroxyethylated Fatty Alcohol	58.0
Methylpyrrolidone	27.0
Sodium Chloride	1.8
Water	2.7

Proc.: The paste from these ingredients is formed into an emulsion with water and good mixing. The viscosity depends on the amount of water added.

Magnetic Ink
(U.S. Patent 3,347,815)

	pts./wt.
Polyvinyl acetate	20-40

Alcohol	80-60
Ferrite (0.5-5 microns)	100-300

Proc.: Mix all ingredients until dissolved. This adheres well to paper, wood and waxed coatings.

Rapid Photographic Developer
(U.S. Patent 2,877,116)

Water	1 l.
Sodium Sulfite	37.8 g.
Pyrogallol	25.2 g.
Amidol	2.07 g.
Sodium Carbonate (Mono)	49.6 g.
Potassium Bromide	3.0 g.

Proc.: Heat all ingredients and stir until uniform.

D-8 Developer

For High Contrast on Films and Plates
Stock Solution

	pts./wt.
Water	24
Sodium Sulphite, Desiccated	3
Hydroquinone	1½
Sodium Hydroxide	1¼
Potassium Bromide	1
Water, To make	32

Proc.: Dissolve chemicals in the order given. Stir the solution thoroughly before use.

For use, take 2 parts of stock solution and 1 part of water. Develop about 2 minutes in a tray at 65°F. (18°C.).

For general use, a developer which is slightly less alkaline and gives almost as much density can be obtained by using 410 grains of sodium hydroxide per 32 ounces of stock solution instead of the quantity given in this formula.

Fine-Grain Developers
I

p-Phenylenediamine	10 g.
Anhydrous Sodium Sulfite	90 g.
Water	To make 1 gal.

Normal exposure time is 15 to 20 minutes.

II

p-Phenylenediamine	10 g.
Glycine	1 g.
Anhydrous Sodium Sulfite	90 g.

Water   To make 1 gal.

This formula requires two times normal exposure. It permits fine-grain development with enlargements beyond average with no graininess.

Stable Single-Powder Developer

"Elon"	115 gr.
Sodium Sulfite	12 oz.
Hydroquinone	1 oz. 30 gr.
Crystalline Sodium Carbonate	7 oz.
Potassium Bromide	290 gr.
Phthalic Anhydride (100 Mesh)	46 gr.

Proc.: Mix and heat to 60°C., cool, and package.

Correcting Photographic Negatives
(U.S. Patent 2,484,091)

	pts./wt.
Gluten	15
Gum Arabic	20
Water	50
Glycerin	5
Sodium Bicarbonate	5
Alcohol	5
Color	To suit

Proc.: The area to be corrected is first bleached and corrections are made with this glycerin-containing ink.

	pts./wt.
Resin Powder	25
Turpentine or Paint Thinner	10
Cuprous Oxide	40
Ground Powdered Glass	15
Spar Varnish	5
Lampblack	5

The bleached area is then covered with this paint.

	pts./wt.
Water	90
Glacial Acetic Acid	5
Glycerin	5

The excess paint is removed by blotting and, while still moist, this solution is applied.

The final solution is allowed to react with the ink and removes the paint over the lines, causing them to appear on the painted background. The correction is completed by polishing the treated portion with a mild abrasive.

Intensifying Baths for Negatives

I

A	Potassium Ferricyanide	100 g.
	Potassium Bromide	100 g.
	Water	1 l.
B	Sodium Sulfide Nonahydrate	40 g.
	Water	100 cc.
	(Dilute 1 to 30 for use.)	

Proc.: Bleach the image in solution A, then bathe in solution B without intermediate washing.

II

A	Potassium Citrate	70 g.
	Copper Sulfate Pentahydrate	6 g.
	Water	1 l.
B	Potassium Ferricyanide	6 g.
	Water	1 l.

Proc.: Mix equal parts of A and B, and bathe the film in the mixture for 5 minutes, then in 10% hypo for 10 minutes, wash for 10 minutes, then bathe for 10 minutes in a bath consisting of:

Basic Aniline Red	3 g.
Acetic Acid	12 cc.
Water	1 l.

Proc.: Finish the processing by a 10 minutes wash in water.

Water-Softening Compound
(U.S. Patent 1,952,408)

A cake for domestic use, formed by pressure when moist, comprises sodium carbonate 62.5, sodium phosphate 30.0, calcium chloride 5.0, and sodium chloride 2.5%.

Latent Fingerprinting

A piece of paper or other material on which one is searching for fingerprints is saturated in a sensitizing solution prepared by dissolving 2 g. of silver nitrate in 1 l. of distilled water. This is stored in a dark place. After having soaked for 2 hours in the silver nitrate bath, the paper is thoroughly washed in distilled water, first by soaking for 30 minutes and then two rinsings. There is left in the paper only the silver chloride which has been formed from the chlorides left by the perspiration and the silver nitrate. *The paper is hung up and allowed to dry thoroughly.* It is then developed, either with a developer of the M.G. type or with others, such as formaldehyde and sodium carbonate. Following the development the paper is again washed in water, then in a bath of hypo, washed, and dried, and is ready for observation.

If kept in a humid atmosphere the migration of the chlorides may be so intensified that in time a gray cloud is formed where the print was originally. In some cases the print goes through the paper. Prints made from the skin of a corpse are very poor and diffuse, although chloride is deposited.

Refrigerator Deodorant

Fill a small muslin bag with a good quality of granular activated carbon. The muslin bag may then be placed in the rear of a lower portion of the ice box and will absorb strong odors which tend to collect.

After six months use, the device may be reactivated by placing in the oven at $350°$F. for about ½ hour.

Fireproof Film Containers

(British Patent 419,249)

The walls are made of a mixture of sawdust 25, calcined magnesite 25, magnesium chloride (as a 25% aqueous solution) 30, potassium alum 10 and a mixture consisting of asbestos flour 4, asbestos fiber 3 and acetic acid 3, 10 parts.

Preserving Fluid for Museum Specimens

Formaldehyde	12-25 oz.
Glycerin	10 oz.
Potassium Nitrite	0.1 oz.
Water	to make 100 oz.

Proc.: Mix all ingredients together.

Tar Asphalt Emulsion for Road Surfacing

		lb.
A	Tar (Coal)	3.8
	Petroleum Asphalt	0.8
	Rosin (Gum)	0.1
	Anthracene Oil	0.1
B	Water	5.0
	Casein	0.1
	Potato Starch	0.1

Proc.: Heat all ingredients in A to $120°$F. until consistency is reached. Heat all ingredients in B to $120°$F. until all are dissolved in the water. Mix B and A together.

If viscosity is too high increase the amount of anthracene oil and decrease the petroleum asphalt.

Equipment: Kettles which can be heated by steam, hot water or hot oil. Stirrers are also necessary but no special kind is needed.

8. Paints and Coatings

This is an area where some of your largest profits can be made, if you choose to market the products you make. Paints are expensive to purchase, but can be made for a fraction of the retail price by you.

It is not to be suggested that this is an easy process. Of all the sections presented in this book, the manufacture of paints will take the most experience and a great deal of care. Do not start your experimenting here unless you are experienced in putting together simpler products.

Sample recipes are given for many different types of paints, oil base, rubber base, whitewash etc.; paints for outdoors, indoors, metal, ceramics, ships, fabrics, you name it. From these sample formulas you can experiment and produce paints for every need and purpose.

In some instance you will need more than kitchen-type equipment. See the "Equipment" chapter for further help. Go to your local industrial equipment dealer. Discuss your needs with him. He is a great source of information. You will not necessarily have to buy expensive new equipment. Many dealers can supply you with good used equipment at greatly reduced prices.

Careful work can give you superior products at small cost. This can be a boon to you and your neighbors, if you choose to turn your experimenting into profit.

House Paint

	oz.
Basic White Lead Carbonate	4.48
Basic White Lead Sulfate	10.88
Zinc Oxide	20.48
Titanium-Magnesium Mixture (Titanium Dioxide, 30, 70% Magnesium Silicate, 70)	28.16
Raw Linseed Oil	14.75
Bodied Soybean Oil	7.76
Bodied Tung Oil	1.94
Mineral Spirits	6.12
Drier	5.43

The bodied tung oil used here has been thickened by heat. Unlike the raw tung oil, it dries in a clear smooth film without wrinkling. Total vehicle, 36% Percentage of drying oil in vehicle, 68%. Percentage of drying, oil in vehicle, 68%. Percentage of thinner and drier, 32%.

Proc.: Mix the oils with the thinner and drier, then stir in the four pigments.

Paints for the Exterior

I

Cement Base

	lb.
Portland Cement	35.0
Hydrated Magnesium Lime	15.0
Water	q.s.

Proc.: Tumble the dry powders in a pebble mill—without pebbles in the jar—for several hours. Stir about 7 lb. of the mixed powder into 1 gallon of water. Stir a small portion of the water into the dry powder until a smooth paste has been formed, add the remainder of the water, and continue to stir until the resultant product is homogeneous. Use a stiff-bristled brush, and paint over damp surface. When the paint sets up, wet it down with water. Use two coats of cement-water paint, allowing 24 hours of drying time between coats. Store the dry powdered paint in moisture-proof containers.

II

Oil Base

	lb.
Pigment:	
White Lead	39.00

Zinc Oxide	16.38
Titanium Dioxide	1.42
Silica and Magnesium Silicate	14.20
Vehicle:	
Pure Linseed and Chinawood Oil	17.11
Mineral Spirits	7.54
Japan Drier	4.35

Proc.: Mix the pigments thoroughly with a small quantity of the oil in a mixing machine until a paste is formed. Finally, stir in the rest of the oil, mineral spirits, liquid drier, and coloring pigments if desired. Continue to mix until the product is uniform. Two coats are necessary for good coverage and durability. The first coat seals the surface to prevent "spotting" of the second coat. Allow 24 hours or more of drying time before applying the second coat. Use a 3-inch paint brush.

Acrylic Exterior Latex Paint
I

	pts./wt.
Water	174.0
"Super AD-IT"	10.0
"Natrosol" 250HR	2.5
"Nuodex" AF-100	3.0
"Tamol" 731	11.0
"Triton" CF10	2.0
Ethylene Glycol	20.0
TiO_2 (rutile)	240.0
TiO_2 (anatase)	10.0
"Nytal" 300	100.0
"Duramite"	90.0
"Rhoplex" AC-34	512.0

Proc.: Grind and mix all ingredients together until uniform.

II
Exterior Latex Paint

	pts./wt.
A Mill Paste	
"Ti-Pure" R-610, Rutile Titanium Dioxide	200.00
"Ti-Pure" FF, Anatase Titanium Dioxide	50.00
"Vi-Cron" 25-11, Calcite	121.80
Mica, Water Ground, 325 Mesh	13.50
"Tamol" 731 (25% Solids)	1.54
Potassium Tripolyphosphate (10% Solids)	15.40
"Triton" CF-10 (25% solids)	7.70
"Balab" 648 Bubble Breaker or	
"Nopco" NXZ	0.90

Phenylmercuric Acetate (6% solids)	10.50
2% "Cellusize" WP-4400	50.00
Water	75.00

B Let Down:

VV 10-VA Latex (52% solids)	433.50
Phenylmercuric Acetate (6% solids)	10.50
Butyl "Dioxitol" /Ethylene Glycol (1/1 by wt.)	45.40
2% "Cellosize" WP-4400	96.60
Water	2.20
Ammonium Hydroxide to pH 7.6	—

Proc.: Grind A, mix B. Mix together A and B.

House-Paint Primer

	pts./wt.
Basic Carbonate White Lead	267.2
Basic Sulfate White Lead	267.2
Rutile Nonchalking Titanium Dioxide	105.8
Barytes	244.4
Magnesium Silicate	173.4
Litharge	10.6
Raw Linseed Oil	159.8
Bodied Linseed Oil (Z-4)	143.3

Proc.: Mix all and grind until uniform.

One-Coat House Paint, White

One-coat house paint has become an important factor in reducing the cost of repainting homes. It is designed to give good hiding in one coat. It should be used only on previously painted or well-primed surfaces.

I

	pts./wt.
Rutile Titanium Dioxide	100.00
Anatase Titanium Dioxide	200.00
35% Leaded Zinc Oxide	400.00
Micronized Magnesium Silicate	150.00
T-1215 Z2 Heat Bodied Linseed Oil	144.00
A-R Varnish and Grinding Linseed Oil	266.00
Mineral Spirits	149.00
24% Lead Naphthenate (0.4%)	7.00
6% Manganese Naphthenate (0.03%)	2.00

Proc.: Mix all and grind through a Buhr stone mill.

II

"Titanox" A160 ALO	250.0 lb.

"Titanox" RCHT	209.0 lb.
Zinc Oxide ("Aztol" 33)	260.0 lb.
Low-Oil-Absorption Manganese Silicate	126.0 lb.
Refined Linseed Oil (Acid No. 2–3)	43.0 gal.
Bodied Linseed Oil (Z2 Viscosity; Acid No. 9–11)	10.0 gal.
24% Lead Naphthenate	0.9 gal.
6% Manganese Naphthenate	0.2 gal.
6% Cobalt Naphthenate	0.1 gal.
Mineral Spirits	19.7 gal.

Proc.: Mix all ingredients and grind until uniform. Instead of Lead Naphthenate use Manganese Naphthenate.

Low-Cost Shingle and Shake Paint

	pts./wt.
Rutile Titanium Dioxide	150.0
Calcium Carbonate	400.0
Diatomaceous Silica	75.0
Aluminum Stearate	7.5
Carbon Black	2.0
"Cykelin" 70	320.0
"Kelecin" F	6.0
Mineral Spirits	200.0
6% Cobalt Naphthenate (0.15%)	0.5
6% Manganese Naphthenate (0.015%)	3.5
4% Calcium Naphthenate (0.1%)	6.0

Proc.: Grind and mix all ingredients together until uniform.

Barn and Roof Paints
I
(Red)

	pts./wt.
Red Oxide No. R-3098	200
F.T. "Asbestine" Pulp	100
"Falkovar" K Heavy Oil (Y Body)	200
Raw Linseed Oil	400
Mineral Spirits	103
6% Cobalt Naphthenate	5
24% Lead Naphthenate	12

Proc.: Grind in a ball mill overnight or 16 hours.

No. 2
(Green)

Medium Chrome Green	46.00 lb.
Aluminum Stearate	2.00 lb.
Blown Fish Oil	4.50 gal.

Gilsonite Varnish	80.00 gal.
Mineral Spirits	5.00 gal.
6% Cobalt Naphthenate	0.06 gal.

Proc.: See Formula I.

Scrub-Resistant White Flat Wall Paint

	pts./wt.
"TiPure" R-901	55
"Satintone" No. 1	195
"Camel" Carb	242
Water	210
25% Soln. "Tamol" 731	5
"Nopco" NDW	2
"Troysan" CMP Acetate	2
Ethylene Glycol	15
"Piccolloid" Emulsion No. 15	88
"Flexbond" 315	83
2% Solution "Cellosize" QP-4400	266

Proc.: Grind with high-speed mixer.

Polyvinyl Acetate Latex Interior Paint

	pts./wt.
Water	325
"Super AD-IT"	1
"Tamol" 731	6
KTPP	1
Lecithin wd	4
"Nuodex" AF-100	3
Ethylene Glycol	30
TiO_2 (rutile)	225
Clay Hydrite	200
Wet Ground Limestone	100
"Methocel"	3
"Resyn" 1251	310

Proc.: Grind and mix all ingredients together until uniform.

Casein Water Paints
I

	pts./wt.
Exterior White	
Heavy Casein Solution*	370
10% Gum Karaya Solution	8
Water	60
Lithopone	500
"Dicalite" L	90

Barytes	45

*Heavy Casein Solution

	lbs.
Casein, oiled	75.00
Water	300.00
Sodium Fluoride	8.40
Borax	8.25
"Preventol" Solution 15% in Water	1.00

II

	pts./wt.
Washable White Casein, Oiled	10.50
Pine Oil	1.00
GC Sodium Silicate	4.00
Univ. Spray Lime	1.00
"Cryptone" No. 19	35.00
No. 232 Talc	11.00
"Pyrax TES" (1)	15.00
No. 1291 Clay	8.00
Lion Clay Powder	9.30
"Pyrax TES" (2)	5.00
"Dowicide" A	0.20

Proc.: The pine oil and talc are mixed together. The silicate and "Pyrax TES" (2) are mixed and passed through the mill. The casein, No. 1291 clay, lion clay, "Pyrax" TES (1) and "Dowicide" A are weighed out.

Whitewash

Casein	2½ lb.
Trisodium Phosphate	1½ lb.
Lime Paste	4 gal.
Formaldehyde	1½ pt.

Proc.: Prepare the lime paste by mixing 25 lb. of hydrated lime with 3 gallons of water. Dissolve the trisodium phosphate in ½ gallon of water. Soak the casein in 1 gallon of water for 2 hours, or until it is soft, add it to the first solution, and dissolve. Add the casein-trisoda solution to the lime paste, stirring continuously. Dissolve the formaldehyde in 1½ gallons of water, and just before use, add it slowly to the white mixture, stirring vigorously. Use two coats of whitewash. Dampen the walls with water before brushing on the first coat. Allow 24 hours of drying time between coats. Use a special whitewash brush.

Exterior Weatherproof Whitewash

I

1. Mix about 120 lb. of spent carbide residue with water to a creamy consistency.

2. Dissolve 2 lb. of common salt and 1 lb. of zinc sulphate in 2 gal. of boiling water.
3. Provide 2 gal. of skimmed milk. Pour (2) into (1), then add (3), and stir well.

II

1. Mix about 15 lb. of spent carbide residue to a creamy consistency with water.
2. Dissolve 1 lb. of carbonate of soda in ¼ gal. of boiling water.
3. Soak in cold water for at least 8 hr. ¼ lb. of common glue and 1 lb. of rice flour; and then thoroughly dissolve the glue mixture in ¾ gal. more water in a double boiler. Mix (1) with (2), then add (3).

III

1. Mix about 12 lb. of carbide residue to a creamy consistency with water.
2. Dissolve 4 oz. of white rosin in 12 fluid oz. of boiled linseed oil.
3. Beat 6 lb. of whiting in 1 gal. of skimmed milk. Mix (2) with (1) while hot, add (3).

Hints for Special Uses

Alum added to whitewash prevents its rubbing off. Flour paste will also prevent rubbing off, but when this is used, zinc sulphate must be added as a preservative.

Molasses causes lime to penetrate wood and plaster better. One pint of molasses to 5 gallons of whitewash is generally considered sufficient. A solution of silicate of soda or water glass, one part to ten parts of whitewash, makes what is commonly referred to as a "fireproof cement" of whitewash.

By adding 1 pound of cheap bar soap dissolved in 1 gallon of boiling water, to every 5 gallons of whitewash, a more or less gloss finish can be obtained.

Extra Fine Quality Calcimine

	pts./wt.
Factory Whiting	75.00
Paragon Clay	12.80
U.S. Gypsum Plaster	7.75
Dry Glue	4.27
Zinc Sulphate	0.18

Proc.: Grind on a Buhr stone mill.

Spackle

	pts./wt.
Powdered Base	
U.S. Gypsum Plaster	83.95

Paragon Clay	12.61
Dry Glue	4.26
Zinc Sulphate	0.18
	100.00 76.00
Regular Silica	6.50
"Pyrax" A	17.00
Terra Alba No. 1	0.50
	100.00

Proc.: Mix and grind on Buhr stone mill.

Quick-Drying Interior Trim Enamel—White

		pts./wt.
A	Rutile Titanium Dioxide	300.0
	"Spenkel" F48-50MS	600.0
	Mineral Spirits	60.0
B	6% Cobalt Naphthenate (0.03%)	1.5
	24% Lead Naphthenate (0.03%)	3.8
	"Exkin" No. 2 (0.02%)	0.6

Proc.: Grind A. Mix B. Stir A and B together until uniform.

White Baking Enamel

	pts./wt.
"Cabot" RF-30	140
"Beetle" 227-8	52
Butyl Alcohol	8

Proc.: Premix and then disperse on a roll mill to a fineness of grind of 7 ns.

"Beetle" 227-8	44
"Rezyl" 7365-51	100
"Rezyl" 99-5	80
VM&P Naphtha	56
Xylol	15
Total Weight	495 lb.
Total Yield (approx.)	50 gal.

This coating should be reduced 4 to 1 with xylol for spray application. Bake at 325°F. for 15 min.

White Four-Hour Exterior Enamel, Lead Free

	pts./wt.
"Titanox" RC	83
"Titanox" R-610	250
Aluminum Stearate	2
Soya Lecithin	4
"Aroplaz" 1085-M50	172

Proc.: Premix and then disperse on a three-roll mill to a fineness of grind of 7ns.

"Aroplaz" 1085 M50	490
"Synthenol" G-H	45
No. 460 Solvent	56
Nuodex Cobalt Octoate 6%	5
Nuodex Calcium Naphthenate 4%	13
"Exkin" No. 2	2
Total weight	1122 lb.
Total yield (approx.)	110 gal.

This coating should be brushed full body.

White "Thumb Tack" Roller Enamel

	pts./wt.
"Cabot" RF-30	150.0
Aluminum Stearate	1.0
"Uformite" MM-55	18.0
"Rezyl" 99-5	60.0

Proc.: Premix and then disperse on a three-roll mill to a fineness of grind of 7 ns.

"Rezyl" 99-5	240.0
"Uformite" MM-55	84.0
Petrolatum	1.0
Nuodex Cobalt Naphthenate 6%	0.3
Pine Oil	5.0
Triethanolamine	1.0
Total Weight	560.3 lb.
Total Yield (approx.)	55.0 gal.

Proc.: This enamel should be applied full body by roller. Bake at 300°F. for 10 min.

Air-Drying Blue Enamel

	pts./wt.
"Stepanyl" 324-57	64.0
Rutile TiO_2	77.0
Blue Pigment	8.6
Deionized Water	89.0
n-Hexanol	3.0

Proc.: Charge pebble mill, grind 46 hr. Let down:

"Stepanyl" 324-57	363.0
24% Pb (drier)	0.2
6% Mn (drier)	1.1
Deionized Water	394.1

Air-Drying Green Enamel

	pts./wt.
"Stepanyl" 324-57	56.0
Rutile TiO_2 (1)	67.5
Water Dispersed Green (26.1% N.V.)	29.8
Deionized H_2O	77.6
n-Hexanol	2.6

Proc.: Charge pebble mill, grind 48 hr. Let down:

"Stepanyl" 324-57	465.0
24% Pb Drier	0.2
6% Mn Drier	1.3
Deionized Water	300.0

Gray Baking Enamel

	pts./wt.
"Stepanyl" 324-57E	165.0
Barytes	70.3
Rutile TiO_2	77.0
Carbon Black	2.7
Aluminum Silicate	26.4
Basic Lead Silico Chromate	26.4
"Cellosolve"	6.4
Deionized Water	194.3

Proc.: Charge pebble mill, grind 22 hr. Let down:

"Advacar" 6% Mn Drier	2.2
"Stepanyl" 324-57E	238.0
Deionized Water	107.5
n-Hexanol	5.5
10% Solution, Water-Soluble Silicone (6)	3.1
Water-Soluble Melamine (80% N.V.) (7)	75.2

Gray Hammered Finish Enamel

	pts./wt.
1593 "Albron" Aluminum Paste	9
"Cycopol" S-102-5	224
"Uformite" MM-55	40
VM&P Naphtha	148
Triethanolamine	1
Total Weight	422 lb.
Total Yield (approx.)	55 gal.

Proc.: Carefully break up the aluminum paste in a portion of the "Cycopol" resin by slowly adding the resin to the aluminum while maintaining agitation. Once the aluminum paste is fully broken up and the mixture is homogeneous, the remaining ingredients are added to complete the batch. This coating should be sprayed full body. The baking schedule is 300°F., 15 min.

Gloss Black Baking Dip or Spray Coat

	pts./wt.
"Stepanyl"324-57E	201.0
Carbon Black	19.5
Supermicrofine Barytes	32.5
Basic Silicate White Lead	6.5
Deionized Water	200.0
Pine Oil	4.9

Proc.: Charge pebble mill, grind 30 hr. Let down:

6% Mn Drier	2.4
"Stepanyl" 324-57E	268.0
Water-Soluble Melamine Resin (80% n.v.) (4)	103.2
Deionized Water	162.0

Architectural High-Gloss Enamel

	pts./wt.
Rutile Titanium Dioxide	300.0
Zinc Oxide	25.0
"Kelsol" DV-1695	515.0
6% Cobalt Naphthenate (0.06%)	2.5
6% Manganese Naphthenate (0.02%)	1.0
24% Lead Naphthenate (0.1%)	1.0
Bubble Breaker-SK	5.0
Water	250.0

Proc.: Grind all ingredients together.

Tile-Like Enamel

	pts./wt.
Component I (Pebble mill grind):	
Rutile Titanium Dioxide	300.0
Phthalo Blue	1.0
CASTOR 1066	240.0
"Cellosolve" Acetate	230.0
"Kelecin" F	6.0

Proc.: Tumble overnight, adjust, filter, and package.

Component II	
"Spenkel" P49-75S	335.0

Add Component II to Component I and mix thoroughly 10 to 20 min before use. For spray application, approximately half of the cellosolve acetate should be replaced with xylol and/or ethyl acetate pot life at least 1 working day.

Brushing Paint, Fluorescent

I

	pts./wt.
"Rezimac" 2420 (40% in Odorless Mineral Spirits)	44.000
"Cyasorb" UV 24	0.400
Pigment, D-Series ("Day-Glo")	50.500
12% Cobalt Octoate Drier	0.075
Antioxidant B	0.200
"Cab-O-Sil"	0.200
Mineral Spirits	4.625

Proc.: High Speed Mix or Roller Mill Grind.

II

	pts./wt.
Paste (roller mill)	
"Aroplaz" 2502	36.00
"Santocel" 54	3.00
Pigment B-3500 Series	32.00
"Nuact" Paste	0.20
Let Down	
6% Cobalt Naphthenate	0.15
24% Lead Naphthenate	0.14
6% Manganese Naphthenate	0.80
Antiskinning Agent	0.10
Mineral Spirits	27.61

Proc.: Grind all ingredients together until uniform.

Acrylic Aerosol Concentrate, Fluorescent

	%/wt.
"Acryloid" F-10 (40% in Mineral Thinner/"Amsco" F)	45.5
or	
"Evacite" 2044 (40% dissolved in Mineral Spirits	
Pigment, A or D-Series ("Day-Glo")	35.0
Xylene	19.5

Proc.: Roller Mill Grind.
Thin 50% with 25% Toluene and 75% Lactol Spirits mixture before packaging with 45% of a propellant such as Freon 12.

Iron Protective Paint

I

	oz.
Lampblack (Ground in Oil)	90.7
*Asphalt Varnish	68.1

Linseed Oil, Raw	68.1
Japan Drier	2.0

II

	oz.
Lampblack	27
Silica	58
Red Lead	10
Graphite	5
*Asphalt Varnish	Sufficient

Proc.: Grind together until smooth.

*Turpentine	1 part
Asphalt in Linseed Oil	1 part

Aluminum Powder Paste

(U.S. Patent 2,002,891)

	oz.
Aluminum, Flaked	58
Stearic Acid, Powdered	1
Aluminum Stearate	1
Naphtha	40

Proc.: Grind together until homogenous.

Paints for Copper

Copper, bronze, or brass gutters and flashings, as well as copper or bronze screening, are apt to cause bad yellowish-green stains on light-or white-painted houses, owing to the washing off of corrosion products. Exposure tests indicate that one of the best ways to paint copper or bronze surfaces is to wash off any grease, using gasoline or turpentine. The surface should be roughened slightly with sandpaper, and a priming coat composed of 1½ to 2 pounds of aluminum powder to 1 gallon of aluminum mixing varnish applied, followed by the desired color coat. Weathered copper or bronze screening should be thoroughly dusted, and then given two coats of a thin black paint. Carbon black enamels, make excellent screen enamel.

"Tornesit" Paints

First, a base solution is prepared, consisting of 33-1/3 per cent Tornesit and 66 2/3 per cent high-flash naphtha. To effect solution is a matter of a very few minutes, if the "Tornesit" is added to the solvent.

Second, a concentrated gum solution is made when required.

Third, the pigments are ground in the plasticizer, or if it is insufficient, some of the "Tornesit" base solution is used.

Fourth, if a brushing paint is required the base solution is thinned to a "Tornesit" content of 21 per cent to 22 per cent by the addition of a solvent mixture consisting of two parts high-flash naphtha and one part xylol. If a spraying composition is desired, the base solution is thinned with toluol to a "Tornesit" content of 11 per cent to 12 per cent.

Finally, the gum solution and pigment paste are added to the reduced solution and the mixture is stirred.

"Tornesit" paints may be applied by spraying, dipping, flowing, or brushing. A good film can be obtained by any of these methods.

Following is a brief outline of procedure to be followed, to obtain most satisfactory results in spraying and brushing:

Spraying

"Tornesit" Solutions can be sprayed, producing a hard, durable, even distributed film. With present equipment, the spraying viscosity is 40 centipoises, which is somewhat lower than the 75 centipoise spraying viscosity of lacquers.

If the "Tornesit" concentration is kept below 12% no difficulty will be encountered from "spider-webbing." By the addition of softening agents, gums, and pigments, the solids content will be increased 30-40 per cent, depending, of course, on choice of ingredients.

Brushing

Brushing paints with as high as 57 per cent solids have been applied successfully. For this purpose, a working viscosity of about 250 centipoises is recommended. In brushing "Tornesit" paint, the surface should be well covered with a full brush, avoiding going over the painted area any more than necessary because of the rapid drying of the product. When bodied tung oil is used as the plasticizer in the priming coat, a second coat may be applied to an interior surface after six to eight hours. On exterior work, three to four hours is an ample drying period with the same priming coat.

"Aquarell" Colors

Pigments
  White:
    Whiting Finest, or China Clay.
  Pale Yellow:
    Pale Yellow Lake, or Yellow Lake, Blended.
  Yellow:
    Yellow Lake, Martius Yellow, Ochre.
  Pale Orange:
    Orange Lake, Blended to get Lighter Colors.
  Orange:
    Orange Lake.

Rosa (Pink):
  Alizarin Lake, or "Echt-Rot" (Genuine-Red),
  Blended to Obtain Lighter Colors.
Red:
  Alizarin Red, Martius Red.
Pale Brown:
  Terra di Siena, Blended
Brown:
  Caput Mortuum (Iron Oxide).
Dark Brown:
  Umbra, or Cassel Brown.
Violet:
  Brilliant Violet Lake.
Pale Blue:
  Blue Violet Lake, Blended.
Blue:
  Blue Lake
Dark Blue:
  Dark Blue Lake.
Pale Green:
  Green Lake, Blended.
Green:
  Green Lake
Gray:
  Black Lake, Blended.
Black:
  Black Lakes.

The blending, to get paler shades, is done by mixing the lake or pigment with white chalk.

Manufacture of "Aquarell" Colors
(Water soluble applied with brush)

Solution for binding of the pigments in the color-paste:

I

	g.
Gum Arabic	26
Water, Distilled	51.9
Glycerin (28° Bé.)	8
Glucose Solution (1 : 1)	10
Beef-Gall, Prepared	4
Moldex or Other Preservative	0.1

Proc.: Dissolve gum powder in cold water, stir, then heat to get complete solution. Add preservative, then glycerin, glucose solution, beef-gall. Filter, when cooled, through a percolator-cloth. (See No. 2)

II

	g.
Dextrin, White	40
Water, Soft or Distilled	41.8
Borax, Crystallized	2
Glycerin (28° Bé.)	6
Glucose Solution (1 : 1)	10
Moldex or Other Preservative	0.2

Proc.: Make dextrin paste in cold water, then warm to get clear solution, add preservative and borax, then glucose-solution and glycerin.

Add the amount of water lost by evaporation (also in No. 1).

Copper Color

A rich copper color can be produced by mixing a chrome yellow paint or enamel with a small quantity of burnt sienna color-in-oil.

Fabric Paint

	pts./wt.
½ Second Nitrocotton	20
Tricresyl Phosphate	24
Cellosolve	20
Toluol	90
Basic Dye	5

Proc.: Mix all ingredients until uniform.

Marble-Effect Dipping Paint

Beautiful, marble-like effects are obtained by dipping objects into many colored paints floating upon the surface of water. In order to float on water, the paints used have to weigh less than 8.33 pounds per gallon. Assuming that a varnish is used which weighs 7 pounds per gallon, the following table gives the number of pounds of pigment which, when ground into 1 gallon of varnish, will yield a paint of sufficiently low weight to float on water, and have good hiding.

Chrome Yellow	1.25
Chrome Green	1.00
Prussian Blue	0.50
Para Red	0.50
Aluminum Bronze Powder	1.50
Gold Bronze Powder	1.50
Carbon Black (High Strength)	0.50

The procedure is important. Select a container which is wide enough and deep enough to hold the largest object to be dipped. Fill the container with water at room temperature. By means of a rod or dropper place a few drops of a colored paint here and there on the surface of the water. Near these drops or upon them place drops of a contrasting colored paint. Three, four or even five different colors may be used, but an excess of paint should be avoided. The colors will spread about, mingling with each other. They may also be blown gently. Hold the object to be decorated in such fashion that the entire outside surface is exposed. Immerse it slowly into the colors and into the water, turning it a bit at the same time. Blow the remaining colors aside in order to withdraw the object without having it traverse the colors again. The designs produced in this manner will always be different from each other, and are almost impossible to reproduce by hand painting.

Colored Asphalt Paint

	%/wt.
85 to 100 Penetration Steam-Vacuum Asphalt	17.5
Gilsonite Selects	17.5
Mineral Spirits (35 KB)	40.0
Metallic Oxide Pigment, Green, Red, Brown, yellow	25.0

Proc.: The pigment is incorporated in ball mills or other dispersing equipment. Again the intensity depends upon the amount of pigment used. Percentages as low as 15% by weight do give an appreciable color but ordinarily approximately 25% is required.

Ship-Botton Paints
(Anticorrosion Primer)

	pts./wt.
"Parlon" S20	2.4
Coumarone Indene Resin, Hard	16.5
Coumarone Indene Resin, Soft	4.2
Coal Tar	4.6
Zinc Oxide	14.0
Zinc Chromate	9.5
Mica	3.7
Indian Red	1.2
Magnesium Silicate	15.2
Aluminum Stearate	0.8
Hi-Flash Naphtha	17.2
Mineral Spirits	10.7

Proc.: Dissolve all ingredients in naphtha and mineral spirits. If necessary use grinder.

Nonflammable Paint for Ship Interiors

	%/wt.
'Parlon" S10	6.57

Long Oil Soybean Alkyd (60% in petroleum solvent)	21.08
Titanium Calcium Pigment	17.28
Titanium Dioxide	14.34
Antimony Oxide	8.86
Yellow Iron Oxide	0.59
Zinc Yellow	0.93
Lampblack	0.08
Xylene	23.78
Petroleum Spirits	6.33
Cobalt Naphthenate	0.13
Epichlorohydrin	0.03
Antisag Agent (Surfactant)	as required

Proc.: Grind all ingredients until uniform.

Salt-Spray Resistant Primer Additive

The addition of 1-2% "Aerosil" R-972 (based on pigments) greatly improves salt spray resistance of primers and other coating systems.

Lead-Free Ceramic Glaze

French Patent 830,994

I

	pts./wt.
Zinc Oxide	37
Borax	55
Boric Acid	6
Calcium Carbonate	7
Sodium Nitrate	5

II

	pts./wt.
Zinc Oxide	41
Borax	55
Boric Acid	6
Barium Carbonate	14
Sodium Nitrate	5

Proc.: Grind all ingredients together until smooth.

Emulsion-Type Terrazzo Paint
(Nondiscoloring)

	pts./wt.
A "Efton" D Super	6.6
"Duroxon" R-21	5.4
"Durmont" E	6.4
"Aristowax" 165	5.6

Stearic Acid, double-pressed	5.0
Mineral Spirits	15.0
B Triethanolamine	3.0
Borax	1.0
Water 210°F.	89.0

Proc.: Melt the waxes together with the stearic acid and slowly add the mineral spirit. If the solution is cloudy, heat the batch slightly until it is clear. In a separate vessel dissolve the triethanolamine and borax in the indicated amount of water. Adjust the temperature to approximately 190 to 205°F., start high-speed agitation, and slowly but steadily pour A into B. Cool with agitation to approx. 100°F. max. Store the product in a covered container.

Air-Drying Black Iron Oxide Primer

	pts./wt.
Charge pebble mill:	
"Stepanyl" 324-57	114.0
Rutile TiO_2	25.4
Black Iron Oxide	84.5
Basic Lead Silicachromate	67.5
Barytes	84.5
Magnesium Silicate	42.2
Ben-A-Gel'E	4.2
Odorless Mineral Spirits	4.2
Deionized Water	170.0

Proc.: Grind 18 hr., let down:

6% Mn Drier	0.9
24% Pb Drier	0.2
"Stepanyl" 324-57	240.0
Deionized H_2O	144.3
n-Butanol	18.1

Red Oxide Baking Primer

	pts./wt.
Charge pebble mill:	
"Stepanyl" 324-57E	163.0
Red Iron Oxide	77.5
Super Microfine Barytes	170.0
Aluminum Silicate	56.5
Basic Lead Silicochromate	56.5
Deionized Water	92.5
Pine Oil	66.2

Proc.: Grind 18 hr., let down:

6% Mn Drier	1.0

"Stepanyl" 324-57E	155.0
Deionized Water	154.5
n-Hexyl Alcohol	5.2
10% Solution, Water-Soluble Silicone	2.1
Water-Soluble Melamine (80% n.v.)	59.5
Antifoaming Agent	0.5

Metal Primers

	I %/wt.	II %/wt.
Chlorinated Polyolefin		
310-6 Resin	10.0	9.85
"Elvax" 40 Wax	—	5.01
"Aroclor" 1260 Plasticizer	—	1.50
"Aroclor" 1254 Plasticizer	5.6	—
"Aroclor" 5460 Resin	2.7	—
Stabilizer A-5	0.2	0.20
Epichlorohydrin	0.1	0.10
Red Pigment	21.7	19.30
"Bentone" 38 Gelling Agent	1.3	1.08
Xylene	47.8	53.60
Mica	3.1	2.69
Talc	7.5	6.67

Proc.: Grind all ingredients together.

Preparing Magnesium Alloys for Painting

To prepare the surface of magnesium alloys so that paint will adhere, it is recommended that the alloy be first immersed in the following:

Sodium Dichromate	1.5 lb.
Concentrated Nitric Acid	1.5 pt.
Water	1 gal.

In a new solution, only 15 seconds are needed. This time increases to two minutes for an old solution.

After rinsing and drying, the proper primer should be used, containing inert pigments or, for example, zinc chromate. For interior work, a minimum of two coats (total) paint should be used; for exterior work, a minimum of four coats.

Wash for Galvanized Iron before Painting

A	Denatured Alcohol	60 fl. oz.
	Toluol	30 fl. oz.
	Carbon Tetrachloride	5 fl. oz.

	Commercial Concentrated	
	Hydrochloric Acid	5 fl. oz.
B	Copper Acetate	6 oz.
	Water	1 gal.
C	Copper Nitrate Crystals	2 oz.
	Copper Chloride Crystals	2 oz.
	Ammonium Chloride Crystals	2 oz.
	Commercial Concentrated	
	Hydrochloric Acid	1/6 pt.
	Water	1 gal.

Proc.: Solution A will cut grease as well as etch. If the metal is not free from grease, solutions B and C must be preceded by a grease-removing operation.

Preparing Aluminum for Enamel

The best method of cleaning aluminum castings, so the finish will adhere tenaciously, is to use the sandblast. Smooth aluminum surfaces are of such character that an ordinary first coat of finishing material will not adhere to them satisfactorily, even when they are clean. The sandblast will leave the surface slightly etched and will aid the first coat in sticking to the metal permanently.

If sandblasting is impractical, about all that can be done is to thoroughly wash the castings with naphtha or some other solvent for grease, and dry them thoroughly with clean cloths.

In other instances it may be satisfactory to bake the castings for a short time at 400 or 500°F., just before finishing them, to burn off any oil or grease. It is not advisable to use caustic cleaning solutions with aluminum, because the metal is so easily attacked and dissolved by this chemical.

Another method is as follows: Immerse them in a 20% solution of acetic acid until all oil and grease is removed or neutralized. Then rinse in a vat of clear hot water and allow castings to drain and dry. Do not wipe them. Spray or brush as soon as the moisture has dissappeared.

Styrene Lacquer
(U.S. Patent 2,603,618)

	pts./wt.
Polystyrene (M.W. 5000-20000)	255
Butylphthalylbutyl Glycolate	45
Isophorone	300
Butyl "Carbitol" Acetate	150

Proc.: Mix all ingredients until uniform.
This can be used for transfers to be applied to styrene plastics.

Coating Lacquer

	pts./wt.
Nitrocellulose (½ Second)	14
Alkyd Resin	4
"Kronisol" or "Kronitex"	5
Blown Castor Oil	2
Toluene	45
Butanol	8
Butyl Acetate	22

Proc.: Mix all ingredients until uniform.

Add loading pigment, such as whiting, or coloring pigments, if desired. To maintain flexibility, increase the ratio of plasticizer if pigment is added to the formula.

Clear Nitrocellulose Lacquer

	(dry basis) %/wt.
Nitrocellulose, RS, ½ sec.	14.4
"Santolite" MHP	5.7
Camphor	2.9
Dibutyl Phthalate	4.4
Isopropyl Alcohol, Anhydrous	10.1
n-Butyl Acetate	29.2
Ethyl Acetate	17.9
Toluene	15.4

Proc.: Mix all ingredients until uniform.

Water-Vaporproof Heat-Sealing Lacquers

	I %/wt.	II %/wt.
"Parlon" S125	58.5	58.5
"Lewisol" 28	15.0	15.0
Dibutyl Phthalate	23.0	23.0
Ceresin Wax, White	3.5	—
Beeswax, Crude	—	3.5

Proc.: Mix all ingredients until uniform.

Lacquer Paste

Titanium Dioxide	264 lb.
Barium-Base Titanium Pigment	264 lb.
"Hercolyn"	24¾ gal.
Ester-Gum Solution (10 lb. Ester Gum to 1 gal. Toluol)	14¼ gal.
½ Second Nitrocellulose Solution (25% Nitrocellulose, 75% Lacquer Thinner)	12 gal.
Lacquer Thinner	14¾ gal.

Proc.: Mix all ingredients until uniform.

This will yield 80 gal. paste of semipaste consistency. The grinding time will be 16 hours.

Wood Floor Lacquer

	%/wt.
EHEC-Low	7.5
"Pentalyn" G	7.5
Ceresin Wax	0.2
Mineral Spirits	76.3
n-Propanol	8.5

Proc.: Mix all ingredients until uniform.

Antislip Lacquer
(U.S. Patent 2,564,735)

	pts./wt.
Thiokol Latex WD-2	906.0
Calcium Stearate	4.5
Red Iron Oxide	11.6
Chrome Yellow	16.5
Carbon Black	1.5
Butyl Acetate	580.0
Acetone	320.0
Ground Cork	136.0
30% Nitrocellulose Solution in Butyl Acetate	1050.0

Proc.: Mix all ingredients until uniform.

Metallized Plastic Lacquer

	%/wt.
"EHEC"-Low	8.50
"Amsco"-Solv D	41.25
V.M. & P. Naphtha	41.25
Butanol	9.00

Proc.: Mix all ingredients until uniform.

Rubber Repairing Lacquer
(For Galoshes)

A	Alcohol	240 cc.
	Nigrosin (Alcohol-Soluble)	2 cc.
B	Nigrosin-Base BT	50 cc.
	Benzol (90%)	180 cc.
	Acetone, Technical	200 cc.

Proc.: Mix A, then B. Combine.

To 350 cc. of this dyestuff solution add

Xylene, Technical	350 cc.
Vinapas B.P. 50T	300 g.

Mix thoroughly, filter through a gauze filter.

Pavement Lacquers
I

	%/wt.
Rosin, Pale	14
Manila Copal	30
Linseed Oil	22
Cobalt Linoleate Drier	1
Benzoline	33

Proc.: Mix all ingredients together until uniform.

II

		pts./wt.
A	Alcohol	40 g.
	Manila Copal	40 g.
B	"Galipot" in Alcoholic Solution (1.5:1)	20 cc.
C	Rosin in Alcoholic Solution (2:1)	20 g.

Proc.: Mix solutions A, B, C.

Pigmented Outdoor Lacquer

	pts./wt.
Cellulose Acetate LL-1	37.0
"Durez" 500	14.8
"Santicizer" M-17	18.7
Zinc Oxide	29.5
Methyl Acetate	15.0
Nitromethane	32.0
"Cellosolve" Acetate	8.0
Ethanol	18.0
Toluene	27.0

Proc.: Mix all ingredients until uniform.

Lacquer Thinners
I

	%/wt.
Ethyl Acetate	15
Butyl Propionate	25
Toluol	60

II

	oz.
Ethyl Acetate	5
Butyl Propionate	10
Fusel Oil	20
Toluol	55
Xylol	10

Proc.: Mix all ingredients until uniform.

Photochemically Nonreactive Thinners
I
Spray Thinner

	pts./wt.
Xylene	6
Toluene	11
2-Nitropropane	30
Isopropanol	14
VM and P Naphtha	25
Butanol	14

II
Brush Thinner

	pts./wt.
MIBK	10
"Solvesso" 150	6
Mineral Spirits	25
"Cellosolve"	29
Butyl "Cellosolve"	10
"Cellosolve" Acetate	10
Butanol	10

Proc.: Mix all ingredients together until uniform.

Wrinkle Finish Varnish

	oz.
Tung Oil	100
Rosin	5-10

Proc.: Heat for 2 to 8 hours at 177-290°C. Cool and dissolve in an equal weight of high-flash naphtha.

Floor Finish

(Permanent, Scratch-free) Clear (Natural) Finish:
I

Castor Oil	1 qt.
Boiled Linseed Oil	½ gal.
Paraffin Wax	3¼ lb.

High-Flash Naphtha	3 qt.
Gasoline	1½ gal.
Varnolene	1 gal.

Proc.: Mix the oils and wax and heat until the wax is molten. Add the varnolene, naphtha and gasoline slowly in the order mentioned.

II
Dark Finish

Castor Oil	1 qt.
*Gilsonite Cook	1 gal.
Paraffin Wax	3 lb.
High-Flash Naphtha	1 qt.
Gasoline	1½ gal.
Varnolene	1 gal.

Proc.: Heat oil and wax until molten, add the gilsonite cook and proceed as above.

*Gilsonite Cook:	
Gilsonite	5 lb.
Kellogg Varnish Oil	1½ gal.
High Flash Naphtha	1¼ gal.

Heat gilsonite and oil to 270°C. (520°F.). Let cool and thin with naphtha.

Any shade may be obtained by intermixing clear and dark finish. Apply by flowing on the freshly scraped floors, distribute and rub in lightly with rags. Permit to dry for at least 48 hours. This finish actually impregnates the floor and will not wear off. It has a velvet sheen and a slight slip, is easy to keep clean and is very resistant to moisture.

Floor Sealer

| | I | II |
	16% solids	40% solids
"RWL"-201 Latex (40%)	340	850
N-Methyl-2-Pyrrolidone	16	40
"Igepal" CO-430	8	20
Water	636	90
	1000	1000
Ammonium Hydroxide (28%)	2	2
Formalin	1.5	4

Proc.: Dilute RWL-201 Latex with water and adjust to a pH of 7.0 to 7.5 with 28% ammonium hydroxide (Precaution: Do not raise pH of high-solids RWL-201 latex above 7.5). Add other ingredients and stir for ½ hr. The pH is adjusted to 8.0 to 8.5 with 28% ammonium hydroxide.

Shiny Black Spirit Stain

	pts./wt.
Alcohol-Soluble Nigrosine	2
Methyl Alcohol	50
Shellac (5-lb. Cut)	50

Proc.: Grind all ingredients until uniform.

Oil Stains
I
Penetrating Oil Stain

	pts./wt.
Oil Red	126
Sudan I	14
Alizarine Cyanine Green Base	60
Toluol	6300
V.M. & P. Naphtha	2500-1000
Varnish	1000-2500

Proc.: Grind all ingredients together until uniform.

II
(Nonpenetrating)

	pts./wt.
Golden Ochre	9.0
Toluidine Toner	1.0
Boiled Linseed Oil	89.2
Concentrated Drier	0.2

Proc.: Grind all ingredients until uniform.

Liquid Oil Graining Color

Raw Linseed Oil	2/5 gal.
Turpentine	3/5 gal.
Drier, Liquid	1/2 pt.
Beeswax, Yellow (Shavings)	2/3 oz.

Proc.: Warm together and mix until clear.

American Walnut Graining Color

Ivory Black	2 oz.
Van Dyke Brown	4 oz.
Burnt Umber	2 oz.
Bolted Whiting	1 oz.
Water	½ gal.

Proc.: Grind all ingredients until uniform.

Dipping Stain for Thermoplastics

	%/wt.
"Celliton" Fast Pine B	15
"Calcosyn" Orange 6R Concentrated	55
"Celliton" Fast Blue FFR	30

Use 3 to 5% in the following solvent:

Pine Oil	50
Acetone	25
Alcohol	25

Proc.: Grind all ingredients until uniform.

Imitating Old Copper Finish

After application of priming coat use

White Lead	6 lb.
Chrome Yellow, Medium,	12 oz.
Venetian Red	1½ lb.
Burnt Umber	4 oz.
Linseed Oil	4¼ lb.
Turpentine	4¼ lb.
Drier	to suit

Proc.: Grind all ingredients until uniform.

After applying above paint, allow to dry and use a coating of copper bronze powder thinned with equal parts of spar varnish and turpentine. When this coat is dry apply a glaze made from chrome green, medium, thinned with equal parts of raw linseed oil and turpentine plus a small amount of drier. While the glaze is still damp wipe it here and there to produce a mottled effect.

Masonry Paint and Waterproofer

	%/wt.
Portland Cement	70
Hydrated Lime	30

Proc.: Grind until uniform.

For a white color, white Portland cement can be used. This, however, has a tendency to darken when wet. If this is objectionable, then 5% of white pigments, such as "Titanox" or lithopone, can be added.

Water repellents, such as calcium or aluminum stearates, may be added up to about 1%. These, however, are of questionable value.

For colors other than white, normal Portland cement with sufficient tinting colors can be added.

To use as a paint, add water to brushing consistency and apply.

To use as a waterproofer, add water to a paste consistency and apply with a stiff-bristle scrub brush. Cracks and open joints should first be filled with a mortar grout.

This paint can be used on porous surfaces of masonry, concrete, stucco, common bricks, and masonry blocks.

Brick and Cement Waterproofing Paint
(U.S. Patent 2,588,438)

	%/wt.
White Portland Cement	50-70
Fibrous Talc	15-22
Precipitated Calcium Carbonate	5-10
Salt	5-10
Calcium Stearate	1-3

Proc.: Grind until uniform. Mix with water before use.

Waterproofing Brick Walls

Walls can be waterproofed by applying a coat of solution made by dissolving 1¾ lb. of paraffin in each gal. of mineral spirits used as a solvent. Use steam to melt rather than a free flame.

Waterproofing Compound and Paint Vehicle
(U.S. Patent 1,965,042)

Three gallons of china-wood oil is raised to a temperature of about 240°C.; at this temperature 12 grams of manganese borate is added with rapid stirring. The temperature is maintained for a period not exceeding about fifteen minutes, but preferably from one to two minutes. In order to quickly cool the oil and also to partially dilute it, about 1 gallon of water white kerosene is added.

The temperature of the mass will thus be reduced to about 175° C. and when this temperature is attained 1½ pints of carbon tetrachloride is gradually added by introducing the same preferably near the bottom of the vessel. The rate of introduction of carbon tetrachloride is such that from 1–2 minutes are required for this step of the process. When the carbon tetrachloride has been introduced and the temperature has been reduced sufficiently, for example, to about 100°C., any desired quantity of diluent such as kerosene or solvent naphtha is added.

This forms a solution of waterproofing material which when applied to stone, brick, masonry and the like penetrates the pores of the same and coats the surface of the material to which it is applied, efficiently protecting it from the elements such as rain, sea water, salt water air, heat and frost. The coating is not substantially acted upon by alkalies or acids and forms a colorless waterproofing material which remains effective for many years.

9. Laboratory Equipment

This chapter consists of a concise listing of equipment important for carrying out experimentation with formulas in the preceding chapters.

Test Tubes

Test tubes are important in testing out sample batches. They are mechanically strong, highly resistant to heat, and chemically stable.

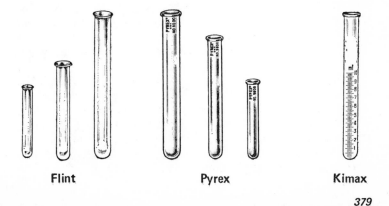

Flint   **Pyrex**   **Kimax**

Beakers

Beakers are important measuring, heating, and pouring laboratory tools.

Pyrex   Kimax

Centrifuges

The centrifuge is used to separate substances from each other. It can separate milk from cream, liquids of different specific gravities, and solids from liquids when they are held in suspension in such a way that they cannot be filtered.

Thermometer

Used for determining the temperature of a solution, it can be obtained in the centigrade or Fahrenheit scales.

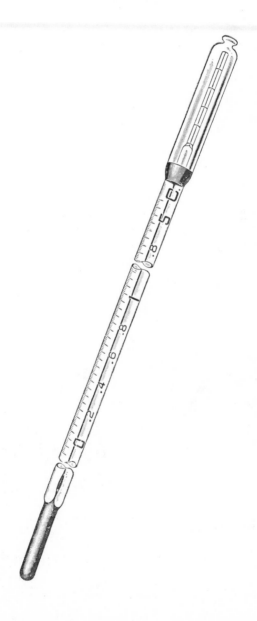

Filter Funnel and Filtering Paper

These materials are used to separate suspended solids from a liquid by forcing the mixture through a porous barrier (filter barrier). The liquid passes through the paper into the flask, while the solids are retained as a residue on the filter paper.

Paper—

Holder—

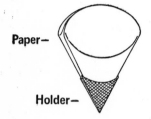

Pre-folded Filter Paper

Filter Funnel   **Flasks**

Mortar and Pestle

These are used for grinding solid materials down to the appropriate sizes called for in the formulas.

Blender

Electric appliance used for quick-mixing of ingredients. Speed of mixing can be controlled by buttons or switch.

Homogenizer

This piece of equipment is used to make a mixture uniform throughout in texture, mixture and quality by breaking down and blending the particles.

pH Indicators

These are used to determine the hydrogen ion concentration in a solution. The pH range is 1 through 14—ranging from acidic to basic. 7 is neutral (e.g. water).

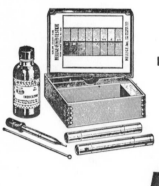

Hydrogen Ion Comparator Kit

pH Meter

Scales

The materials used in the preparations must be the exact weight required in the formula.

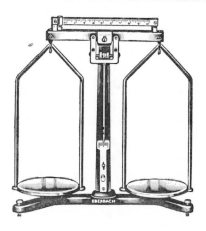

Utility Balance

Direct Reading Balance

Colloid Mill

Colloid mills produce stable emulsions to the sub-micron range—Gifford-Wood Colloid Mills are specifically designed to disperse solids, liquids or plastic materials into a carrying vehicle rapidly and economically. Any emulsion or suspension—liquids in liquids or solids in liquids—can be reproduced repeatedly with unvarying particle-size accuracy.

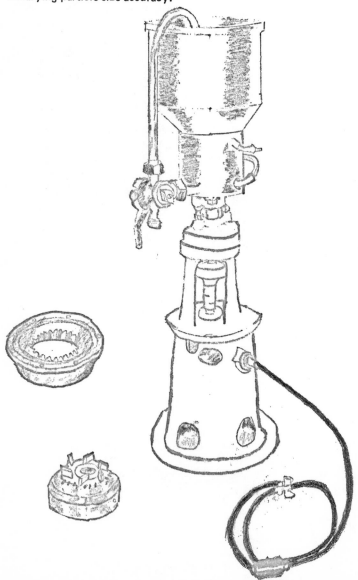

Ball Mill and Balls

This is a motor driven device containing pebbles for grinding small amounts of pigments or other materials.

Ball Mill

Flint Grinding Pebbles

Distilling Apparatus

This equipment is used for purifying water and other liquids and also for the separation of liquids of different boiling points from each other.

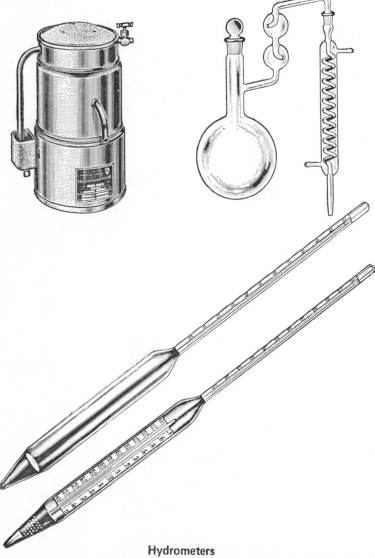

Hydrometers

This is an instrument used for measuring the specific gravity or density of liquids. It is a graduated weighted tube that sinks in a liquid up to the point determined by the density of the liquid.

Stormer Viscosimeter

The Stormer Viscosimeter is designed for determining viscosities by measurement of the time required for a definite number of revolutions of a rotating cylinder—or other type rotor—immersed in sample in the test cup, maintained at a desired temperature with a water or oil bath. Driven by a falling weight through a series of gears. A revolution counter is attached to the spindle of the rotating cylinder or other rotor.

Relative viscosity is obtained by dividing the time required for the rotor to make a specified number of revolutions in the material under examination, by the time required for the cylinder to make the same number of revolutions in distilled water or other reference, under identical conditions.

Viscosities can be determined and recorded in the absolute unit, i.e., the centipoise, with a calibration table easily prepared by the user. It is well adapted for such use, as its readings are independent of the specific gravity of the fluid.

The weight box regularly supplied with the instrument is filled with lead shot and weighs approximately 160 grams. The weight of the box empty is approximately 35 grams, so that the operating weight can be adjusted within these limits by adding to or removing shot from the weight box.

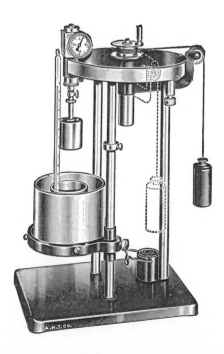

10. Start Your Own Business

The history of this nation is filled with examples of successes made from small beginnings. Many of the largest enterprises began in a small way in small communities. Woolworth's for example, originated from a "five and dime" store in a small town and spread its operation throughout the United States. There are countless opportunities for developing a new business today. The large resevoir of savings, changes in buying habits, higher levels of income, increase in population, and the introduction of new products are opening further channels for enterprises.

This chapter takes off from the point that you've found a product or products (from your experimentation with the formulas) that you feel you can sell successfully. You have already experimented with its saleability on a small scale and you feel you can benefit financially by selling it on a large scale. This chapter has been structured to give you a well-rounded perspective on what is involved in setting up your own business.

Small business has particular characteristics which make it an essential and enduring part of our country's economic life. It is one of the chief outlets for individual initiative and expression, for the operator of a small business like a pioneer farmer, must be a jack-of-all-trades.

The small businessman obviously must know and meet his customers' needs. He must select his location with an eye to market centers or neighborhoods. He must train his own help, arrange for banking services and insurance, do his own buying and record keeping, make his own store layout and window displays and promote customer goodwill while maintaining proper credit control. In a very real sense he is a *businessman,* instead of a specialist in some particular business activity within a large organization.

You cannot become an expert in banking, real estate, insurance, purchasing and inventory control, record keeping, regulation, advertising and credit. But there are certain things you should know about each of these subjects so that you may better make use of such specialists as are available in your community.

A person's interests, experience and knowledge of trends in different trades will determine his selection of the kind of business to operate. Assuming you have already made that decision, how much capital—cash and credit—do you need to start a small retail service or other business?

Because of changes in price levels the amounts required today are substantially above those of a few years ago. Prices may or may not stabilize around current levels. Therefore, any dollar figures given at a specific time may be substantially different two or three years later. The State Departments of Commerce keep posted on changes of this order and has material written expressly for the purpose of guiding prospective businessmen.

How much money does it take to keep a store running? One way of answering this is to show operating expenses for typical stores as a percentage of their sales. Operating expenses include wages (including average salary drawn by proprietors), rent, advertising, supplies (not stock), heat, light, power, taxes, insurance, interest on borrowed money, repairs, improvements, delivery services and so forth. These are commonly lumped together as "overhead."

Do you have to locate in a big city to enjoy a large sales volume? The figures do not indicate it. For all independent stores taken together, the average sales per store in the United States do not change much except for extremes of population. In places below 2,500 population, sales drop off sharply, because of diminishing turnover in such small communities and because of the drawing power and wider shopping opportunities available in large surburban shopping centers and in cities.

How long may you expect to stay in business—what are your chances of survival?

The answer to these questions depends less on general economic conditions than on your own business acumen and personal good fortune. Although most casualties are due primarily to inexperience, only 2.7 percent of the firms that failed in 1965 were in their first year of operation. The majority of fatalities occurred among establishments two to five years old. After a firm, has survived for five years, its life expectancy increases in each succeeding year of its existence.

The main causes of failure in retail trade are lack of experience, inadequate sales and competitive weakness. Business depressions are not the chief cause, although they do contribute to the mortality rate.

Borrowing too heavily is another danger. The evidence in the retail field indicates that at least 50 percent of the capital—75 percent, if possible—should be put up by the owner.

How much money should you expect to make on your investment of time, energy and financial resources? This question is the one of most interest, and yet the most difficult to answer.

It is important in appraising your financial picture to allow a reasonable sum for your personal compensation or salary as a cost of doing business. Too many proprietors "live off their shelves" and fail to appreciate the difference between actual profit and the amount received from sales minus the cost of goods and overhead. Profit is what is left over when you have paid for the goods sold and all other costs normally met, and have deducted a reasonable amount for your own personal services. You can determine this deduction best by inquiring how much managers of businesses like yours normally receive in your area. Your records should reflect true costs including your own services. Otherwise you may some day awaken to the fact that you have been operating at a loss for years. It stands to reason that you need good records when the margin between profit and loss on your sales is so narrow.

However, not all of the prifit in self-employment is measurable in dollars and cents. Running your own business means careful planning, hard work and long hours; it also means an opportunity to profit from your own best efforts and to correct your own mistakes rather than suffer from the mistakes of others.

Begin your business career by saying to yourself: "Let's sit down and talk this over from all sides." Make sure you realize what you are getting into. Don't be afraid to ask questions about things you do not understand. Don't accept every statement at its face value. The big thing is to know what you are doing. "Be sure you're right, then go ahead," is still a good motto. Remember that a business—large or small—is a big responsibility.

As a starter, here are a few questions to ask yourself:

1. Do I get along with people and inspire confidence?
2. Am I willing to shoulder the responsibility of meeting a payroll and paying debts on time?

3. Do I like the business I want to enter so that I won't mind working long hours and making other personal sacrifices?
4. Do I understand that business is a speculation, and am I willing to take the risk involved?
5. Do I like to sell?
6. Can I make decisions and can I weather wrong ones?
7. Am I resourceful in emergencies?
8. Am I a good organizer?

If you can answer most of the above questions affirmatively, you can feel quite certain you have the personal aptitude you need to start a business.

Business is not a bed of roses. Basically, it is a great adventure and calls for initiative, integrity, good judgment, courage and determination. You probably have some or all of these qualities or you wouldn't consider going into business for yourself. You may have to grit your teeth many times and stick through disappointments. But if you win, you will reap ample rewards. Is it worth the try? You are the best judge.

Anybody can start a business--the trick is to stay in business. There are several necessary qualifications for making a business succeed.

Good business practice indicates that the primary requirement for a new venture is to offer something that satisfies a public need. The customer can afford to spend only so much and he wants to get the most for his money. In that respect the world of commerce and industry is cold and impersonal. You must have something to sell that people really want, and you must know how to sell—and keep selling.

No matter what business you select, you must have a market for your product or service. If you plan to sell to manufacturers or distributors, locate where you can reach them easily and where they can find you. If you plan to sell directly to consumers, locate where they are. People make markets. They may beat a path to your door if you can offer then something good enough, but you'll get more customers if you locate where they can reach you easily. Go where the market is.

Financing a new enterprise involves a great deal more than drawing out savings or applying for a loan. You will need experienced counsel, for too little capital invested can cause trouble while too large a loan can break the back of a young business. There is plenty of money in this country to be invested, and loans at a reasonable rate of interest are available for sound propositions. Do not go to loan sharks who charge excessive interest rates. Do not give a fee to a promoter who promises to get new capital for you.

Visit a banker. You will be met with courtesy and consideration and will get sound advice concerning your business.

In the final analysis, success or failure depends upon the businessman himself. If you try to run a $20,000 business on $1,000, if you take too much money from the business for personal expenses, if you give credit with an easy hand, if you put too much money into plant, store or staff in relation to the amount of business you are likely to do, you may get into serious difficulty.

There are plenty of advantages in being in business for yourself, but the extent of your success may depend on your patience and adaptability. You may make a good living and may build a fortune, but you will have plenty of responsibility—perhaps more than you will relish. You will have to make crucial decisions and stand by them. You may—and it is important to keep this in mind—lose every cent you have. If you are hoping to embark on a life of ease, forget it. You have to work for the big prize.

Don't make a hasty decision. Think it over. Discuss it with your friends and others whose opinions you respect. This country grew to greatness because men with ideas, some capital and plenty of courage could start small businesses of their own and make them grow. That is the opportunity you have if you are self-reliant, eager to get off somebody's payroll and make your own way.

Borrowed Money

Very few businesses can be run without permanent capital—money put into the business by owners, partners or shareholders. This money is invested in plant and equipment or used as working capital, year in and year out, without being removed from the business. No legal obligation exists to pay a return on such funds or to repay the amount invested. The risk involved is substantial. But few businesses confine their financing to permanent capital alone. They borrow money and make it work for them at a profit. In this way they can expand their profit-making resources beyond what would otherwise be possible.

Your banker has an impartial, practical point of view. He will tell you whether your proposition seems sound. He knows the financial structure of a great many businesses, including some that are much like your own. He can generally tell when a good opportunity presents itself, or when you are in danger of overexpanding. He will lend you the money you need, or tell you where you can get it.

Even the best managed business can be burglarized, destroyed by fire, or sued for heavy damages as a result of the negligence of its employees. These represent possible loss of capital or interruption of earnings which can lead to failure.

Fortunately you can insure against such losses even if you cannot prevent their occurrence. To do so is a matter of good business judgment. It is difficult, for instance, to conceive of a sound business not carrying fire insurance. The word "accident" covers a multitude of possible losses: an errand boy crushed in an elevator, a patron poisoned by apparently fresh food, a pedestrian slipping on the sidewalk or injured by a truck. The hazard of dishonesty may also spell disaster: a figure-juggling bookkeeper, a thieving clerk or a "stickup" man.

While these and many other dangers face every business, it is not necessary that you bear the losses. Even though these perils are largely beyond your control, you can protect yourself against financial loss by transferring your risks to a professional risk bearer—an insurance company.

1. Property insurance, which protects the owner of the property (or the mortgagee) against loss caused by the actual destruction of a part or all of the property by fire, windstorm, explosion, falling aircraft, riot and other perils.

2. Business interruption insurance, and other Time Element coverages, which protects a business against loss of earnings resulting from an interruption caused by damage to or destruction of the physical property.

3. Liability insurance, which protects a person (or persons) against loss arising out of his legal liability for death, injury or damage to the person or property of others caused by his negligence, including his obligation to pay medical, hospital, surgical and disability benefits to injured persons, and funeral and death benefits to dependents, beneficiaries or personal representatives of persons who are killed, irrespective of his legal liability.

4. Fidelity bonds, which guarantee against loss due to the dishonesty of employees, and surety bonds, which guarantee the performance of various types of obligations assumed by contract or imposed by law.

5. Workmen's compensation insurance, which provides for payment of compensation benefits, as established by State law, to injured employees of a business. As later described, this coverage is compulsory under New York State law under certain conditions.

6. Disability benefits insurance, which provides for payment of benefits to employees who are disabled by reason of injury or illness occurring outside of their employment. This coverage became compulsory on July 1, 1950, under certain conditions as described later.

Keeping Records

You may have the idea that you do not need to keep accurate and up-to-date records to run your business; that you carry your office in your hat. If you have that idea—a mistaken one as you will find out—your bank and the tax collector will correct your false impression. They have the power to enforce their ideas on the subject whenever you need a loan or have to file your income tax return. You will have to give them information about your business which you can get only from an adequate set of records. Banks cannot lend money on guesswork and the tax collector may make an arbitrary assessment if you cannot support your own figures with concrete evidence from day-to-day records.

Standard account books for a small business can be obtained at any business stationery store and you may even be able to find an account book especially designed for your particular line of business. If your business is large enough, you will probably hire a clerk bookkeeper who has been trained in keeping business records.

First you will need a record of cash receipts and payments. You must determine the degree of detail in which they are to be kept. On the payment side you will probably want to record such items as wages, rent, interest and taxes. The account books which are available at stationery stores will suggest additional items and, of course, your bookkeeper or accountant will have suggestions. The main thing to remember is that books are your servants—they should never become your master. Keep bookkeeping down to the minimum required to meet bank, government and your own requirements.

There should also be a record of credit sales and purchases and these should be arranged so that you can easily tell when a customer is delinquent and when your own accounts are due to be paid. Details regarding all sorts of property and equipment bought by the business should also be recorded.

The number and complexity of annual summaries and statements will also be a matter for you to decide. But there is one annual statement, known as *profit and loss statement,* which is absolutely necessary for income tax purposes. Before making a loan, the bank will also want to examine another statement known as the *balance sheet*. These two statements also contain most of the information you will need to make important business decisions.

Here are some questions your banker, the income tax collector or the company that supplies you with goods on credit might ask about your business:

1. How much do you own, how much do you owe and how much are you worth?
2. What was your income last year?
3. How much of your sales are for cash, how much for credit and what is your collection record?
4. What is the total of your "overhead" expenses and what percentage of gross sales do they represent?
5. What items of expense do you have?
6. What is the present value (after allowance for depreciation) of

your building, your machinery, your truck, your furniture and fixtures?

7. What are your best and worst sellers?
8. What are the most profitable and the least profitable departments of your business?
9. Are you taking full advantage of cash discounts given for prompt payment for purchases?
10. What is the normal size of your inventory and what does it cost you to carry it?

These are only a few of the things that *others* may want to know about your business.

Balance Sheet

You might compare the balance sheet to a "still" picture. It shows the condition of the business at a specific time, usually the close of business December 31. The profit and loss statement is a "moving" picture. It gives a brief review of what happened during the course of the year.

Balance Sheet
December 31, 19

Assets

Current Assets

Cash on Hand ....$	54.95	
Cash in Bank	548.63	
Merchandise Inventory	3,250.00	
Accounts Receivable	325.00	
		$ 4,178.58

Fixed Assets

Real Estate—Land ....$	950.00	
Real Estate—Buildings		
Original Cost ....$ 5,000.00		
Less Reserve for 1,000.00		
Depreciation	4,000.00	
Furniture and Fixtures		
Original Cost ....$ 600.00		
Less Reserve for 100.00		
Depreciation	500.00	

Delivery Truck
    Original Cost ....$ 985.00
    Less Reserve for   200.00
        Depreciation
                                              785.00

                                                          6,235.00

Total Assets ....$ 10,413.58

Liabilities

Current Liabilities
    Accounts Payable ....$ 2,050.75
    Notes Payable 1,000.00

Total Liabilities ....$ 3,050.75

Capital or Net Worth
    Capital, January 1, 19____ ....$ 7,015.68
    Profit After Withdrawals
        January 1-December 31, 19____ 347.15   7,362.83

Total Liabilities and Capital ....$ 10,413.58

The balance sheet is a list of Assets (what you own), Liabilities (what you owe) and Capital (what you are worth). Capital is the difference between Assets and Liabilities. In other words, the amount left over after you deduct what you owe from what you own is what you are worth. This is sometimes called "proprietorship," "equity" or "net worth."

Liabilities are also divided into "current" and "fixed." Goods which have to be paid for during the course of the coming year, bank loans and promissory notes are among items included under current liabilities. A long term loan such as a mortgage loan is classified as a fixed liability.

The difference between assets and liabilities is the "net worth" or "proprietorship" or "capital." It is the capital originally invested in the business with additions or subtractions. Additions may come about through new money invested in the business or through retention of some of the profits. Subtractions may come through withdrawal of some of the money from the business or through losses.

In the case of a partnership the share of capital belonging to each of the partners is shown separately on the balance sheet. This may be important in deciding the distribution of profits year after year and it is absolutely necessary in case the partnership is dissolved. Records must, therefore, be kept of each partner's equity. This involves recording his original investment and the share of the profit to which he is entitled each year minus his withdrawals.

If the business is organized as a corporation—and many small businesses are—the capital section of the balance sheet will list the capital stock outstanding at the original issue price. Any profits which have been allowed to accumulate in the business will be shown in a separate subdivision of the capital section called "surplus."

There are some cases where a balance sheet is required. This is true where an application is made for a bank loan. For your own purposes, however, the balance sheet is useful because of the wealth of information it contains. A few of its uses are considered here.

Growth of business can be measured in various ways but the most important item that concerns you is growth in capital. How much more is your business worth than a year ago or five years ago or when you first started? The capital section of the balance sheet will show precisely that. It will help you decide whether the work and worry connected with the business have been worthwhile. It will let you know whether the business is stagnant or declining. It will disclose the soundness of your financial position. In a small business, there is always a danger of "eating into" capital. If you take out of the business from week to week whatever money is considered necessary as living expenses, you do not know at the time whether you really have earned that money. A comparison of capital figures in the balance sheet over a period of years will show up any tendency to dissipate the company's capital.

The total of assets is also an indication of what progress you have been making. It represents the total amount of resources under your control and even though they are partly offset by liabilities, they are still a good measure of your economic strength. For this reason the size of banks, for instance, is usually measured in terms of total resources—that is, their assets rather than their capital alone.

In order to stay in business you will have to watch both your "liquidity" and your "solvency." You are insolvent as soon as your liabilities exceed your assets. If a condition like that persists for very long you are almost certain to go bankrupt. Even though you are solvent and your total assets exceed your total liabilities, you may still be in serious difficulty if you are short of funds. If you cannot meet your current liabilities when they come due, you may so impair your credit standing as to make it necessary for you to close down. For this reason it is very important to check the size of current liabilities and compare them with current assets. The balance sheet provides this information. Frequent and periodic estimates of current assets and current liabilities from day-to-day records can help you arrange your finances so as to avoid temporary embarrassment, and at the same time make the best use of your funds.

Profit and Loss Statement

The profit and loss statement is a brief summary of what has taken place during the course of the year. The various sections of this statement provide important information in making everyday decisions. The final figure, net profit, is of the greatest significance.

Gross profit is determined first, then expenses must be deducted to arrive at net profit.

To determine gross profit, you must first find the cost of goods sold. The total of goods handled during the year is obtained by adding the inventory of goods at the beginning of the year to the cost of goods purchased during the year (after deducting any returns and allowances). From this total of beginning inventory and purchases is subtracted the inventory of goods at the end of the year. The difference between the two items is the cost of the goods sold during the year. Freight paid on purchases is included in the cost of these goods.

The difference between net sales for the year (after deducting returns and allowances on sales) and cost of goods sold is the gross profit on sales.

All of the expenses incurred in carrying on business during the course of the year are deducted from gross profit. These include such items as wages and commissions; freight, trucking and express; printing and stationery supplies; heat, light and power; rent; insurance and certain taxes; telephone and telegraph charges.

Expenditures in purchasing property or equipment must not be included as expenses. They are capital items which will stay in the business for a period of years. It would not be fair to deduct their cost from the gross profit of any one year. Depreciation on property and equipment is, however, an annual operating expense.

There are a few additional items and adjustments which may not be self-explanatory. Among expenses are included supplies which are used up during the course of the year. That is why it is necessary to inventory supplies on hand at the year's end. If fire insurance has been paid in advance for three years, only one year's cost is considered an expense of this year. Loss from bad debts is another expense that is deducted from gross profit.

Daily Records

You have seen how the balance sheet and the profit and loss statement—aside from being practical requirements of the bank and the tax collector—can be of real use to you in running your business. If you plan to keep your own records and have had no experience in bookkeeping, you should buy one of the account books on the market especially designed for a small business. The column headings are already printed for you, and by exercising discretion you can make such changes as will make these account books ideal for your business.

If you hire a trained bookkeeper or accountant, he should be able to present you with a balance sheet and a profit and loss statement at your request. Some account books include forms for the balance sheet and for the profit and loss statement with space to fill in figures derived from daily records.

Your most important daily records are those which record your cash balance, your sales and your purchases. From cash records you keep track not only of your balance at any moment, but also of the amount of cash you need on hand to handle a typical day's sales and purchases. Your sales records go further and show—in any detail you desire—the kinds of merchandise or service most or least in demand and also which clerks are apparently most "on the job." Your purchase records are basic to your knowledge of expenses for equipment, supplies, merchandise and overhead.

Advertising

The first step in determining your advertising requirements is to do a "situation analysis." You can begin by asking yourself these questions:

1. Who are my potential customers?
2. How many are there?
3. Where do they live?
4. Can they get to me conveniently?
5. Are they the kind of people who want charge accounts and delivery service?
6. Where do they now buy the goods or services I want to sell them?
7. Can I offer them anything they are not now getting? What? How?
8. How can I convince them that they should do business with me?

Other questions will emerge in the process of finding the answers to these. Honest appraisal of answers to these questions will go far towards determining how much a business should spend in advertising.

Mail advertising is rifle, not shotgun shooting. Thus, in any of the various ways it can be used effectively, there should be a specific thing to say which is of interest to specific people. Among the many ways mail advertising can be used more effectively is by inserting enclosures with bills (don't send bills alone; make the stamp do its full job). Tell your regular customers of sales ahead of time (they appreciate the chance to make their choices before everyone hears of the sale). Announce a new service or product (sometimes the manufacturer or distributor will help pay for this). Send circulars or self-mailing folders cheaply at bulk mailing rates.

Handbills, business cards, circulars and other "throw-aways" are effective, especially for announcements of openings or sales. Like other forms of advertising, they should be used with care. Avoid such pitfalls as placing them where they will litter premises and irritate possible customers, putting them in home mail boxes, or distributing them in communities where ordinances forbid it.

Credit

Credit is permission to pay in the future for goods or services received now. It is used by business to facilitate trade and to increase sales.

Credit allows an individual of limited means to engage in a business of his own and it encourages the growth and development of all business.

Retail or consumer credit plays a most important part in retail distribution. It is not only a convenience to the customer, but is also a sound method of sales promotion.

A credit policy that operates efficiently for merchants extending either open account, installment, revolving or other credit must include certain well-established principles:

1. Establish terms that are clearly defined and fair to both merchant and customer.
2. Obtain good accounts by careful selection and investigation.
3. Commence immediate collection efforts according to procedure previously established.
4. Freeze or close delinquent accounts until arrangements for settlement have been made.
5. Maintain contact with delinquents in order to arrange mutually satisfactory measures for correcting the situation.
6. Follow through even to a legal conclusion when investigation reveals that the debtor is financially able to meet his obligations.

The following pages outline only those broad phases of regulation which are common to all business, an understanding of which is essential to successful operation. These include regulation of the forms of business organization with brief mention of business taxes, licensing, and the regulation of labor, competition and trade practices.

There are three main forms of business organization, each subject to different legal requirements. These are the individual proprietorship, the partnership and the corporation. While theoretically the tax laws and other aspects of regulation are not designed to influence your choice of the form in which you do business, the fact remains that the state of the law influences choice.

The individual proprietorship is the oldest and most common form of business organization, where one man owns and operates the business and hires all other help for pay. In organizing an enterprise under the individual or single proprietorship form, there are no special legal requirements to be met. Of course, you may have to obtain a license if you engage in certain trades (see the next section on licensing); but this is unrelated to the form of business organization.

However, you may wish to conduct your business under a name other than your own, such as the "Fix-It Shop." In such cases, to enable creditors and other interested parties to determine the actual ownership of any business conducted under a trade name, that name must be filed with the county clerk of each county in which the business is conducted or transacted.

Men and women who are planning to start a new business of their own, and owners of existing enterprises have a common interest in the laws and governmental regulations affecting such establishments.

Under the individual proprietorship form you would be liable for Federal and State personal income taxes, just as though you were on a salary. As a self-employer your income will also be subject to Federal old age and survivor insurance (social security).

Another common form of business organization is the partnership, which results from a contract or agreement in which two or more persons agree to combine property or labor, or both for a common undertaking and the acquisition of a common profit. The key to the rights, duties and obligations of the partners is in the "partnership agreement," and you should, therefore, obtain the services of a lawyer in its preparation.

Persons organizing a *general partnership* must file a certificate of doing business as partners and file the firm's name with the county clerk of each county in which the business is conducted or transacted. You need not file the *partnership agreement,* but it is desirable to have this in written form, and containing specific provisions with respect to all organizational and operational phases of the proposed enterprise.

Persons organizing under the *limited partnership* form have additional obligations. A limited partnership is a firm in which one or more of the partners are relieved from liability beyond the amount of the capital contributed by them, under provisions of special legislation.

The third most common form of business organization is the corporation. Unlike proprietorships and partnerships, the corporation is wholly a creation of the State. Incorporation is a *privilege* rather than a *right*; and since the State can grant or withhold the privilege of doing business under this form, it can and does subject it to certain special conditions and regulations. The major distinguishing characteristics of a corporation, which also indicate the special privileges attaching to this form of organization, are threefold: (1) unlike proprietorships and partnerships it can have rights of its own differing from those of its members, and can survive those members and preserve those rights as well as its own identity; (2) the liability of the members is generally limited to the value of their particular investment; and (3) it can increase or decrease the number of owners without altering fundamentally the nature of the corporation or its rights and obligations.

There is a fourth "privilege" or "advantage" which the corporate form has over the proprietorship or general partnership: the stockholder is an owner without assuming the responsibilities of management, while the managers are employees on salary instead of depending wholly on the profits of the enterprise. All of these features mentioned have made the corporation the normal form for large-scale operations involving long-term commitments or large amounts of capital.

Licensing

The license is a tool of regulation to control a course of conduct or a line of business having some particularly close relationship to the public health, safety or morals. It is also a familiar method of municipal regulation and prospective businessmen should consult their local ordinances which may have been passed under State enabling laws. Check with local authorities if your proposed business requires a license.

Protection of Employees

The man or woman starting a small business either has one or more employees or probably will have in the course of time. It is, therefore, important to know the major State and Federal laws and regulations designed to provide safe and healthful working conditions, a floor to wages and a ceiling on hours, prohibiting certain employments, restricting industrial homework, providing workmen's compensation, unemployment insurance, disability insurance, social security, governing industrial relations and fair employment practices.

Workmen's compensation insurance must be carried by every business employer of one or more workers. The hazardous nature of employment is no longer a determining factor for coverage.

As personal injuries incurred in the course of employment and deaths resulting from such injuries are compensable under the Workmen's Compensation Law, insurance for workmen's compensation must be had before putting employees to work for the first time. Rates for workmen's compensation insurance vary according to the nature of the employment and hazards involved.

APPENDIX

pH Values

Acids	pH Value	Bases	pH Value
Hydrochloric Acid	1.0	Sodium Bicarbonate	8.4
Sulfuric Acid	1.2	Borax	9.2
Phosphoric Acid	1.5	Ammonia	11.1
Sulfuric Acid	1.5	Sodium Carbonate	11.6
Acetic Acid	2.9	Trisodium Phosphate	12.0
Alum	3.2	Sodium Metasilicate	12.2
Carbonic Acid	3.8	Lime, Saturated	12.3
Boric Acid	5.2	Sodium Hydroxide	13.0

pH Ranges of Common Indicators

	Useful pH Range
Thymol Blue	1.2-2.8
Bromphenol Green	2.8-4.6
Methyl Orange	3.1-4.4
Bromcresol Green	4.0-5.6
Methyl Red	4.4-6.0
Propyl Red	4.8-6.4
Brom Cresol Purple	5.2-6.8
Brom Thymol Blue	6.0-7.6
Phenol Red	6.8-8.4
Litmus	7.2-8.8
Cresol Red	7.2-8.8
Cresolphthalein	8.2-9.8
Phenolphthalein	8.6-10.2
Nitro Yellow	10.0-11.6
Alizarin Yellow R	10.1-12.1
Sulfo Orange	11.2-12.6

INTERNATIONAL ATOMIC WEIGHTS

	Symbol	Atomic Number	Atomic Weight		Symbol	Atomic Number	Atomic Weight
Actinium	Ac	89	227	Neodymium	Nd	60	144.27
Aluminum	Al	13	26.98	Neptunium	Np	93	237.00
Americium	Am	95	243	Neon	Ne	10	20.183
Antimony	Sb	51	121.76	Nickel	Ni	28	58.71
Argon	A	18	39.944	Niobium	Nb	41	92.91
Arsenic	As	33	74.91	Nitrogen	N	7	14.008
Astatine	At	85	210	Osmium	Os	76	190.20
Barium	Ba	56	137.36	Oxygen	O	8	16
Berkelium	Bk	97	249	Palladium	Pd	46	106.4
Beryllium	Be	4	9.013	Phosphorus	P	15	30.975
Bismuth	Bi	83	209.00	Platinum	Pt	78	195.09
Boron	B	5	10.82	Plutonium	Pu	94	242.00
Bromine	Br	35	79.916	Potassium	K	19	39.100
Cadmium	Cd	48	112.41	Praseodymium	Pr	59	140.92
Calcium	Ca	20	40.08	Promethium	Pm	61	145
Californium	Cf	98	249.00	Protactinium	Pa	91	231
Carbon	C	6	12.011	Radium	Ra	88	226.05
Cerium	Ce	58	140.13	Radon	Rn	86	222
Cesium	Cs	55	132.91	Rhenium	Re	75	186.22
Chlorine	Cl	17	35.457	Rhodium	Rh	45	102.91
Chromium	Cr	24	52.01	Rubidium	Rb	37	85.48
Cobalt	Co	27	58.94	Ruthenium	Ru	44	101.10
Copper	Cu	29	63.54	Samarium	Sm	62	150.35
Curium	Cm	96	245	Scandium	Sc	21	44.96
Dysprosium	Dy	66	162.51	Selenium	Se	34	78.96
Erbium	Er	68	167.27	Silicon	Si	14	28.09
Europium	Eu	63	152	Silver	Ag	47	107.880
Fluorine	F	9	19	Sodium	Na	11	22.991
Francium	Fr	87	223	Strontium	Sr	38	87.63
Gadolinium	Gd	64	157.26	Sulfur	S	16	32.066
Gallium	Ga	31	69.72	Tantalum	Ta	73	180.95
Germanium	Ge	32	72.60	Technetium	Tc	43	99
Gold	Au	79	197.00	Tellurium	Te	52	127.61
Hafnium	Hf	72	178.50	Terbium	Tb	65	158.93
Helium	He	2	4.003	Thallium	Tl	81	204.39
Holmium	Ho	67	164.94	Thorium	Th	90	232.05
Hydrogen	H	1	1.0080	Thulium	Tm	69	168.94
Indium	In	49	114.82	Tin	Sn	50	118.70
Iodine	I	53	126.91	Titanium	Ti	22	47.90
Iridium	Ir	77	192.20	Tungsten	W	74	183.86
Iron	Fe	26	55.85	Uranium	U	92	238.07
Krypton	Kr	36	83.80	Vanadium	V	23	50.95
Lanthanum	La	57	138.92	Xenon	Xe	54	131.30
Lead	Pb	82	207.21	Ytterbium	Yb	70	173.04
Lithium	Li	3	6.940	Yttrium	Y	39	88.92
Lutetium	Lu	71	174.99	Zinc	Zn	30	65.38
Magnesium	Mg	12	24.32	Zirconium	Zr	40	91.22
Manganese	Mn	25	54.94				
Mendelevium	Mv	101	256.00				
Mercury	Hg	80	200.61				
Molybdenum	Mo	42	95.95				

Temperature Conversion Tables

°F.	°C.	°F.	°C.	°F.	°C.	°F.	°C.	°F.	°C.
−40	−40.0	9	−12.8	58	14.4	107	41.7	156	68.9
−39	−39.4	10	−12.2	59	15.0	108	42.2	157	69.4
−38	−38.9	11	−11.7	60	15.6	109	42.8	158	70.0
−37	−38.3	12	−11.1	61	16.1	110	43.3	159	70.6
−36	−37.8	13	−10.6	62	16.7	111	43.9	160	71.1
−35	−37.2	14	−10.0	63	17.2	112	44.4	161	71.7
−34	−36.7	15	−9.4	64	17.8	113	45.0	162	72.2
−33	−36.1	16	−8.9	65	18.3	114	45.6	163	72.8
−32	−35.6	17	−8.3	66	18.9	115	46.1	164	73.3
−31	−35.0	18	−7.8	67	19.4	116	46.7	165	73.9
−30	−34.4	19	−7.2	68	20.0	117	47.2	166	74.4
−29	−33.9	20	−6.7	69	20.6	118	47.8	167	75.0
−28	−33.3	21	−6.1	70	21.1	119	48.3	168	75.6
−27	−32.8	22	−5.6	71	21.7	120	48.9	169	76.1
−26	−32.2	23	−5.0	72	22.2	121	49.4	170	76.7
−25	−31.7	24	−4.4	73	22.8	122	50.0	171	77.2
−24	−31.1	25	−3.9	74	23.3	123	50.6	172	77.8
−23	−30.6	26	−3.3	75	23.9	124	51.1	173	78.3
−22	−30.0	27	−2.8	76	24.4	125	51.7	174	78.9
−21	−29.4	28	−2.2	77	25.0	126	52.2	175	79.4
−20	−28.9	29	−1.7	78	25.6	127	52.8	176	80.0
−19	−28.3	30	−1.1	79	26.1	128	53.3	177	80.6
−18	−27.8	31	−0.6	80	26.7	129	53.9	178	81.1
−17	−27.2	32	0.0	81	27.2	130	54.4	179	81.7
−16	−26.7	33	0.6	82	27.8	131	55.0	180	82.2
−15	−26.1	34	1.1	83	28.3	132	55.6	181	82.8
−14	−25.6	35	1.7	84	28.9	133	56.1	182	83.3
−13	−25.0	36	2.2	85	29.4	134	56.7	183	83.9
−12	−24.4	37	2.8	86	30.0	135	57.2	184	84.4
−11	−23.9	38	3.3	87	30.6	136	57.8	185	85.0
−10	−23.3	39	3.9	88	31.1	137	58.3	186	85.6
−9	−22.8	40	4.4	89	31.7	138	58.9	187	86.1
−8	−22.2	41	5.0	90	32.2	139	59.4	188	86.7
−7	−21.7	42	5.6	91	32.8	140	60.0	189	87.2
−6	−21.1	43	6.1	92	33.3	141	60.6	190	87.8
−5	−20.6	44	6.7	93	33.9	142	61.1	191	88.3
−4	−20.0	45	7.2	94	34.4	143	61.7	192	88.9
−3	−19.4	46	7.8	95	35.0	144	62.2	193	89.4
−2	−18.9	47	8.3	96	35.6	145	62.8	194	90.0
−1	−18.3	48	8.9	97	36.1	146	63.3	195	90.6
0	−17.8	49	9.4	98	36.7	147	63.9	196	91.1
1	−17.2	50	10.0	99	37.2	148	64.4	197	91.7
2	−16.7	51	10.6	100	37.8	149	65.0	198	92.2
3	−16.1	52	11.1	101	38.3	150	65.6	199	92.8
4	−15.6	53	11.7	102	38.9	151	66.1	200	93.3
5	−15.0	54	12.2	103	39.4	152	66.7	201	93.9
6	−14.4	55	12.8	104	40.0	153	67.2	202	94.4
7	−13.9	56	13.3	105	40.6	154	67.8	203	95.0
8	−13.3	57	13.9	106	41.1	155	68.3	204	95.6

Temperature Conversion Tables (Con.)

°C.	°F.	°C.	°F.	°C.	°F.	°C.	°F.	°C.
96.1	254	123.3	303	150.6	352	177.8	401	205.0
96.7	255	123.9	304	151.1	353	178.3	402	205.6
97.2	256	124.4	305	151.7	354	178.9	403	206.1
97.8	257	125.0	306	152.2	355	179.4	404	206.7
98.3	258	125.6	307	152.8	356	180.0	405	207.2
98.9	259	126.1	308	153.3	357	180.6	406	207.8
99.4	260	126.7	309	153.9	358	181.1	407	208.3
100.0	261	127.2	310	154.4	359	181.7	408	208.9
100.6	262	127.8	311	155.0	360	182.2	409	209.4
101.1	263	128.3	312	155.6	361	182.8	410	210.0
101.7	264	128.9	313	156.1	362	183.3	411	210.6
102.2	265	129.4	314	156.7	363	183.9	412	211.1
102.8	266	130.0	315	157.2	364	184.4	413	211.7
103.3	267	130.6	316	157.8	365	185.0	414	212.2
103.9	268	131.1	317	158.3	366	185.6	415	212.8
104.4	269	131.7	318	158.9	367	186.1	416	213.3
105.0	270	132.2	319	159.4	368	186.7	417	213.9
105.6	271	132.8	320	160.0	369	187.2	418	214.4
106.1	272	133.3	321	160.6	370	187.8	419	215.0
106.7	273	133.9	322	161.1	371	188.3	420	215.6
107.2	274	134.4	323	161.7	372	188.9	421	216.1
107.8	275	135.0	324	162.2	373	189.4	422	216.7
108.3	276	135.6	325	162.8	374	190.0	423	217.2
108.9	277	136.1	326	163.3	375	190.6	424	217.8
109.4	278	136.7	327	163.9	376	191.1	425	218.3
110.0	279	137.2	328	164.4	377	191.7	426	218.9
110.6	280	137.8	329	165.0	378	192.2	427	219.4
111.1	281	138.3	330	165.6	379	192.8	428	220.0
111.7	282	138.9	331	166.1	380	193.3	429	220.6
112.2	283	139.4	332	166.7	381	193.9	430	221.1
112.8	284	140.0	333	167.2	382	194.4	431	221.7
113.3	285	140.6	334	167.8	383	195.0	432	222.2
113.9	286	141.1	335	168.3	384	195.6	433	222.8
114.4	287	141.7	336	168.9	385	196.1	434	223.3
115.0	288	142.2	337	169.4	386	196.7	435	223.9
115.6	289	142.8	338	170.0	387	197.2	436	224.4
116.1	290	143.3	339	170.6	388	197.8	437	225.0
116.7	291	143.9	340	171.1	389	198.3	438	225.6
117.2	292	144.4	341	171.7	390	198.9	439	226.1
117.8	293	145.0	342	172.2	391	199.4	440	226.7
118.3	294	145.6	343	172.8	392	200.0	441	227.2
118.9	295	146.1	344	173.3	393	200.6	442	227.8
119.4	296	146.7	345	173.9	394	201.1	443	228.3
120.0	297	147.2	346	174.4	395	201.7	444	228.9
120.6	298	147.8	347	175.0	396	202.2	445	229.4
121.1	299	148.3	348	175.6	397	202.8	446	230.0
121.7	300	148.9	349	176.1	398	203.3	447	230.6
122.2	301	149.4	350	176.7	399	203.9	448	231.1
122.8	302	150.0	351	177.2	400	204.4	449	231.7

Equivalents
Volumes

1 barrel (bbl), liquid	31 to 42 gallons
1 barrel (bbl), dry commodites such as fruits, vegetables, etc.	7056 cubic inches 105 dry quarts 3.281 bushels
1 barrel, standard, cranberry	5826 cubic inches 86.703 dry quarts 2.709 bushels
1 bushel (bu) (U.S.)	2150.42 cubic inches 35.238 liters
1 bushel, British Imperial	1.032 U.S. bushels 2219.36 cubic inches
1 cord (cd) (firewood)	128 cubic feet
1 cubic centimeter (cm^3)	0.061 cubic inch
1 cubic decimeter (dm^3)	61.023 cubic inches
1 cubic foot (cu ft)	7.481 gallons 28.317 cubic decimeters
1 cubic inch (cu in)	0.554 fluid ounce 4.433 fluid drams 16.387 cubic centimeters
1 cubic meter (m^3)	1.308 cubic yards
1 cubic yard (cu yd)	0.765 cubic meter
1 cup, measuring	8 fluid ounces ½ liquid pint
1 dram, fluid (fl dr) U.S.	1/8 fluid ounce 0.226 cubic inch 3.697 milliliters 1.041 British fluid drams
1 drachm, fluid (fl dr) British	0.961 U.S. fluid dram 0.217 cubic inch 3.552 milliliters

1 dekaliter (dkl)	2.642 gallons 1.135 pecks
1 gallon (gal) U.S.	231 cubic inches 3.785 liters 0.833 British gallon 128 U.S. fluid ounces
1 gallon (British Imperial)	277.42 cubic inches 1.201 U.S. gallons 4.546 liters 160 British fluid ounces
1 gill (gi)	7.219 cubic inches 4 fluid ounces 0.118 liter
1 hectoliter (hl)	26.418 gallons 2.838 bushels
1 liter	1.057 liquid quarts 0.908 dry quart 61.025 cubic inches
1 milliliter (ml)	0.271 fluid dram 16.231 minims 0.061 cubic inch
1 ounce, fluid (fl oz.) U.S.	1.805 cubic inches 29.573 milliliters 1.041 British fluid ounces
1 ounce, fluid, British	0.961 U.S. fluid ounce 1.734 cubic inches 28.412 milliliters
1 peck (pk)	8.810 liters 33.600 cubic inches
1 pint (pt), dry	0.551 liter 28.875 cubic inches
1 pint (pt), liquid	0.473 liters 67.201 cubic inches

1 quart (qt), dry (U.S.)	1.101 liters
	0.969 British quart
	57.75 cubic inches
1 quart (qt), liquid (U.S.)	0.946 liter
	0.833 British quart
	69.354 cubic inches
1 quart (qt) (British)	1.032 U.S. dry quarts
	1.201 U.S. liquid quarts
1 tablespoon	3 teaspoons
	4 fluid drams
	½ fluid ounce
1 teaspoon	1/3 tablespoon
	1-1/3 fluid drams
1 water ton (English)	270.91 U.S. gallons
	224 British Imperial gallons

Weights

1 assay ton (AT)	29.167 grams
1 carat (C)	200 milligrams
	3.086 grains
1 dram, apothecaries (dr ap)	60 grains
	3.888 grams
1 dram, avoirdupois (dr avdp)	27.344 grains
	1.772 grams
1 grain	64.799 milligrams
1 gram (g)	15.432 grains
	0.035 ounce, avoirdupois
1 hundredweight, gross (gross cwt)	112 pounds
	50.802 kilograms

1 hundredweight, net (net cwt)	100 pounds
	45.359 kilograms
1 kilogram (kg)	2.205 pounds
1 microgram (μg)	0.000001 gram
1 milligram (mg)	0.015 grain
1 minim	0.0616 cc.
1 ounce, avoirdupois (oz avdp)	437.5 grains
	0.911 troy ounce
	28.350 grams
1 ounce, apothecaries or troy (oz ap)	1.097 avdp ounces
	31.103 grams
1 pennyweight (dwt)	1.555 grams
1 point	0.01 carat
	2 milligrams
1 pound, avoirdupois (lb avdp)	7000 grains
	1.215 apothecaries pounds
	453.592 grams
1 pound, apothecaries (lb ap)	5760 grains
	0.823 avoirdupois pound
	373.242 grams
1 scruple (S ap)	20 grains
	1.296 grams
1 ton, long	2240 pounds
	1.12 short tons
	1.016 metric tons
1 ton, short	2000 pounds
	0.893 long ton
	0.907 metric ton
1 ton (t) metric	2204.622 pounds
	0.984 long ton
	1.102 short tons

EQUIVALENTS OF TWADDELL, BAUME AND SPECIFIC GRAVITY SCALES

Twaddell	Baumé	Specific Gravity	Twaddell	Baumé	Specific Gravity	Twaddell	Baumé	Specific Gravity	Twaddell	Baumé	Specific Gravity
0	0	1.000	44	26.0	1.220	88	44.1	1.440	131	57.1	1.655
1	0.7	1.005	45	26.4	1.225	89	44.4	1.445	132	57.4	1.660
2	1.4	1.010	46	26.9	1.230	90	44.8	1.450	133	57.7	1.665
3	2.1	1.015	47	27.4	1.235	91	45.1	1.455	134	57.9	1.670
4	2.7	1.020	48	27.9	1.240	92	45.4	1.460	135	58.2	1.675
5	3.4	1.025	49	28.4	1.245	93	45.8	1.465	136	58.4	1.680
6	4.1	1.030	50	28.8	1.250	94	46.1	1.470	137	58.7	1.685
7	4.7	1.035	51	29.3	1.255	95	46.4	1.475	138	58.9	1.690
8	5.4	1.040	52	29.7	1.260	96	46.8	1.480	139	59.2	1.695
9	6.0	1.045	53	30.2	1.265	97	47.1	1.485	140	59.5	1.700
10	6.7	1.050	54	30.6	1.270	98	47.4	1.490	141	59.7	1.705
11	7.4	1.055	55	31.1	1.275	99	47.8	1.495	142	60.0	1.710
12	8.0	1.060	56	31.5	1.280	100	48.1	1.500	143	60.2	1.715
13	8.7	1.065	57	32.0	1.285	101	48.4	1.505	144	60.4	1.720
14	9.4	1.070	58	32.4	1.290	102	48.7	1.510	145	60.6	1.725
15	10.0	1.075	59	32.8	1.295	103	49.0	1.515	146	60.9	1.730
16	10.6	1.080	60	33.3	1.300	104	49.4	1.520	147	61.1	1.735
17	11.2	1.085	61	33.7	1.305	105	49.7	1.525	148	61.4	1.740
18	11.9	1.090	62	34.2	1.310	106	50.0	1.530	149	61.6	1.745
19	12.4	1.095	63	34.6	1.315	107	50.3	1.535	150	61.8	1.750
20	13.0	1.100	64	35.0	1.320	108	50.6	1.540	151	62.1	1.755
21	13.6	1.105	65	35.4	1.325	109	50.9	1.545	152	62.3	1.760
22	14.2	1.110	66	35.8	1.330	110	51.2	1.550	153	62.5	1.765
23	14.9	1.115	67	36.2	1.335	111	51.5	1.555	154	62.8	1.770
24	15.4	1.120	68	36.6	1.340	112	51.8	1.560	155	63.0	1.775
25	16.0	1.125	69	37.0	1.345	113	52.1	1.565	156	63.2	1.780
26	16.5	1.130	70	37.4	1.350	114	52.4	1.570	157	63.5	1.785
27	17.1	1.135	71	37.8	1.355	115	52.7	1.575	158	63.7	1.790
28	17.7	1.140	72	38.2	1.360	116	53.0	1.580	159	64.0	1.795
29	18.3	1.145	73	38.6	1.365	117	53.3	1.585	160	64.2	1.800
30	18.8	1.150	74	39.0	1.370	118	53.6	1.590	161	64.4	1.805
31	19.3	1.155	75	39.4	1.375	119	53.9	1.595	162	64.6	1.810
32	19.8	1.160	76	39.8	1.380	120	54.1	1.600	163	64.8	1.815
33	20.3	1.165	77	40.1	1.385	121	54.4	1.605	164	65.0	1.820
34	20.9	1.170	78	40.5	1.390	122	54.7	1.610	165	65.2	1.825
35	21.4	1.175	79	40.8	1.395	123	55.0	1.615	166	65.5	1.830
36	22.0	1.180	80	41.2	1.400	124	55.2	1.620	167	65.7	1.835
37	22.5	1.185	81	41.6	1.405	125	55.5	1.625	168	65.9	1.840
38	23.0	1.190	82	42.0	1.410	126	55.8	1.630	169	66.1	1.845
39	23.5	1.195	83	42.3	1.415	127	56.0	1.635	170	66.3	1.850
40	24.0	1.200	84	42.7	1.420	128	56.3	1.640	171	66.5	1.855
41	24.5	1.205	85	43.1	1.425	129	56.6	1.645	172	66.7	1.860
42	25.0	1.210	86	43.4	1.430	130	56.9	1.650	173	67.0	1.865
43	25.5	1.215	87	43.8	1.435						

Relation of Capacity, Volume and Weight

1 pint	=28.875 cubic inches
1 quart	=57.75 cubic inches
1 gallon (U. S.)	=231 cubic inches
1 gallon (English)	=277.274 cubic inches
7.4805 gallons	=1 cubic foot

1 gallon water at 62° Fahr. weighs 8.3356 lbs.

Chemical Substitutes

Product	Substitute or Alternative
Accroides, Gum	Rosin
	Seed-Lac
	"Vinsol"
Acetaldehyde	Aldol
	Formaldehyde
	Furfuraldehyde
	Glyoxal
Acetamide	Ammonium Acetate
	Ethanolamine Acetate
	Formamide
	Urea
Acetanilide	Alum
	Pyramidone
	Zinc Sulphocarbolate
Acetic Acid	Ammonium Sulphate with
	dilute sulphuric acid
	Boric acid
	Citric acid
	Formic acid
	Glycollic acid
	Lactic acid
	Levulinic acid
	Phosphoric acid
	Propionic acid
	Pyroligneous acid
Acetic Acid	Saccharic acid
	Salt
	Sodium diacetate
Albumen	Agar
	Alum, potash
	Casein
	Emulsifiers
	Protein, fish
	Protein, soybean
	Resin, natural
	Resins, synthetic
	Thickeners
Alcohol	See Ethyl alcohol
Alkalies	Amines, primary, secondary,
	tertiary, quaternary
	Aminoalcohols
	Ammonium hydroxide
	Barium hydroxide
	Borax

Product	Substitute or Alternative
	Calcium hydroxide
	Calcium oxide
	Lithium hydroxide
	Magnesium hydroxide
	Magnesium oxide
	Nephelin
	Potassium carbonate
	Potassium hydroxide
	Potassium silicate
	Sodium carbonate
	Sodium hydroxide
	Sodium metasilicate
	Sodium orthosilicate
	Sodium pyrophosphate
	Sodium silicate
	Trisodium phosphate
Alkyd Resins	"Flexoresins"
	"Piccolyte" resins
	Resins, synthetic
Almond Oil	Apricot kernel oil
	Benzaldehyde
	Cherry kernel oil
	Mineral oil, refined
	Peach kernel oil
	"Persic" oil
	Vegetable oils
"Alperox"	Hydrogen peroxide
Alpha Protein	Casein
	"G"-protein
Alum	Acetanilide
	Alum, potash
	Aluminum chloride
	Aluminum hydrate
Aluminum Hydrate	Alum
	Ammonia alum
	Copperas
	Ferric sulphate
	Lime
	Potassium alum
	Sodium aluminate
Aluminum Oleate	Calcium oleate
	Lead oleate
	Magnesium oleate
Aluminum Phosphate	Calcium phosphate

Product	Substitute or Alternative
Aluminum Powder	Graphite
	Mica
	Pearl essence
	Sericite
	Slate powder
Aluminum Resinate	Calcium resinate
	Lead resinate
	Magnesium resinate
Aluminum Silicate	Alum, potash
	Talc
Ammonium Compounds	Alkalies
	Amides
	Amines
	Ammonium thiocyanate
	Cyanamide
	Dicyandiamide
	Urea
Ammonium Chloride	Manganese chloride
	Zinc chloride
Ammonium Hydroxide	See Alkalies
Ammonium Lactate	Ammonium glycollate
	Glycerin
Ammonium Phosphate	Calcium cyanamid
	Guano
	Sodium nitrate
Ammonium Sulphamate	"Abopon"
	Borax with boric acid
	Sodium chlorate
Ammonium Sulphate	Ammonium phosphate
Ammonium Sulphite	Potassium bisulphite with ammonia
Ammonium Thiocyanate	Sodium bisulphite with ammonia
	Ammonium compounds
Amyl Acetate	Butylacetate
	Fusel oil
Amyl Alcohol	Capryl alcohol
	Fusel oil
	Hexyl alcohol
	Octyl alcohol
	Solvents
	Tetrahydrofurfuryl alcohol
Aniline	o-Aminodiphenyl
	Furfural
	Pyridin
Anise Oil	"Annol"
Annatto	Dyes, aniline

Product	Substitute or Alternative
Antimony ....,	Cadmium
	Calcium
	Selenium
	Tellurium
Antimony Lactate ......,	Tartar emetic
Antimony Oxide.........	Tin oxide
	Titanium oxide
Apricot Kernel Oil	Almond oil
Beeswax ....,	"B-Z Wax"
	Ceresin with soap
	Coffee wax
	Flax wax
	"Flexo Wax"
	"Isco 662, 663"
	"Norco Wax 36"
	Sugar cane wax
	Wax, synthetic
Belladonna ....,	Stramonium
Bentonite ....,,....	Alum, potash
	Clay, colloidal
	Emulsifiers
	Fillers
	Gums, water dispersible
	Thickeners
Benzaldehyde ...,.,....,,	Bitter almond oil
	Nitrobenzol
Benzene Sulphonic Acid ...	Phenolsulphonic acid
Benzine..............,,,	Petroleum ether
Calcium Sulphide	Barium sulphide
Camel's Hair	"Nylon" fleece
Camphor ....,,,	Benzyl benzoate
	Camphene
	Dibutyl tartrate
	"Dehydranone"
	Diethyl phthalate
	Esparto wax
	Hexachloroethane
	Menthol
	Naphthalene
	Phenol
	Plasticizers
	Resorcinol
	"Tetralin"
	Triphenyl phosphate

Product	Substitute or Alternative
Camphor Oil	"Japp-O"
	"Terpesol"
	Turpentine
Candelilla Wax	"Norcowax 72"
Cane Sugar	See Sugar
Capric Acid	"Alox" acids
	Cocoanut oil fatty acids
Capryl Alcohol	See Octyl alcohol, normal
Carbon Dioxide	Ammonium bicarbonate
	Carbon tetrachloride
	Methyl chloride
	Nitrogen
	Sodium bicarbonate
Carbon Tetrachloride	Carbon dioxide
	Chloroform
	Ether, petroleum
	Ethylene dichloride with sulphur dioxide
	Methyl bromide
	Methyl chloride
	Solvents
	Trichlorethylene
"Carbowax"	Glycerin
	Polymerized glycol stearate
Cardamom Oil	"Card-O-Mar"
Carnauba Wax	"Acrawax"
	Candelilla wax
	Cotton wax, green
	Esparto wax
	Hydrogenated castor oil
	"Norcowax 350"
	Ouricouri wax
	"Rezowax"
	"Santowax M"
	Stearamides, substituted
	"Stroba Wax"
	Sugar cane wax
Carob Gum	Sodium alginate
Carragheen	Sodium alginate
Casein	Albumen
	Alum, potash
	Alkyd resins
	"Alpha" protein
	Cellulose esters

Product	Substitute or Alternative
	Emulsifiers
	Gluten
	Gums, water dispersible
	"Proflex"
	"Prosein"
	Resins, natural
	Resins, synthetic
	Shellac
	Thickeners
	Zein
Castor Oil	"Dipolymer"
	"Flexoresin L 1"
	Glycol hexaricinoleate
	Grapeseed oil
	Vegetable oils
Cetyl Alcohol	Lanolin alcohols
	Monostearin
	Oleyl alcohol
	Stearyl alcohol
Cherry Kernel Oil	Almond oil
China Clay	Barytes
	Talc
	Whiting
China Wood Oil	See Tung oil
Chinese Blue	See Iron blue
Chinese Wax	See Insect wax
Chloramin	Hydrogen peroxide
	Sodium chlorite
Chlorine	Bleaching powder
	Bromine
	Catalysts
	Hydrogen peroxide
	Iodine
	Nitric acid
	Sulphur dioxide
Chloroform	Carbon tetrachloride
Chlorophyll	Dyes, aniline
Chloropicrin	"Dry-Ice" with 10% ethylene oxide
	"Ethide"
	Furoylchloride
	Insecticides
	Methyl bromide
Chlororubbers	"Halowax"
	Rubbers, synthetic

Product	Substitute or Alternative
Cholesterol	Lanolin alcohols
	Phytosterols
Chondrus	Agar
Chrome Alum	Alum, potash
Chrome Orange	Ochre
	Orange mineral
	Sienna
Chromic Acid	Nitric acid
Chromium Acetate	Aluminum sulphate
Chromium Plating	Cadmium plating
	Nickel and silver plating
	Pearl lacquer
Chromium Sulphate	Ferric sulphate
Cinnamon	Cinnamaldehyde and eugenol with powdered nut shells
Citral	Lemongrass oil, Florida
Citric Acid	Acetic acid
	Gluconic acid
	Glycollic acid
	Lactic acid
	Levulinic acid
	Malic acid
	Phosphoric acid
	Propionic acid
	Saccharic acid
	Sodium acid sulphate
	Sodium bisulphite
	Sodium diacetate
	Tartaric acid
	Sulphuric acid, dilute
	"Tartex"
	Vinegar
Citronella Oil	"Andro"
	"Javonella"
Clay	Alum, potash
	Whiting
Clay, Colloidal	Bentonite
	Kaolin
Clove Oil	"Clovel"
	Eugenol
Cobalt	Lead
	Manganese
	Tantalum

Product	Substitute or Alternative
Cobalt Chloride	Barium chloride
Cochineal	Dyes, aniline
Cocoanut Oil	Babassu oil
	Castor oil
	Castor with cottonseed oils
	Cocoanut oil fatty acids
	Cohune oil fatty acids
	Confectioners' oil, "Crystal"
	Carozo oil
	Coyol oil
	Glyceryl myristate with castor oil
	Hydrogenated vegetable oils, partially
	Lard oil
	Macanilla oil
	Mineral oil with lard oil
	Murumuru oil
	Myristic with ricinoleic acid
	Neatsfoot oil
	"Neo-Fat 13"
	Oleic with ricinoleic acid
	Olive oil with mineral oil
	Oxidized paraffin wax with red oil
	Palm kernel oil
	Peanut oil, blown
	Polyhydric alcohol, fatty acid esters, e.g., Diglycol ricinoleate
	Rosin with linseed oil
	Tucum oil
	Vegetable oils
	Vegetable oils, blown
Cod Liver Oil	Rice bran oil, purified
	Sardine oil
	Shark liver oil
	Sterols, irradiated animal
	Tuna liver oil
Cod Oil	Degras
	Herring oil, blown
	Menhaden oil, blown
	Pilchard oil
	Sardine oil, blown
	Whale oil

Product	Substitute or Alternative
Copper Naphthenate	Copper carbonate, basic
	Copper "mahogany" sulphonate
	Copper oleate
	Creosote
Copper Oxide . . . . , ... , . .	Manganese dioxide
	Mercuric chloride
Copper Sulphate . . . . , . .	Aluminum sulphate
Cork	Asbestos fiber with asphalt
	or resin binder
	Bark fiber with asphalt
	or resin binder
	Bran fiber with asphalt
	or resin binder
	"Cushiontone"
	Felt, hair or wool, impregnated
	"Fiberglas"
	"Foamglass"
	"Joinrite"
	Linseed meal
	Millboard soft
	Mineral wool
	"Naturazone"
	Oatmeal
	Palmetto wood
	Paper pulp
	Peat moss
Cream of Tartar	Adipic acid
	Ammonium sulphate
	Mucic acid
	Saccharolactic acid
Creosote . . . , ... , . . .	Coal tar
	Copper chromate
	Copper naphthenate
	Copper oleate
	Copper phosphate
	Copper sulphate
	Cresylic acid
	Pentachlorphenol
	Tar oils
	Zinc chloride
Cresol . . . . , ...	Coal tar acids
	Creosote
	Furfural
	Phenol

Product	Substitute or Alternative
Cresylic Acid	Creosote
	See Cresol
Degras	Cod oil
	Petrolatum
	"Sublan"
Derris	Devil's shoe-string root
	"Thanite"
Dextrin	"Abopon"
	Adhesives
	Glycerin
	Malt extract
	Sodium silicate
	Sodium sulphate
	Sugar
	Urea
Dextrose	See Sugar
Diacetone	Acetone
	Solvents
Diamond, Industrial	Abrasives
	Boron
	Boron carbide
	"Corundum"
	Silicon Carbide
Diatomaceous Earth	Carbon, activated
	"Dicalite"
Dibutyl Phthalate	Butyl oleate
	Castor oil
	Castor oil blown
	"Dipolymer"
	"Glaurin"
	Glycol hexaricinoleate
	2, 5 Hexanediol
	Monoglycollin
	Plasticizers
	"Theop"
Dichloramine	Chlorine
	Hydrogen peroxide
Dichlorethylene	Methyl chloride
Dicresyl Carbonate	Glycerin
Dyes, Aniline	Amaranth
	Annatto
	Caramel coloring
	Chlorophyll
	Cochineal
	Coffee grounds

Product	Substitute or Alternative
	Cudbear
	Cutch
	Dragon's blood
	Fustic
	Hypernic
	Indigo
	Logwood
	Madder
	Orchil extract
	Osage Orange extract
	Pigments, mineral, e.g. sienna
	Precipitates, chemical e.g. antimony sulphide
	Quercitron bark
	Saffron
	Tannin
	Turmeric
Dyes, Vat	"A.A.P. Naphthols"
	Dyes, aniline
	Precipitate, chemical e.g. lead chromate
Ether, Petroleum	Benzol
	Carbon tetrachloride
	Ether, ethyl
	Ethyl chloride
	Isopropyl ether
	Pentane
Ethyl Acetate	Acetone
	Isopropyl acetate
	Methyl acetate
Ethyl Alcohol	Methyl alcohol
	Pentane
	Propylene glycol
	Rum
	Solubilizers or emulsifiers
	"Glycol S533"
	"Stago CS"
	Sulphonated oils
	Solvents
	Tetrahydrofurfuryl alcohol
	Wine
Fluorine	Iodine
Fluorspar	Ammonium bifluoride
	Cryolite
	Sodium silicofluoride

Product	Substitute or Alternative
Formaldehyde	Acetaldehyde
	Aluminum chloride
	Aluminum sulphate
	Furfural
	Glyoxal
	Potassium bichromate
	Sodium bichromate
	Tannin
Gluconic Acid	Citric acid
	Lactic acid
	Mucic acid
Glucose	See Sugar
Glue	Casein
	Emulsifiers
	Gelatin
	Gums, water dispersible
	Latex
	Resins, synthetic
	Rosin soap
Gluten	Casein
Glycerin	Aminoalcohols
	Ammonium lactate
	Apple syrup
	"Aquaresin"
	Butylene glycol
	Calcium chloride
	"Carbowax"
	Corn syrup
	Dextrin
	Dicresyl carbonate
	Diglycol oleate
	Ethylammonium phosphate
	Glycols
	"Glycopon"
	"Glucarine B"
	Glucose
Glycerin	Invert sugar
	Kerosene
	Lactic acid
	Magnesium chloride
	Methyl cellulose
	Methyl sodium potassium phosphate
	Mineral oil

Product	Substitute or Alternative
	Nonaethylene glycol ricinoleate
	Polymerized glycol oleate
	Sorbitol syrup
	Sugar
	Sulphonated castor oil
	"Yumidol"
Glyceryl Chlorhydrin	Ethylene chlorhydrin
Glyceryl Phthalate	Glyceryl maleate
Glycol Diacetate	Monoglycollin
Glycol Monoacetate	Monoglycollin
Glycollic Acid	Acetic acid
	Citric acid
	Lactic acid
Glycols	Ethyl potassium phosphate
	Glycerin
	Sorbitol
Glyoxal	Formaldehyde
Grapeseed Oil	Castor oil
Graphite	"Acrawax"
	Bone black with talc
	Iron oxide
	Metals, powdered
	Mica, powdered
	Paraffin wax
	Red lead
	Silica black
	Talc
Gum Arabic	See Gum Acacia
Gum Benzoin	See Benzoin
Gum Karaya	Gums, water dispersible
Gum, Locust Bean	Gums, water dispersible
Gum Tragacanth	Gums, water dispersible
	Thickeners
Gums, Water Dispersible . . .	"Abopon"
	Agar
	Algin
	"Algaloid"
	Ammonium alginate
	Carragheen
	Casein
	Cherry gum
	Dextrin
	"Diglycol" stearate
	Emulsifiers

Product	Substitute or Alternative
	"G" protein
	Gum acacia
	Gum karaya
	Gum tragacanth
	Locust bean gum
	Methyl cellulose
	Pectin
	Quince seed
	Sodium alginate
	Sodium borophosphate
	Soap
Gut	Fiber
	Metal wire
	"Nylon"
	Protein
	Resins, synthetic
Gutta Percha	Balata
	Resins, synthetic
	Rubbers, synthetic
"Halowax"	"Arochlor"
	Chlorinated mineral oil
	Chlorinated paraffin wax
	Chlororubbers
	Resins, synthetic
Hydrofluoric Acid	Aluminum chloride, anhydrous
	Ammonium bifluoride
	Phosphoric with chromic acid
Hydrofuramide	Hexamethylenetetramine
Hydrogen	Acetylene
	Helium
Hydrogen Peroxide	"Alperox"
	Benzoyl peroxide
	Bleaching powder
	Calcium peroxide
	Chloramine
	Chlorine
	Dichloramine
	Magnesium peroxide
	Oxalic acid
	Oxygen
	Ozone
	Potassium bichromate
	Potassium chlorate
	Potassium chromate

Product	Substitute or Alternative
	Potassium perchlorate
	Potassium permanganate
	Selenium dioxide
	Sodium chlorate
	Sodium chlorite
	Sodium hypochlorite
	Sodium hydrosulphite
	Sodium perborate
	Sodium perchlorate
	Sodium peroxide
	Sulphur dioxide
	Zinc hydrosulphite
	Zinc peroxide
Hydroquinone	Maleic acid
	Naphthol, beta
	Pyrogallol
	Resorcinol
	Selenium
Hydroxycitronellal	Cyclamen aldehyde
Hypernic	Dyes, aniline
Iceland Moss	Gums, water dispersible
	Thickeners
Indigo	Dyes, aniline
Indol	"Indolene"
Infusorial Earth	See Diatomaceous Earth
Insecticides	Amines, higher fatty
	Bordeaux mixture
	Calcium arsenate
	Castor leaf extract
	Chloropicrin
	Cryolite
	"Derex"
	"Ethide"
	Hydrocyanic acid
	Ketones, higher fatty
	Lead arsenate
	"Lethane"
	Methyl bromide
	Naphthalene
	Nicotine
	Paradichlorobenzene
	Paris green
	Phenothiazin
	Phthalonitrile
	Pyrethrum

Product	Substitute or Alternative
	Rotenone
	Sodium fluoride
	Sodium silicofluoride
	Sulphur
	Tetrahydrofurfuryl lactate
	"Thanite"
	Tobacco dust
Invert Sugar . . . . . , ...	Glycerin
Iodine	Bromine
	Chlorine
	Fluorine
Irish Moss	Gum, water dispersible
	Thickeners
Lard . . . . ,	Hydrogenated vegetable or fish oils
	Tallow, refined
Lard Oil.	Fish oil
	Mineral oil
	Mustard seed oil
	Polyhydric alcohol fatty acid esters, e.g. Diglycol oleate
	Rosin oil
	Vegetable oil
Latex	Blood albumen
	"Dispersite"
	"Emulsion 58-8"
	Gelatin
	Glue
	Methyl cellulose
	Resin emulsions
	"Seatex"
	Vinyl copolymer emulsions
Monoacetin	Monoglycollin
Monoglycollin	"Carbitol"
	Glycol diacetate
Montan Wax	Lignite wax
	"Monten" wax
	"Norcowax 12A"
	Peat wax
	"Rezo Wax"
	"Santowax"
Morpholine	Ammonia
	Ethylenediamine
	Methylamine

Product	Substitute or Alternative
Mucic Acid	Adipic acid
	Gluconic acid
	Saccharic acid
Mucin	Agar
"Nipagen"	"Moldex"
	"Parasept"
	Preservatives
Nitre Cake	Hydrochloric acid
Nitric Acid	Chlorine
	Chrome acid
	Hydrochloric acid
	Hydrogen peroxide
Nitrocellulose	Cellulose esters
	Plastics
	Resins, synthetic
	Vinyl copolymers
Nitrogen	Ammonia, anhydrous
	Carbon dioxide
Nitromannite	Mercury fulminate
Nutgalls	Gall apples
	Oak galls (oak apples)
	Tannin
Nux Vomica	Strychnine hydrochloride
Ochre, French	"Witco Yellow"
Octyl Alcohol, Normal	Hexyl alcohol
	Tributyl phosphate
Oleic Acid	Fatty acids
	"Indusoil"
	Talloil
Oleyl Alcohol	Cetyl alcohol
Olive Oil	Apricot kernel oil
	"Lenolene"
	Corn oil with crushed green olives
	Diglycol laurate
	"Glaurin"
	Grapeseed oil
	Lard oil with mineral oil
	Mineral oil with cocoanut oil
	"Nopco C.P."
	"Olev-ol"
	Peach kernel oil
	Peanut oil, destearinated
	Rice oil
	Vegetable oils

Product	Substitute or Alternative
Platinum	Iron containing 42-50% nickel
"Pliofilm"	"Cellophane"
	"Ethocel"
	Plastic films
	Parchment paper
Polystyrene	Cellulose acetopropionate
	Glass, tempered
Polyvinyl Alcohol	"Abopon"
	Gums, water dispersible
	"Hevealac"
	Methyl cellulose
	Synthetic resin emulsion
	Urea-formaldehyde resins
Pyrogallic Acid	Hydroquinone
Pyrogallol	Hydroquinone
	Naphthol, beta
	Pyrogallic acid
	Resorcinol
	"S A 326"
Quartz	Garnet
	Silica, fused
	"Vycor"
Quercitron Bark	Dyes, aniline
Quince Seed	Gums, water dispersible
	Psyllium seed
Quinine	"Atabrin"
	Pamaquine naphthoate
	"Promin"
	Quinarine hydrochloride
	Salicin
	Sulphadiazin
Rubber	"Foamglas"
	"Jointite"
	Lead
	Lead oleate with carbon black
	Polyvinyl butyral resin with 1% Acrawax C
	Resins, synthetic
	"Resistoflex PVA"
	Rubbers, synthetic
	"Saflex"
	Shellac
	"Tygon"
	Varnished cambric

Product	Substitute or Alternative
	Vinyl acetate and chloride copolymers
	"Vinylite"
Rubber Cement	Blood albumen
	"Cement E C-226"
	Polyvinyl acetate and copolymer emulsions
	Resins, synthetic, solutions of
	Rubber, synthetic solutions
	"Vinylite" solutions
Rubber, Chlorinated	Cumarone resin
Rubber, Hard	Ceramics, pressed
	"Densite"
	"Electrite"
	Vulcanized fiber
Saponin	"Foamapin"
	Soap bark
	"Virifoam"
Sapphire	Glass, fused hard
Sardine Oil	Cod oil
	Rice bran oil
	Vegetable oils
Sassafras Oil	"Cam-O-Sass'
	"S-O-Frass"
Sebacic Acid	Maleic acid
Seed-Lac	Accroides, gum
	Ester gum, alcohol soluble
	Resins, synthetic
Selenium	Antimony
	Hydroquinone
	Manganese dioxide
	Sulphur
	Tellurium
Selenium Dioxide	Hydrogen peroxide
Sesame Oil	Diglycol dilaurate
	Peanut oil
	Sunflower seed oil
Shellac	Alkyd resins
	"Bullzite"
	Batavia gum
	Casein
	Copal, alcohol soluble
	"Elastolac"
	Ester gum with plasticizer

Product	Substitute or Alternative
	Gelatin
	Glass with "Vinylite" coating
	Glyceryl phthalate
	Gum accroides
	Gum kauri
	Polyvinyl chloride
	"Protoflax"
	Resins, synthetic
Silica Gel	Agar
	Aluminum hydroxide
	Calcium chloride
	Carbon, activated
	Sodium alumino-silicate
Sodium Alginate	Agar
	Emulsifiers
	"G" protein
	Gelatin
	Gums, water dispersible
	Thickeners
Sodium Alkyl Sulphate	"Wetanol"
Sodium Aluminate	Alum, potash
	Aluminum hydrate
	Copperas with slaked lime
Sodium Antimony Fluoride	Tartar emetic
Sodium Benzoate	Preservatives
Sodium Bichromate	Aluminum sulphate
	Formaldehyde
	Hydrogen peroxide
	Tannin
Sodium Bisulphite	Acetic acid
	Catalysts
	Potassium metabisulphite
	Sodium hyposulphite
Suet	Tallow, edible
Sugar	Apple juice, concentrated
	Calcium chloride
	Dextrin
	"Diglycol" stearate with water and saccharin
	Glycerin
	Glycols
	Glucose
	Gum, water dispersible
	Honey
	Invert sugar

Product	Substitute or Alternative
	Lactose
	Magnesium chloride
	Malted barley
	Malt syrup
	Molasses
	"Nulomoline"
	Saccharin
	Sorghum
	"Sweetose"
	Urea
Sugar Coloring, Burnt	See Caramel coloring
Sulphated Fatty Alcohol . .	Emulsifiers
	Wetting agents ("Wetanol")
Sulphonated Castor Oil	Emulsifiers
	Glycerin
	Naphthenate soaps
	Polyglycol fatty acid esters
	with or without wetting
Sugar	agents, e.g. nonaethylene
	glycol oleate
	Sulphonaphthenic soaps
	Sulphonated olive oil
	Sulphonated tall oil
	Sulphonated vegetable oil
Sulphonated Cocoanut Oil .	Sulphonated castor oil
Sulphonated Olive Oil . . . . ,	Diglycol monoricinoleate
	Diglycol oleate
	Emulsifiers
	Glyceryl mono-oleate
	Sulphonated castor oil
Sulphonated Red Oil	Sulphonated castor oil
Sulphonated Tallow	Sulphonated castor oil
Sulphur Dioxide	Chlorine
	Hydrogen peroxide
	Methyl chloride
Talc	Soapstone
	"Stroba" wax
	Wax, synthetic
	Zinc, stearate
Tallow	Fatty acids
	Garbage grease
	Glyceryl oleo-stearate
	Hydrogenated vegetable oil
	Lard
	Lubricating grease

Product	Substitute or Alternative
	Petrolatum
	Soap
	Stearin
	Vegetable oils
	Whale oil with stearin
Tallow Oil	Menhaden oil
Tannic Acid	Alum
	Ammonium bichromate
	Dyewoods
	Formaldehyde
	Potassium bichromate
	Sodium bichromate
	"Syntans"
Tannin	Dyes, aniline
	Formaldehyde
	Lignin sulphonates
	"Maratan"
Tartaric Acid	Acetic acid
	Citric acid
	Saccharolactic acid
	Sodium acid sulphate
	Sulphuric acid
Tellurium	Antimony
	Sulphur
Tetrachlorethylene	Trichlorethylene
Tetrahydrofurfuryl Alcohol	Ethyl alcohol
	Trichlorethylene
"Tetralin"	Turpentine
Thickeners	Agar
	Albumen
	Ammonium caseinate
	Ammonium stearate
	Arrowroot
	Bentonite
	Blood, dried
	Casein
	Clay
	"G" protein
	Glue
Titanium Dioxide	"Celite No. 340"
	Tin oxide
Titanium Tetrachloride	Silicon tetrachloride

Product	Substitute or Alternative
Toluol	"Enn Jay" solvents
	Hydrogenated petroleum fractions
	"Nevsol"
	"Notol 1"
	Solvent
	"Solvesso 1"
	Tollac solvent
Tonka Beans	Coumarin
	"Tonka-Mel"
Triacetin	Butyl "Carbitol"
	"Carbitol"
	Plasticizers
	Triglycollin
Trichloracetic Acid	Salicylic acid
Trichlorethylene	"Dresinate"
	Mineral spirits
	Naphtha, petroleum (340-410°F.)
	Soap with solvent
	Solvents
	Tetrahydrofurfuryl alcohol
Tricresyl Phosphate	Diglycol oleate
	"Glaurin"
	Sperm oil
Triethanolamine	Alkalies
	Alkyl amines e.g. amylamine
	Amino alcohols e.g. aminomethyl propanol
	Emulsifiers
	Glycerin
	"Glyco S489"
	"Trigamine"
Tripoli	Diatomaceous earth
Trisodium Phosphate	Alkalies
Tritolyl Phosphate	See Tricresyl phosphate
Tung Oil	Castor oil, dehydrated
	"Kellsey"
	"Kellin"
	Linseed oil, polymerized
Tung Oil	Synthetic resin solutions
	Vegetable oil
Zein	Casein
	"G" protein
Zinc Oleate	Calcium oleate
	Lead oleate
	Magnesium oleate

Product	Substitute or Alternative
Zinc Oxide	Barium sulphate
	"Bolted King White"
	Titanium dioxide with
	talc or kaolin
	Whiting
Zinc Perborate	Sodium perborate
Zinc Peroxide	Hydrogen peroxide
	Sodium peroxide
	Zinc perborate
Zinc Stearate	Aluminum stearate
	Barium sulphate, purified
	Graphite
	Magnesium stearate
	Stearic acid
	Talc
Zinc Sulphate	Alum, potash

Incompatible Chemicals

The substances in the lefthand column must be stored and handled so that they cannot come into any contact with the substances in the right-hand column.

Alkaline and alkaline-earth metals, such as sodium, potassium, cesium, lithium, magnesium, calcium, aluminum	Carbon dioxide, carbon tetrachloride, and other chlorinated hydrocarbons. (Also prohibit water, foam, and dry chemical on fires involving these metals.)
Acetic acid	Chromic acid, nitric acid, hydroxyl-containing compounds, ethylene glycol, perchloric acid, peroxides, and permanganates.
Acetone	Concentrated nitric and sulfuric acid mixtures.
Acetylene	Chlorine, bromine, copper, silver, fluorine, and mercury.
Ammonia (Anhydr)	Mercury, chlorine, calcium hypochlorite, iodine, bromine, and hydrogen fluoride.
Ammonium Nitrate	Acids, metal powders, flammable liquids, chlorates, nitrites, sulfur, finely divided organics or combustibles.
Aniline	Nitric acid, hydrogen peroxide.

Bromine	Ammonia, acetylene, butadiene, butane and other petroleum gases, sodium carbide, turpentine, benzene, and finely divided metals.
Calcium Carbide	Water (See also acetylene.)
Calcium Oxide	Water.
Carbon, Activated	Calcium Hypochlorite.
Copper	Acetylene, hydrogen peroxide.
Chlorates	Ammonium salts, acids, metal powders, sulfur, finely divided organics or combustibles.
Chromic Acid	Acetic acid, naphthalene, camphor, glycerol, turpentine, alcohol, and other flammable liquids.
Chlorine	Ammonia, acetylene, butadiene, butane and other petroleum gases, hydrogen, sodium carbide, turpentine, benzene, and finely divided metals.
Chlorine Dioxide	Ammonia, methane, phosphine, and hydrogen sulfide.
Fluorine	Isolate from everything.
Hydrocyanic Acid	Nitric acid, alkalis.
Hydrogen Peroxide	Copper, chromium, iron, most metals or their salts, any flammable liquid, combustible aniline, nitromethane.
Hydrofluoric Acid, Anhydrous (Hydrogen Fluoride)	Aqueous or anhydrous ammonia
Hydrogen Sulfide	Fuming nitric acid, oxidizing gases.
Hydrocarbons (Benzene, Butane, Propane, Gasoline, Turpentine, Etc.)	Fluorine, chlorine, bromine, chromic acid, sodium peroxide.
Iodine	Acetylene, anhydrous or aqueous ammonia.
Mercury	Acetylene, fulminic acid, ammonia.
Nitric Acid (Conc)	Acetic acid, aniline, chromic acid, hydrocyanic acid, hydrogen sulfide, flammable liquids, flammable gases, and nitritable substances.
Nitroparaffins	Inorganic bases.

Oxygen	Oils, grease, hydrogen, flammable liquids, solids or gases.
Oxalic Acid	Silver, mercury.
Perchloric Acid	Acetic anhydride, bismuth and its alloys, alcohol, paper, wood, grease, oils.
Peroxides, Organic	Organic or mineral acids; avoid friction.
Phosphorus (White)	Air, oxygen.
Potassium Chlorate	Acids (See also chlorate.)
Potassium Perchlorates	Acids (See also perchloric acid.)
Potassium Permanganate	Glycerol, ethylene glycol, benzaldehyde, sulfuric acid.
Silver	Acetylene, oxalic acid, tartaric acid, fulminic acid, ammonium compounds.
Sodium	See alkaline metals.
Sodium Nitrate	Ammonium nitrate and other ammonium salts.
Sodium Oxide	Water.
Sodium Peroxide	Any oxidizable substance, such as ethanol, methanol, glacial acetic acid, acetic anhydride, benzaldehyde, carbon disulfide, glycerol, ethylene glycol, ethyl acetate, methyl acetate, and furfural.
Sulfuric Acid	Chlorates, perchlorates, permanganates.
Zirconium	Prohibit water, carbon tetrachloride, foam, and dry chemical on zirconium fires.

Determination of Physical Constants

Melting-point.—The purity of a solid synthetic may be determined by its melting-point, for it is well known that the presence of
a foreign substance will lower the melting-point. The apparatus required is not complicated:—

A good centrigrade thermometer.

A thin-walled glass tube of about 1 mm. diameter, closed at one
end.

A large glass tube having a bulb at one end.

A liquid having a higher boiling-point than the melting-point of
the solid, such as paraffin oil, castor oil, glycerine, etc.

A small quantity of the dried and powdered substance is introduced into the capillary tube, closed at one end. This tube is attached to the thermometer so that the substance is on the same level
as the bulb containing the mercury (the rubber ring being at the top
of the capillary tube and not immersed). The liquid is poured into
the bulb and the thermometer bulb and tube immersed. It is held in
place by a rubber cork inserted in the top. This is fitted on a metal
stand and a bunsen flame placed below the bulb which is heated very
gradually. When a certain temperature is reached the synthetic, if
pure, melts suddenly within a range of one or two degrees. The
process should be repeated several times with fresh portions of the

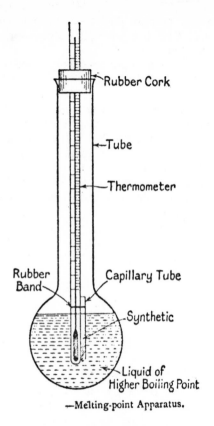

Rubber Cork

Tube

Thermometer

Rubber Band

Capillary Tube

Synthetic

Liquid of Higher Boiling Point

—Melting-point Apparatus.

synthetic, and the mean temperature taken as the melting-point of the substance.

Boiling-point.—Pure synthetics have a constant and definite boiling-point when the liquid is boiled under the same pressure and similar conditions. The apparatus required for its determination is as follows:—

A centigrade thermometer.

A distilling flask with a side tube.

A condenser.

A receiving flask.

The synthetic is placed in the distilling flask and the thermometer, held by a rubber cork on the top, is placed so that the just opposite the side tube. The condenser is connected with the side tube and the receiving flask placed at the bottom orifice.

Heat is applied to the flask very gradually until the liquid boils, the vapour being carried away by the side tube and condensed. When the reading of the thermometer remains constant, the temperature is that of the boiling-point of the liquid.[1]

The atmospheric pressure should be constant and that of 760 mm. of mercury. When it is below this a correction of 0.043° is necessary or every 1 mm.

[1] When the thread of mercury extends above the cork the following correction must be made according to the formula:—

$$N(T - t) \times 0.000156,$$

where T = the apparent temperature in degrees centigrade,

t = the temperature of a second thermometer, the bulb of which is placed at half the length N above the cork,

N = the length of the thread of mercury in degrees above the vessel to T.

0.000156 = the apparent expansion of mercury in glass.

Some liquids undergo decomposition near the boiling-point, and the boiling-point is therefore taken under reduced pressure and is in consequence lower. The only alteration to the apparatus is the use of a second distilling flask at lower end of the condenser (sometimes this may be dispensed with) when the side tube is attached to a gauge and water pump

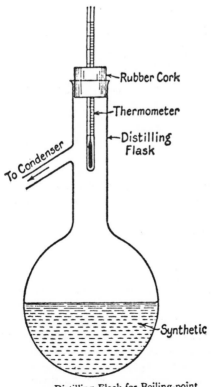

—Distilling Flask for Boiling-point Determination.

Specific Gravity.—The density of a liquid is an important indication of its purity. It is obtained by comparing the weight of a given volume of the liquid with the weight of an equal volume of water—both at the same temperature usually 15°C. All that is required is—

A specific gravity bottle of known tare.

A balance.

The best type of specific gravity bottle is fitted with a perforated stopper which enables the precise filling of the bottle with the liquid and with the exclusion of air bubbles. The bottle should be absolutely dry and is then filled with the liquid at the necessary temperature. The stopper is fitted and any liquid escaping from the perforation is wiped off with the dry hand. The bottle is then wiped dry and weighed. By subtracting the tare of the bottle from the total weight, the weight of the liquid is obtained.

This is repeated with distilled water at the same temperature in a perfectly dry and clean bottle.

In view of the fact that the volumes of the liquid and the water which has been weighed are equal, the specific gravity is obtained by dividing the former by the latter.

—Specific Gravity
Bottle.

The *hydrometer* furnishes a quick way of taking the specific gravity of a liquid and dispenses with the necessity for weighing.[1]

[1] Other indications of the purity of substances are obtained by taking the refractive index and the optical activity (rotation), but since a refractometer and a polarimeter are necessary and are not in the possession of the average chemist, these methods are not detailed here.

SAFETY IN THE LABORATORY OR HOME WORKSHOP.

It is necessary to learn:

Use of laboratory fume hoods

Handling flammable solvents

Mixing acids

Glass blowing

Common electrical hazards

First aid (for four situations only)

Stoppage of breathing

Profuse bleeding

Chemical burns (water only)

Fire in clothing

Use of portable fire extinguishers

Compressed or flammable gases

Disposal of hazardous chemicals

Handling and storing dangerous chemicals, including
alkali metals.

An outstanding deficiency pertaining to laboratory safety seems to be a lack of awareness of hazards among nontechnical personnel. It is conceivable that increased emphasis on "briefing" custodial workers about the dangers of the laboratories in which they work, and periodic review of these conditions could substantially reduce the hazard of ignorance.

Third, a more universal use of safety glasses, reaction shields, and other personal protective devices seems to be needed. From the responses received, an increased program of education on the hazards of common laboratory procedures and the use of personal protective equipment to lessen these hazards would be helpful.

Chemical Hazards

All laboratories, whether they be biological, chemical, or radiological, utilize hazardous chemicals. The hazard may result from utilizing the "raw" product or from products of a chemical reaction between two or more substances or breakdown products developed through heating or aging. Laboratory personnel should have an acquaintance at least with the modes of entry, the physiological responses, both acute and chronic, and methods of roughly assessing the hazards potential of chemicals they are using.

Nonionizing Radiation Hazards

Nonionizing radiation includes ultraviolet, infrared, noise, microwaves, ultrasonics, and lasers and masers. All require special protective measures which must be understood and used. Lasers and masers are gaining in prominence and can be extremely dangerous if not handled properly.

Biological Hazards

Biological hazards are numerous, especially within the health sciences. These hazards include contact with the microorganisms with the hazards potential ranging from low to extremely high.

Electrical Hazards and Management

The problem of handling electricity is probably one of the most ignored facets of safety, yet each year many needless deaths and injuries are caused through carelessness in handling even low voltages. It is also of importance to recognize that electrical equipment can act as an ignition source to activate a fire or explosion. Static electricity should be considered in this category.

Pressure-Hazards

Pressure equipment, either high or low (vacuum), is a part of most laboratories. High-pressure apparatus such as gas cylinders, if improperly handled, can be very dangerous. This is especially true of oxygen. Precautions are necessary in handling, transporting, and in storing. Vacuum equipment, through implosions, can be every bit as dangerous as high-pressure explosions.

Cryogenic Hazards

Cryogenics or the use of low-temperature refrigerants require a knowledge of the behavior of these materials under laboratory uses. It is impossible to understand the design of a piece of cryogenic equipment or cryogenic experiment without an appreciation for the principles of insulation or the significance of extremely low temperatures. Misuse can result in severe injury.

Flammable Chemicals-Hazards

Fires and explosions account for the most dangerous and the most expensive types of laboratory accidents. A knowledge of the flammable properties of chemicals along with an understanding of potential sources of ignition is extremely vital. Storage and handling of these materials also requires special attention.

General Safety Considerations

A number of accidents and injuries in laboratories could very well result from improper lifting, falls, and lacerations from improper handling of glassware. Preventive measures in these areas are worthy of mention.

Ventilation

The principal method of hazards control in laboratory involves the effective use of ventilation, both general and exhaust. An example of exhaust ventilation is the fume hood which if improperly designed or used fails to give the desired protection. Observations indicate that the function of this equipment is not entirely understood and a number of misuses have been witnessed.

Laboratory Sanitation

Poor laboratory sanitation practices may be the cause of contaminating potable water supplies through temporary cross connections. At times, poor housekeeping practices may be dangerous because of blocking passages or by providing tripping hazards. Many chemicals are kept well beyond their usefulness, causing containers to deteriorate and leak, or chemicals to become unstable. Disposal of flammable and toxic chemicals also presents a problem.

Protective Equipment

All laboratories require that protective equipment of one type or another be immediately available. These devices may include eye wash, emergency shower, safety glasses, eye shields, protective clothing, and respiratory protection. Knowledge of the proper usage and limitations of such equipment is extremely important. At times, injury or death may result from improper selection and application of protective equipment.

Reports and Records

Reports and records are necessary adjuncts to any safety program and should be complete, accurate, and disseminated to the appropriate administrators. Accident reports are of little value unless periodically examined and tabulated in order to obtain a picture of local and overall problems.

Emergencies

The initial procedures one follows in an emergency oftentimes determine the ultimate outcome of the accident, both to the individuals and to the installation. The rudiments of first aid, fire fighting and reporting are vital. Personnel have to be continually instructed on procedures for medical and fire emergencies and how and where to make these initial contacts. Such procedures are critical, especially when working alone.

Contact Lenses

It is important to wear eye protection in the chemical laboratory and especially when wearing contact lenses.

Danger in Handling Acid

The heat evolution of the solution could have provided sufficient thermal shock to the glass to permit it to crack when lifted free of the counter, or setting it on a cold counter top, or a shock in setting it down, could have contributed to the bottom separating when the jug was lifted.

Written procedures for handling of acids should always be followed. Personal protective equipment consisting of face protection, rubber apron, and gloves are a necessity for this operation.

You Must Have Fire Extinguishers

If a fire breaks out in your office or apartment, get out fast. Many people are killed because they don't realize how fast a small fire can spread.

If you are caught in smoke take short breaths, breathe through your nose, and crawl to escape. The air is better near the floor.

Head for stairs-not elevator. A bad fire can cut off the power to elevators. Close all doors and windows behind you.

If you are trapped in a smoke-filled room, stay near the floor, where the air is better. If possible, sit by a window where you can call for help.

Feel every door with your hand. If it's hot don't open. If it's cool, make this test: open slowly and stay behind the door. If you feel heat or pressure coming through the open door, slam it shut.

If you can't get out, stay behind a closed door. Any door serves as a shield. Pick a room with a window. Open the window at the top and bottom. Heat and smoke will go out the top. You can breathe out the bottom.

DON'T fight a fire yourself.

DON'T jump. Many people have jumped and died-without realizing rescue was just a few minutes away.

If there is a panic for the main exit, get away from the mob. Try to find another way out. Once you are safely out, *DON'T* go back in. Call the Fire Department immediately. Use alarm box or telephone. DIAL 911

If you find smoke in an open stairway or open hall, use another preplanned way out.

REMEMBER: Get out fast. Don't underestimate how fast a small fire can spread. Use stairs, not the elevator. Close all doors behind you. Don't panic. Once you are safely out, call the Fire Department. Dial 911 or use alarm box. Don't go back in.

SAMPLE LABEL

(your brand name here)

Insect Repellent Lotion

Contains DEET* for longer protection - resistant to perspiration, water and wear. Not greasy.

Repels mosquitoes, biting flies, gnats, chiggers, ticks and fleas.

Active ingredient* N, N-diethyltoluamide	50%
*meta isomer	47.5%
*other isomers	2.5%
Inert ingredients	50%

Directions For Use

Shake into hands and rub lightly over all exposed skin, stockings and all other thin clothing. Reapply as necessary. For chiggers, ticks and fleas, use as above and also apply to socks, pant cuffs and around other openings in outer clothing. Safe to use on nylon, orlon, cotton and wool.

Caution: Harmful if swallowed. Avoid contact with eyes and lips. Keep out of reach of children.

Do not use on acetate, rayon and Dynel. May have plasticizing or softening action on certain synthetic fibers and paint and lacquer surfaces.

Manufactured by:

(customer's name
and address here)   Net Weight   Oz.

CHEMICALS (TRADEMARK)

454

LIST OF SUPPLIERS

1. Abbott Labs .... No. Chicago, Ill.
2. Advance Solvents and Chem. Corp. ..... Jersey City, N. J.
3. Airco Chem. and Plastics .... New York, N. Y.
4. Alco Chem. Corp. .... Philadelphia, Pa.
5. Alcolac Chem. Corp. .... Baltimore, Md.
6. Allied Chem. Corp. .... New York, N. Y.
7. American Aniline Prod. Inc. .... Paterson, N. J.
8. American Cholestrol Prod. Inc. . ... Edison, N. J.
9. American Cyanamid Co. .... Bound Brook, N. J.
10. American Hoechst Corp. .... Mountainside, N. J.
11. American Lecithin Co., Inc............ Woodside, N. Y.
12. American Mineral Spirits Co............. Palatine, Ill.
13. American St. Gobain Corp. .... Kingsport, Ind.
14. Apco Oil Corp. .... Oklahoma City, Okla.
15. Archer-Daniels Midland Co. .... Minneapolis, Minn.
16. Arco Chemicals .... Philadelphia, Pa.
17. Arizona Chem. Co. .... New York, N. Y.
18. Armour Indust. Chem. Co. .... Chicago, Ill.
19. Ashland Chem. Co. .... Columbus, Ohio
20. Atlantic Refining Co.................. Philadelphia, Pa.
21. Atlas Chem. Ind. .... Wilmington, Dela.
22. Baker Castor Oil Co. .... Bayonne, N. J.
23. Balab Inc. .... Burlingame, Ca.
24. BASF Wyandotte Corp................ Paramus, N. J.
25. Bayer Chem. Co. .... New York, N. Y.
26. Beck Koller and Co.................... Detroit, Mich.
27. C. H. Boehringer Sohn .. Ingelheim, Fed. Rep. of Germany
28. Bordon Co. .... New York, N. Y.
29. Cabot Corp.......................... Boston, Mass.
30. H. T. Campbell's Sons Corp. .... Baltimore, Md.
31. Carlisle Chem. Works Inc. .... New Brunswick, N. J.
32. Central Soya Co. .... Chicago, Ill.
33. Chemo Puro Mfg. Co. .... L. I. C., N. Y.
34. Ciba Co............................ Fair Lawn, N. J.
35. Clintwood Chem. Co. .... Chicago, Ill.
36. Clovens Ltd. .... England
37. Columbian Carbon Co. Inc. .... New York, N. Y.
38. Commercial Solvents Corp. .... New York, N. Y.
39. Concord Chem. Co..................... Camden, N. J.
40. Continental Oil Co. .... Saddlebrook, N. J.
41. Cowels Chem. Corp. .... Cleveland, Ohio
42. Croda Co. .... New York, N. Y.
42A. Croda Ltd......................... Goole, England
43. Culver Chem. Co.................. Melrose Park, Ill.

44.	Cutler-Tabs .... Berkeley, Ca.
45.	Day Glo Color Corp. .... Cleveland, Ohio
46.	Degusa Inc. .... New York, N. Y.
47.	Diamond Alkali Co. .... Cleveland, Ohio
48.	Dianol Div. of Mills-Pearson Corp. ..... St. Petersburg, Fla.
48A.	Diva Commodities Corp. .... New York, N. Y.
49.	Dodge and Olcott Inc. .... New York, N. Y.
50.	Dover Chem. Corp. .... Dover, Ohio
51.	Dow Chem. Corp. .... Midland, Mich.
52.	Dragoco Inc. .... Totowa, N. J.
53.	Drew Chem. Corp. .... Boonton, N. J.
54.	Duke Labs Inc. .... South Norwalk, Conn.
55.	E. I. Dupont de Nemours & Co., Inc. ..... Wilmington, Del.
56.	Dura Commodities Corp. .... New York, N. Y.
57.	Eastman Chem. Prod. .... Kingsport, Tn.
58.	Eastman Kodak Co. .... Rochester, N. Y.
59.	Emery Industries .... Cincinnati, Ohio
60.	Englehard Min. Chem. Corp. .... Menlo Park, N. J.
61.	English Mica Co. .... Stamford, Conn.
62.	Enjay Chem. Co.
	(Div. of Humble Oil Refining Co.) .... New York, N.Y.
63.	Falk & Co. .... Chicago, Ill.
64.	Falleck Chem. Corp. .... Newark, N. J.
65.	Falleck Prod. Co. .... New York, N. Y.
65A.	Firmenich Co. .... New York, N. Y.
66.	F. M. C. Corp.
	(Inorganic Chem. Div.) .... New York, N. Y.
67.	FREON Prod. Div.
	(E. I. Dupont de Nemours) .... Wilmington, Del.
68.	Geigy Chem. Corp. .... Ardsley, N. Y.
69.	General Aniline and Film Corp. .... New York, N. Y.
70.	General Electric Corp. .... Waterford, N. Y.
71.	General Mills Corp. .... Minneapolis, Minn.
72.	General Tire & Rubber Co. .... Akron, Ohio
73.	Georgia Marble Co. .... New York, N. Y.
74.	Givaudan Corp. .... New York, N. Y.
75.	Givaudan Delawanna Inc. .... New York, N. Y.
76.	Glidden Co. .... Cleveland, Ohio
77.	Glovers Chemicals Ltd. .... Leeds, England
78.	Glyco Chem. Inc. .... Greenwich, Conn.
79.	Goldschmidt Chem. Div. .... New York, N. Y.
80.	B. F. Goodrich Chem. Co. .... Cleveland, Ohio
81.	Goodyear Tire and Rubber Co. .... Akron, Ohio
82.	Gulf Oil Co. .... Houston, Tx.
83.	Hampshire Chem. Corp. .... Nashua, N. H.
84.	Hercules Inc. .... Wilmington, Del.

85.	Hercules Powder Co.	Wilmington, Del.
86.	Hooker Chem. Corp.	Niagara Falls, N. Y.
87.	Hostawax Div.	New York, N. Y.
88.	J. M. Huber Corp.	Edison, N. J.
89.	Humble Oil and Refining Co.	Humble, Tx.
90.	Humko Div.	Memphis, Tn.
91.	ICI Inc. (Atlas Chem. Div.)	Wilmington, Del.
92.	Interchemical Corp.	Bound Brook, N. J.
93.	International Talc. Co.	New York, N.Y.
94.	International Wax Refining Co.	Valley Stream, N. Y.
95.	Jefferson Chem. Co.	Houston, Tx.
96.	Johns-Manville Products Corp.	New York, N. Y.
97.	Kaopolite Inc.	Elizabeth, N. J.
98.	Kelco Co.	San Diego, Ca.
99.	H. Kohnstamm & Co. Inc.	New York, N. Y.
100.	Koppers Co. Inc.	Pittsburgh, Pa.
101.	Krumbhaar Resin Div.	Kearny, N. J.
102.	Lonza Inc.	Fairlawn, N. J.
103.	Lubrizol Corp.	Cleveland, Ohio
104.	3M Co.	St. Paul, Minn.
105.	M & T Chemicals Inc.	Rahway, N. J.
106.	Mallinckrodt Chem. Works	St. Louis, Mo.
107.	Malmstrom Chem. Corp.	Linden, N. J.
108.	Marbon Chem. Div. Borg-Warner Corp.	Washington, W. Va.
109.	McLaughlin Gormley King Co.	Minneapolis, Minn.
110.	McWhorter Chem. Co.	Chicago, Ill.
111.	The Mearl Corp.	New York, N. Y.
112.	Meer Corp.	New York, N. Y.
113.	Merck & Co., Inc.	Rahway, N. J.
114.	Millmaster Onyx Corp.	New York, N. Y.
115.	Minerals & Chem. Division	Edison, N. J.
116.	The Miranol Chem. Co.	Irvington, N. J.
117.	Mobay Chem. Co.	Pittsburgh, Pa.
118.	MONA Industries	Paterson, N. J.
119.	Monsanto Co.	St. Louis, Mo.
120.	Morton Chem. Co.	Chicago, Ill.
121.	Nalco Chem. Corp.	Chicago, Ill.
122.	National Casein Co.	Chicago, Ill.
123.	National Lead Co.	New York, N. Y.
124.	National Starch & Chem. Corp.	New York, N. Y.
125.	Naugatuck Corp.	New York, N. Y.
125A.	N. J. Zinc Co.	New York, N. Y.
126.	Neville Chem. Co.	Pittsburgh, Pa.
127.	Nopco Chem. Co.	Newark, N. J.
128.	Norda Co.	New York, N. Y.

129.	Onyx Chem. Co.	Jersey City, N. J.
130.	Ott Chem. Co.	Muskegon, Mich.
131.	Ozokerite Mining Co.	Grand Rapids, Mich.
132.	Penn. Glass Sand Corp.	Pittsburgh, Pa.
133.	Penn. Indust. Chem. Corp.	Clairton, Pa.
135.	Penn. Refining Co.	Butler, Pa.
136.	Perry Bros.	Woodside, N. Y.
137.	Petrochemicals Co.	New York, N. Y.
138.	Petroleum Specialties Inc.	New York, N. Y.
139.	Chas. Pfizer & Co.	New York, N. Y.
140.	Philadelphia Quartz Co.	Philadelphia, Pa.
141.	Pittsburgh Plate Glass Co. Chem. Corp.	Pittsburgh, Pa.
142.	Plough Inc.	Memphis, Tn.
143.	Procter & Gamble	Cincinnati, Ohio
144.	Refined Products	Lyndhurst, N. J.
145.	Reheis Chem. Co.	Chicago, Ill.
146.	Reichhold Chem. Co. Inc.	White Plains, N. Y.
147.	Rhodes Chem. Corp.	Jenkintown, Pa.
148.	Richardson Chem. Co.	Melrose, Ill.
149.	RITA Chem. Co.	Chicago, Ill.
150.	Robeco Chem. Inc.	New York, N. Y.
151.	Robertet Inc.	New York, N. Y.
152.	Robinson Wagner Co., Inc.	Mamaroneck, N. Y.
153.	Rohm & Haas	Philadelphia, Pa.
154.	Ru-Jac, Inc.	Upper Montclair, N. J.
155.	Sandoz, Inc.	New York, N. Y.
156.	San Yo Chemicals	Japan
157.	Schenectady Chem. Inc.	Schenectady, N. Y.
158.	Schuylkill Chem. Co.	Philadelphia, Pa
159.	Shanco Plastics & Chemical	Tonawanda, N.Y
160.	Sinclair Petrochemicals	New York, N. Y
161.	Sindar Corp.	New York, N. Y.
162.	Socony Mobil Oil Co.	New York, N. Y.
163.	Socony-Vacuum Oil Co.	New York, N. Y.
164.	Sole Chem. Div.	Skokie, Ill.
165.	Sonneborn Div.	New York, N. Y.
166.	Spencer Kellogg Div. of Textron Inc.	Buffalo, N.Y.
167.	Staleg Chem. Co.	Cambridge, Mass.
169.	A. E. Staley Mfg. Corp.	Decatur, Ill.
170.	Standard Chem. Prod. Inc.	Hoboken, N. J.
171.	Stepan Chem. Co.	Northfield, Ill.
172.	Sun Oil Co.	Philadelphia, Pa.
173.	Tamms Ind. Inc.	Chicago, Ill.
174.	Tenneco Chem. Inc. Newport Div.	Pensacola, Fla.
175.	Thompson-Weinman Co.	Cartersville, Ga.
177.	Troy Chem. Corp.	Mt. Kisco, N. Y.

Aerosols

The E. I. DuPont de Nemours & Co. (Inc.) "Freon" Products Division at Wilmington, Delaware, has been kind enough to supply the author with the information which appears in the following pages. The data supplied here is meant to introduce the layman to the workings of the Aerosol industry. It also provides him with access to more technical bulletins and the names of the leading companies involved in the various areas of the industry.

Definition

An aerosol is defined technically as a suspension of fine solid or liquid particles in air or gas. Through usage, the term "aerosol" today includes all those products that are dispensed from a container by a pressure exerting gas, even though no suspension of particles in air may be involved. Shaving lather is an example.

In spray type aerosols, the size of the particles varies with the product. Some idea of the order of size may be obtained from the U.S. Department of Agriculture's definition of insecticide aerosols: ". . . all particles must be less than 50 microns in diameter, and 80% (by weight) less than 30 microns . . ." It would take 847 particles, 30 microns in size and laid side by side, to make one inch. In paints, perfumes and other sprays the particle size may be much larger.

Basic Aerosol Patent

The use of liquefied compressed gas in forming aerosol insecticides is the result of research by L. D. Goodhue and W. N. Sullivan of the U.S. Department of Agriculture. Their work formed the basis of the well-known aerosol insecticides of World War II. U.S. Patent No. 2,321,023 was issued June 8, 1943 and assigned to the Secretary of Agriculture.

Originally, aerosol insecticide formulations were pretested by the Pesticide Chemicals Research Section of the U.S. Dept. of Agriculture at Beltsville, Md. When a manufacturer now has a new formula, unless it is the same as one in the Government files, the sponsor must do his own pretesting until Government requirements are satisfied. Industry representatives with new or altered formulas now deal directly with the Pesticide Regulation Section, Plant Pest Control Branch, Agricultural Research Service, U.S.D.A., Washington, D.C.

A company that has aerosol containers custom-filled does not need a license because the filler is licensed. The same is true of firms that supply concentrates or are concerned with aerosols other than as fillers.

It should be noted that all insecticides must be registered under the Federal Insecticide, Fungicide and Rodenticide Act.

Advantages

Products such as insecticides, sun tan lotions, room deodorants, shaving creams, and many others can be packaged and dispensed in many different ways. They may be put up in squeeze bottles, jars, or tubes, handled in liquid form with a spray gun, or they may be packaged in aerosol containers.

The aerosol-type package has a number of distinct advantages over other methods of packaging. In fact, there are a number of products that could not exist except in aerosol form. These advantages of aerosol packaging vary somewhat with the particular class of product.

Speaking broadly, consumers prefer aerosols because of their convenience, speed, effectiveness, cleanliness, and ease of use.

The aerosol package is a completely sealed unit so that the product is never exposed to the deteriorating effects of air. This is especially important with paints, perfumes, and deodorants, which are very susceptible to oxidation.

Another advantage found only in aerosols is the ability to control the particle size so as to give just the desired properties in a particular product.

The "Freon" propellant in an aerosol changes state from a liquid to a gas during dispensing and expands as much as 260 times. This can be controlled to give a fineness of particle size or bubble structure not practically obtainable in any other way.

The propellant provides the same dispensing pressure at all times, regardless of whether the container is full or practically empty. This means the user always gets a uniform product and a uniform application rate.

These many advantages can be demonstrated readily. The fact that they are recognized and approved by the consuming public is clearly indicated by the amazing growth of the industry and the application of this packaging principle to an ever-increasing list of products.

Aerosol Product Types

Since the first aerosol product insecticides were introduced publicly in 1946, this method of packaging has been applied successfully to a constantly expanding variety of products:

Household—Personal—
Pharmaceutical

Air Sanitizer (glycol)

Anesthetic, Topical
Anti-Fungal (personal use)
Anti-Perspirant Spray
Anti-Static Sprays
Anti-Tarnish Sprays
Asthma Relief Spray
Auto Polishes, Waxes

Bactericides
Bandage (spray-on plastic)
Burn Treatment Preparations

Charcoal Lighter
Cold Relief Sprays
Colognes-Perfumes
Color Sprays (live foliage, styrofoam)
Cleaners

De-Icers (windshields, etc.)
Deodorants: Personal, Space (Air Fresheners)
Dog and Cat Repellents
Dust Mop Sprays

Fabric Soil Repellent
Fire Alarm Devices
Feminine Hygiene products
Fog Horn Devices
Food Specialties
Furniture Polish (and waxes)
Fungicides (for garden plants)
Flameproofers (X-mas trees)
Furniture Touch-up Sprays

Garbage Can (odor and insect control)
Germicides (pharmaceutical)
Grease Removers (wallpaper, motors, engines)

Hand Creams and Lotions (including medicated)
Hair Curl and Set Sprays
Hair Dressings (lanolin, brilliantine, etc.

Ignition Sprays and Sealers
Insecticides
Insect Repellents

Liniment (for athletes, general)
Lubricants
Mothproofers
Motor Degreasers and Cleaners

Oven Cleaners

Paints-Enamels-Lacquers-Varnishes

Pet Sprays
Protective Coatings

Shave Cream
Shave Lotion (after)
Snow, Artificial
Solvents
Starch
Sunburn Creams
Sun Tan Sprays

Tire Inflator
Toilet Bowl Cleaner

Warning Horns and Alarm Devices
Water Repellents
Waxes
Window and Windshield Cleaner

Commercial—Industrial Use Products

Anti-Static Sprays

Battery Protection Spray (terminal posts to control corrosion)

Belt Dressings

Chilling Spray (for use in locating electronic parts that fail on overheating)

Cleaner-Lubricant for electrical contacts

Color Sprays
Corrosion Inhibitor Cutting Oil

Decorating Sprays
De-Icer Spray (for windshields, etc.)
Degreaser for engine cleaner, also for metal, glass and film
Deodorants

Engine Starting Fluid (diesel and gasoline in all types weather)

Fire Alarm Devices
Fog Horn Device

Grease Remover

Ignition Waterproofers
Insecticides for institutional, hotel, restaurant, dairy and food industry

Layout Ink
Lubricants .

Mildew Preventives
Mold Release Compounds and Lubricants

Paints
Protective Coatings

Release Agent (Silicone) for use on heat sealing equipment

Soil and Stain Retardant Fabric Sprays (institutional, hotel, also home)

Tire Inflator
Tuner-Restorer Cleaner-Lubricant

Upholstery Cleaners and Spot Removers

Warning Horns
Water Repellents
Water Displacing Spray (for drying out flooded motors, electrical equipment, etc.)

Waxes (for aluminum, metal, wood, ski, furniture, etc.)

Principles of Operation

An aerosol is, under the present day meaning, any substance that is packaged in a container under pressure. This substance may be discharged as a foam, spray, stream, or drops. The contents of the container may or may not be a homogeneous mixture of the propellant and concentrate, but all aerosols have a propellant gas inside the container.

When an aerosol stands at room temperature, the contents exert a pressure against the walls of its container. The space above the liquid is filled by gas exerting the same pressure against the valve as well as the walls of the container.

When the valve is opened the pressure forces some of the liquid up the dip tube and through the valve. Here the "FREON" propellant flashes into vapor because of the drop in pressure. A small part of the main body of the propellant evaporates into the gas space to occupy the space left by the liquid lost. These actions take place at about the same rate and pressure whenever the valve is opened, until the liquid is gone.

It is not like using a compressed gas as the propellant. In that case there is no reservoir of liquid to produce pressure and so the pressure keeps dropping as the valve is opened and the gas used up.

In the case of the space sprays, and some surface sprays, the active ingredient is dissolved in the liquid propellant. When the solution is blown out of the nozzle it forms extremely fine particles in the air as the liquid propellant evaporates away from them to form a gas.

When the active material is suspended and not dissolved, the tiny solid particles are carried out through the nozzle and left behind in the air or carried to a surface as the liquid evaporates. These are the typical surface or wet sprays, such as paint.

In foams where the propellant is intimately mixed (partially emulsified) with the active ingredient, rather than completely dissolved in it, the mixture is carried through the nozzle. The particles of "FREON" evaporate into a gas, whipping the active ingredient into a foam of countless bubbles. Where the active ingredient and the propellant will not mix, carrying cosolvent can often be used to overcome the difficulty.

It is obvious, that pressure is an important factor in aerosol performance. It can be controlled by: (1) the kind and amount of propellant; (2) the kind and amount of material dissolved in the propellant.

In most aerosols the propellant system contains "FREON-11", "FREON-12" or "FREON-114" blended with each other or with other materials: methylene chloride, vinyl chloride or hydrocarbons such as propane and isobutane. Stability, safety and appearance characteristics determine the composition of propellant blend used.

Space Sprays

These are the aerosols used for space insecticides and room deodorants. The particle size is quite small and must be carefully controlled. A one-second burst from a typical aerosol space spray will produce 120 million particles. Under average conditions a substantial portion of these particles will remain suspended in the air for an hour.

These aerosols usually operate at pressures between 30 and 40 psig at $70°F$. They may contain as much as 85% propellant, usually a combination of "FREON-12" and "FREON-11".

Surface Sprays

As the name indicates, these aerosols carry an active ingredient to a surface rather than suspend it in air as the space sprays do. Surface sprays include, among others, residual insecticides, lubricants, and surface-coating products like paints and paint removers. Particle size is varied with the job the product has to do. The average is somewhat larger than that in a space spray.

Foam Products

Foam aerosols are used principally for personal products such as shave creams. The firmness and density of the foam may be varied by changing the quantity and/or type of "FREON" propellant as well as by changing the active ingredient formulation. Foam products usually operate between 33-55 psig at $70°F$. They may contain 6-10% propellant which is usually a blend of "FREON-12" and "FREON-114" or a hydrocarbon.

Water-In-Oil-System

The water-in-oil system provides another method for dispensing water base products which do not mix with the "FREON" propellants. In this system the water base product is dispersed as tiny individual droplets in the "FREON" propellant or in a mixture of the "FREON" propellant and a solvent. Surface active agents are incorporated into this system to facilitate the dispersion of the water base product in the propellant. When a water-in-oil system is sprayed, the propellant rapidly evaporates and the individual droplets of the water base product remain. An example would be furniture polish.

Three-Phase-System

The three-phase system provides one method for the aerosol dispensing of water base products which do not mix with the "FREON" propellants. In a three-phase system the propellant forms an individual layer, usually in the bottom of the container, and the water layer floats on the top. The lower opening in the standpipe extends into the upper water layer. The vapor pressure exerted by the propellant forces the water phase up the standpipe and through a special nozzle where the spray is formed.

Containers

Drawn or seamed coated-steel cans and drawn or extruded aluminum cans in various sizes are available for use with low pressure aerosols.

Several plastic, plastic-coated glass, and glass bottles can be obtained to withstand pressures up to 25 psig at 70° F. These have many advantages in attractiveness and resistance to corrosion.

It should be kept in mind that the aerosol pressure adds to the hazards that are normally associated with any glass container. All glass bottles should be pressure-tested before filling. Loaders usually are equipped with semi-automatic devices that will test bottles at air pressures up to 150 pounds.

Aerosol containers holding up to 50 pounds can be bought for commercial use, although there is no reason why any size container cannot be made practical. Most of the existing ones have a fixed nozzle, but they can be used with pressure hose and a remote nozzle.

New developments are constantly being studied to add to the variety of containers already available. Plastics are being evaluated and developed for use as molded aerosol containers.

Glass, plastic-coated glass, or plastic bottles are usually used for low-pressure formulations. The chief propellant is "FREON-114", although some "FREON-12" is often used. Depending on the product, the amount of propellant may range from 35 to 90%. Because of the low pressure, the balance between the amount of propellant and solvent must be carefully controlled to give a satisfactory spray. These aerosols are used chiefly for personal products where corrosion problems and aesthetic appeal rule out metal containers.

Valves

Aerosol valves designed to meet the requirements of the various aerosol products are available from many manufacturers. Decorative, protective caps also are sold by valve manufacturers.

The metering valves that deliver a controlled amount of material each time the valve is opened are of particular interest for cosmetic or pharmaceutical aerosols.

The valve manufacturer or contract loader should be consulted before a selection is made, to assure the best valve for the purpose intended. A list of valve manufacturers is given.

Developing A Formulation

A prospective manufacturer may either engage one of the custom packers listed in the Appendix, or conduct his own laboratory experiments with the help of the raw material suppliers.

In the latter case, however, it should be remembered that a satisfactory formula usually involves much more than mixing active ingredients and propellant. The mixture and valve must give the desired spray or foam without corrosion of the container or the valve. But, in addition, the ingredients and the "FREON" propellant must perform well when mixed. The formulation must be stable. It should be shelf-tested for settling out of solid particles, degradation of perfume, crystal formation, lumping together of particles, etc. Custom packers can be of considerable help in making these tests.

An important fact to remember in developing a formula is that aerosols are subject to I.C.C. regulations, which limit the pressure allowed in the containers.

Shelf-Testing

A product should be carefully shelf-tested for chemical break-down, container corrosion, leakage, or other factors which might cause difficulty in use. And it should be kept in mind that there is no substitute for a full life test. A container opened after a six to nine month shelf-test gives a far better indication of package and formulation reliability than any shorter, speeded-up test.

Such accelerated tests may be helpful for screening purposes but the results do not necessarily correlate with full life tests. It cannot be emphasized too strongly that the advantage gained by getting into the market early can be completely destroyed by the difficulties that may arise from marketing a product that gives trouble later.

The experience and research ability of the "FREON" Products Division is available to customers in planning shelf-tests and inter-preting results. Container and valve manufacturers, and experienced custom packers, may also be consulted for recommendations and advice regarding shelf-testing of aerosol products.

Aerosol Production

After a product has been adequately tested, the prospective seller should look for the best advice he can get on packaging costs and distribution.

The custom packers listed will be glad to give the benefit of their advice, experience, and background on regulatory as well as technical problems. They can help to choose the size and type of container, as well as the proper valve. They will load the formula into a container and label with the name of the seller. Some of these firms will accept loading contracts for as few as 100-500 units.

How To Get Your Product Loaded

You should realize that aerosol loading is a highly specialized business and that a large volume may be necessary to make a loading plant pay.

Because of the special equipment and techniques required to package aerosol products, an industry called "Aerosol Custom Packaging" has grown up. Thus, in order for you to market an aerosol product it is unnecessary to invest in production facilities.

The custom packers vary in size and method of operation. Some have been established just to load aerosol containers, and others are parts of old-line production facilities. All have quality control laboratories and are equipped to carry out product evaluation studies.

Packaging services range from filling containers on a fee basis with the customer supplying all materials, to delivering a finished package with the loader responsible for purchasing, scheduling, and shipping ingredients, containers and valves. The packers should be contacted to determine the type of services they offer.

Because "FREON" propellants are compressed liquefied gases, loading must be done either with pressure equipment at room temperature or with all the ingredients at low temperatures so that the propellants may be handled as liquids. The physical properties of a formulation determine to a large extent which method is better.

The "FREON" Products Division and the custom packager will be glad to give their advice in deciding which method of operation should be followed in a particular case.

Cold Filling

When the low pressure aerosol container was devised, a low cost method of loading was needed. The process now known as "cold filling" was developed, and is still used for a few aerosol products.

Cold filling derives its name from the fact that the "FREON" propellant is cooled below its boiling point before it is put into the open aerosol container. The active ingredient is cooled also, but not to a temperature as low as that of the "FREON" propellant.

A typical cold fill line operates as follows:

1. The concentrate is chilled and loaded into the container under careful control.
2. The "FREON" propellant is chilled and loaded to meet all conditions of regulation and use.
3. The valve is crimped onto the container.
4. The container and contents are heated to $130°F$. in a water bath, as required by government regulations.

Although a wide range of products may be filled, the method cannot be used with all types of products. Water base products, such as shaving cream, must be pressure loaded because chilling would freeze the water in the concentrate.

Pressure Filling

To market aerosol shaving cream, it was necessary to develop a method of loading that did not require chilling the ingredients to a low temperature. Pressure filling accomplishes this by loading the concentrate at room temperature, removing the remaining air from the container, capping it, and forcing the "FREON" propellant through the valve into the container. With the use of multiple filling heads or rotary filling equipment, rapid pressure filling line rates can be obtained.

Experimental Loading Requirements

When a manufacturer wants to load a few containers in his own laboratory, it is generally desirable to use the same method as is planned for production. In addition to the loading equipment, there must be a crimper for sealing the valve to the container.

In cold filling, the cylinder of "FREON" propellant should be inverted so liquid is forced through the cooling coil. Care should be taken that ice forming on the tube does not drop into the container.

Pressure loading has advantages for laboratory work. Very small amounts of propellant can be added accurately. Propellant solutions can be added without loss, and the enclosed system excludes moisture.

Details on a pressure-loading assembly used at the "FREON" Products Laboratory are given in Bulletin FA-10 available from the "FREON" Products Division. Manufacturers of laboratory loading equipment are listed.

Effects Of Air In The Aerosols

Pressure must be determined for all aerosol products. This is essential for the protection of all who manufacture and handle aerosol packages as well as for compliance with regulations. The effect of air on pressure in aerosol containers is an important factor in loading operations. A detailed report on this subject is given in Bulletin FA-15. Anyone manufacturing or marketing aerosols should read this bulletin.

Reducing Contamination By Air

There are several methods for eliminating air or reducing it to allowable limits. Which is the best will depend upon how much air is to be removed, the product, and the particular equipment.

Purging

A simple, effective method of removing the bulk of the air is to purge the open container with a small amount of "FREON-12" vapor before capping. The "FREON-12" vapor displaces air readily because it is nearly five times as heavy.

Evacuation Of The Container

Air can also be removed by mechanical evacuation. This can be done best before the valve has been attached. The effectiveness of mechanical evacuation depends upon the initial evacuation and also upon how successful the valve assembly is in preventing outside air from leaking back.

"FREON" Propellants . . .

General Characteristics

"FREON" propellants are extremely stable and have similar chemical properties because they are closely related members of a chemical family. Technically they are fluorinated hydrocarbon derivatives of the short-chain aliphatic series of organic compounds. Those of commercial importance are derived from the methane and ethane series and contain chlorine as well as fluorine. "FREON" propellants are nonflammable, nonirritating, colorless, essentially odorless, and relatively nontoxic.

Toxicity

A propellant must be of relatively low-order toxicity to minimize the health hazard in case of escape from the loading equipment or

Flammability

"FREON" propellants are nonflammable under all conditions of use. These qualities are important to the loader who handles great quantities of material, and they are important to the consumer who uses the product.

In Underwriters' Laboratories Reports MH-2375, MH-3072, and MH-3134, "FREON-11", "FREON-12", "FREON-113", "FREON-114" and "FREON-22" have been classified as non-explosive and nonflammable at ordinary temperatures.

Even with safe propellants, the flammability of active ingredients and complete formulations must be considered in non-aqueous systems, when the concentration of halogenated material, including "FREON", falls much below 75%, the product may be flammable.

leakage from a package. But toxicity should not be confused with suffocation from excess propellant in the air. If the oxygen concentration in air is lowered too much by addition of propellant, the consequences can be serious.

The major points to be watched are accumulation of vapors in loading areas and accidental leakage of larger containers. In the home with normal, proper handling, there need be no concern whatever with the use of "FREON" propellants. A twelve ounce can emptied in a room 8' x 8' x 10' would produce gas only to the extent of about one half of one percent of the total volume.

It is well established the "FREON" fluorocarbons are of a relatively low order toxicity. These data have been summarized in Safety Bulletin S-16.

Generally speaking, if a new formulation of established safe ingredients is proposed, there is little need for further toxicity studies. But if the active ingredient is new, its health effects in aerosol form should be carefully tested. There are indications that even though a propellant and an active ingredient are harmless individually, use of the active ingredient in aerosol form may be irritant or even toxic. This is because the exceedingly fine particles of an aerosol may reach parts of the respiratory system that would not be reached if the particles were larger, as in a spray mist, for example.

Another factor to be kept in mind is the effect of temperatures that might occur in a fire in a factory or elsewhere. "FREON" propellants would decompose to a greater or less extent. The resulting fluorine and chlorine compounds would be actively toxic. Precautions should be taken to meet these contingencies.

In addition to these considerations, manufacturers of drug, food, or cosmetic aerosols must comply with regulations of the Federal and State Food, Drug, and Cosmetic Acts, both with respect to "FREON" and other ingredients. "FREON" compounds are being used in drugs and cosmetics. In fact, several drugs containing "FREON" propellants are now on the market.

Safety In Consumer Use

There is little hazard to the consuming public when aerosol products are used properly. Over ten billion units have been sold in the past ten years and the safety record is excellent.

Nevertheless the operation of an aerosol depends on a gas under pressure and so has certain potential hazards for the user. The pressure inside the aerosol container may rise rapidly as the temperature rises. An aerosol container should never be exposed to temperatures above 120°F. because bursting of the container might result. Products should be labeled accordingly to prevent misuse.

Other Information

Additional requisite cautions for statements as to hazards to person and property, the safe discharge of contents, etc., will depend upon the particular product. All other information required on the label by law should be complied with, e.g., the Federal and State Pesticide Acts, Net Weight Laws, and such local laws as those in the Administrative Code of the City of New York.

Labeling Suggestions

The Precautionary Labeling Committee of the Chemical Specialties Manufacturers Association has prepared a suggested warning label to appear on all aerosol (pressurized) packages.

Corrosion

The ideal propellant should not corrode metals used in loading equipment and in the aerosol dispenser. "FREON" propellants of themselves do not corrode common metals such as steel, cast iron, brass, copper, tin, lead, zinc, stainless steel, and aluminum.

For formulations without water where the ingredients do not react chemically, the chance of corrosion is remote. Although the propellant itself may not be corrosive, the combination of propellant, plus active ingredient, plus water, or alcohol, or both, may be. In some cases this problem can be solved by changing the formulation.

The experience of the "FREON" Products Laboratory is available to all customers with aerosol corrosion problems.

Solvent Properties

"FREON" propellants have solvent qualities similar to those of the chlorinated solvents, like carbon tetrachloride. But there is enough difference so that direct comparison is not too helpful.

More detailed information on solvent properties is available in Technical Bulletin B-7.

Container And Shipping Information

"FREON" propellants are shipped in cylinders containing 10, 25, and 145 pounds, and tanks of approximately 2000 pounds. Larger volumes can be handled economically in specially built tank trucks of approximately 35,000 pounds, or tank cars holding 60,000 to 120,000 pounds.

The following is a list of technical bulletins put out by the "FREON" Products Division which apply to all aspects of the aerosol industry.

474

LIST OF AEROSOL COMPANYS

Royal Aerosol Corp.
175 Express Street
Plainview, L. I., New York

Rudd Paint and Varnish Co.
1608 15th Avenue, West
Seattle 99, Washington

Sargent Paint Manufacturing Co.
323 West 15th Street
Indianapolis, Indiana

Schaefer Paint Company
334 West Marion Street
Lancaster, Pennsylvania

Service Supply Company
115 Seventh Street
Denver, Colorado

Seymour of Sycamore, Inc.
917 Crosby Avenue
Sycamore, Illinois

Shield Aerosol Company
Turner Place
Piscataway, New Jersey

Shield Chemical Company, Inc.
University Road
Canton, Massachusetts

Sifers, Inc.
P. O. Box 274
Iola, Kansas

Southeastern Packaging Company
P. O. Box 8057
St. Petersburg, Florida

Southwest Aerosols
6419 Longwood Road
Little Rock, Arkansas

Spray-Chem Corporation
6651 Salzman Industrial Court
St. Louis, Missouri

Sprayon Products, Inc.
3818 E. Coronado Street
Anaheim, California

Sprayon Products, Inc.
26300 Fargo Avenue
Bedford Heights, Ohio

Stalfort, Inc.
319 West Pratt Street
Baltimore, Maryland

Stanley Home Products, Inc.
116 Pleasant Street
Easthampton, Massachusetts

Star Chemical Co.
9830 Derby Avenue
Westchester, Illinois

Stoner's Ink Company
Quarryville, Pennsylvania

Strobel Products, Inc.
3017 West Main Street
Louisville, Kentucky

Strouse, Inc.
Conshohocken Road
Norristown, Pennsylvania

Talsol Corporation
Cincinnati Aerosol Division
2017 East Kemper Road
Cincinnati 41, Ohio

Technair Packaging Laboratories
414 E. Inman Avenue
Rahway, New Jersey

Texas Phenothiazine Co.
2021 North Grove
Fort Worth, Texas

Trio Chemical Works, Inc.
260 Meserole Street
Brooklyn, New York

Tru-Scale, Inc.
1123 North Mosley
Wichita, Kansas

United States Aviex Company
1056 Huntley Road
Niles, Michigan

Vim Laboratories
Adamstown, Maryland

Virginia Aerosols
P. O. Box 685
Winchester, Virginia

Virginia Chemicals, Inc.
West Norfolk, Virginia

Watkins Products, Inc.
175 Liberty Street
Winona, Minnesota

Wellston Aerosol Manufacturing
Company
105 West "A" Street
Wellston, Ohio

Western Filling Division
Aerosol Techniques, Inc.
6423 Bandini Boulevard
Los Angeles, California

Westwood Chemical Co., Inc.
715 S. Haven Street
Baltimore, Maryland

Whitmire Research
Laboratories, Inc.
339 South Vandeventer Avenue
St. Louis, Missouri

Wilson Aerosol, Inc.
Spring Hope, North Carolina

Workman Electronic Products,
Inc.
P. O. Box 3238
Sarasota, Florida

Zenith Chemical Corporation
175 Mount Pleasant Avenue
Newark, New Jersey

Zoe Chemical Company, Inc.
1801 Falmouth Avenue
New Hyde Park, New York

Zolene Organic Cosmetics, Inc.
616 Canal Street
San Rafael, California

Aerosol Valve Manufacturers

Aerosol Research Company
743 Circle Avenue
Forest Park, Illinois

Avoset Company
5131 Shattuck Avenue
Oakland 18, California

The Caldwell Casting Company
Market Street
Cambridge, Maryland

Clayton Corporation
4205 Forest Park Blvd.
St. Louis, Missouri

Eaton Manufacturing Co.
Dill Division
700 E. 82nd Street
Cleveland, Ohio

Emson Research, Inc.
118 Burr Court
Bridgeport, Connecticut

Newman-Green, Inc.
57 Interstate Road
Addison, Illinois

Oil Equipment Laboratories, Inc.
600 Pearl Street
Elizabeth, New Jersey

Precision Valve Corporation
700 Nepperhan Avenue
Yonkers, New York

Risdon Manufacturing Company
P. O. Box 520
Naugatuck, Connecticut

A. Schrader's Son
Div. of Scovill Mfg. Co., Inc.
470 Vanderbilt Avenue
Brooklyn 17, New York

Seaquist Manufacturing
Div. of Pittsburgh Railway Co.
225 First Street
Carey, Illinois

Sprayon Products Company
26300 Fargo Avenue
Bedford Heights, Ohio

Valve Corporation of America, Inc.
1720 Fairfield Avenue
Bridgeport, Connecticut

Aerosol Container Manufacturers

American Can Company
100 Park Avenue
New York, New York

Colt Plastic Company
North Grosvenordale,
Connecticut

Continental Can Company, Inc.
633 Third Avenue
New York, New York

Crown Cork & Seal Company, Inc.
Can Division
9300 Ashton Road
Philadelphia, Pennsylvania

Emson Research, Inc.
118 Burr Court
Bridgeport, Connecticut

Foster-Forbes Glass Co.
6 E. 45th Street
New York, New York

Impact Container Corporation
11125 Walden Avenue
Alden, New York

National Can Company
5959 Cicero Avenue
Chicago 32, Illinois

Owens-Illinois Glass Company
1300 Adams Street
Toledo, Ohio

Peerless Tube Company
58 Locust Avenue
Bloomfield, New Jersey

Precision Valve Company
700 Nepperhan Avenue
Yonkers, New York

Aerosol Custom Packers

A-M-R Chemical Company, Inc.
115 Jacobus Avenue
South Kearny, New Jersey

Aero-Fill International
501 N. E. 33 Street
Ft. Lauderdale, Florida

Aero-King, Inc.
1530 Stillwell Avenue
New York 6, New York

Aeropak Division of
DeMert & Dougherty
5000 West 41 Street
Chicago, Illinois

Aeropak of New Jersey, Inc.
548 Belleville Turnpike
Kearny, New Jersey

Aerosol Corporation
P. O. Box 501
Camp Hill, Pennsylvania

Aerosol Cosmetic Enterprises
430 Sniffens Lane
Stratford, Connecticut

Aerosol of Georgia, Inc.
4113 Peachtree Road
Atlanta 19, Georgia

Aerosol International, Inc.
3511 8th Avenue
Baltimore, Maryland

Aerosol Techniques, Inc.
Old Gate Lane
Milford, Connecticut

Airosol Co., Inc.
525 North Eleventh Street
Neodesha, Kansas

Allstadt Manufacturing Company
2004 Wall Street
Dallas, Texas

American Aerosols, Inc.
636 E. 40 Street
Holland, Michigan

American Jet Spray Industries, Inc.
1240 Harlan Street
Denver, Colorado

Armstrong Laboratories Div.
Aerosol Techniques, Inc.
423 LaGrange Street, West Roxbury
Boston, Massachusetts

Astoria Manufacturing Company
42-02 Vernon Boulevard
Long Island City, New York

G. Barr Company
Division of Pittsburgh
   Railway Company
6100 W. Howard Street
Niles, Illinois

Beryman Products
350 N. Montgomery St.
San Jose, California

Big "D" Chemical Company
1708 W. Main Street
Oklahoma City, Oklahoma

Boyertown Packaging Service
East 2 Street
Boyertown, Pennsylvania

Capitol Packaging Company
1502 North 25th Avenue
Melrose Park, Illinois

Carbisulphoil Company
2917 Swiss Street
Dallas, Texas

Chase Products Company
19th & Gardner Road
Broadview, Illinois

Chemical Packaging Corporation
800 N. W. 57th Place
Ft. Lauderdale, Florida

Chem-Spray Aerosols, Inc.
5814 Heffernan
Houston, Texas

Chemspray Filling Corporation
16 Commerce Road
Cedar Grove, New Jersey

Chem-Tech, Inc.
P. O. Box 1748
Wilmington, Delaware

Cincinnati Aerosols, Inc.
11536 Gondola Avenue
Cincinnati, Ohio

Claire Manufacturing Company
7620-28 South Harvard Avenue
Chicago, Illinois

Clifton Private Brands
200 Entin Road
Clifton, New Jersey

Clinton Chemical Company
79 Elmwood Street
Leonard, Michigan

Continental Filling Division
Aerosol Techniques, Inc.
800 South Gilbert Street
Danville, Illinois

Cramer Chemical Company
West Warren Street
Gardner, Kansas

Davis Paint Company
1311 Iron Street
Kansas City, Missouri

Delkote, Inc.
Virginia Street and Park Avenue
Pennsgrove, New Jersey

DeMert & Dougherty
1600 South Baker Street
Ontario, California

Denniston Chemical Company
411 South Mercantile Court
Wheeling, Illinois

Dentocide Chemical Company
3437 S. Hanover Street
Baltimore, Maryland

DeSoto Chemical Coatings, Inc.
P. O. Box 239
Chicago Heights, Illinois

Dettlebach Chemical Corp.
4181 Peachtree Road, N.E.
Atlanta, Georgia

Diamond Chemical Company
R. D.
Glen Gardner, New Jersey

Dixie Paint Manufacturing Corp.
Cleveland & 15th Streets, S. W.
Roanoke, Virginia

Douglas Chemical Company
P. O. Box 297
Liberty, Missouri

Dupli-Color Products Company
1601 Nicholas Boulevard
Elk Grove, Illinois

Emko Company
411 East Gano Street
St. Louis 17, Missouri

Eska Chemical Corporation
375 Herzel Street
Brooklyn, New York

Eveready Products, Inc.
1101 Belt Line
Cleveland 9, Ohio

Farris and Company
8401 Chancellor Row
Dallas, Texas

Fluid Chemical Company
878 Mount Prospect Avenue
Newark, New Jersey

Frekote, Inc.
2300 North Emerson Ave.
Indianapolis, Indiana

Fre-Wa Enterprises, Inc.
25739 Van Horn Road
Taylor, Michigan

Fuld Bros., Inc.
702-710 South Wolfe Street
Baltimore 31, Maryland

Funkhouser Chemical Company
Div. of Funkhouser Industries, Inc.
Ranson, West Virginia

Gard Industries, Inc.
1970 Estes Avenue
Elk Grove Village, Illinois

Gebauer Chemical Company
9408-10 St. Catherine Avenue
Cleveland, Ohio

Gem, Inc.
3742 Lamar Avenue
P. O. Box 258
Memphis, Tennessee

Glastonbury Toiletries, Inc.
224 Williams Street, East
Glastonbury, Connecticut

Gulf States Paint
P. O. Box 14267
Houston, Texas

Haas Chemical Corporation
430 North Main Street
Taylor, Pennsylvania

Hampton Products Company
Hester Street
Portland, Pennsylvania

Harris Paint Company
1026 North 19th Street
Tampa, Florida

Hot Shot Quality Products, Inc.
P. O. Box 297
Memphis, Tennessee

Hue Chemical Company
67 Woodland Avenue
Westwood, New Jersey

Hysan Products Company
919 West 38th Street
Chicago, Illinois

I.K.I. Products Company
Maple Court & West Rollins Street
Edgerton, Wisconsin

Illinois Bronze Powder, Inc.
300 East Main Street
Lake Zurich, Illinois

Impact Container Corporation
11125 Walden Avenue
Alden, New York

Jet-Aer, Inc.
165 Third Avenue
Paterson 4, New Jersey

Jet-Air Products Company
1250 Majestic Drive
Dallas, Texas

Kerr Chemicals, Inc.
1001 N. W. Highway
DesPlaines, Illinois

Kerr Chemicals, Inc.
306 Industrial Way
San Carlos, California

Kerr Chemicals, Inc.
4647 Hugh Howell Road
Tucker, Georgia

Kolmar Laboratories, Inc.
Skyline Drive
Port Jervis, New York

Krylon Division
The Borden Company
Ford & Washington Streets
Norristown, Pennsylvania

LaMaur, Inc.
110 North Fifth Street
Minneapolis 3, Minnesota

Lawson Chemical Products Co
19500 South Normandie
   Avenue
Torrance, California

A. G. Leclair, Inc.
250 North Avenue
Bridgeport, Connecticut

Leeming/Pacquin Division
Charles Pfizer Co.
100 Jefferson Road
Parsippany, New Jersey

Lenk Manufacturing Company
P. O. Box 324
Franklin, Kentucky

Luster Leaf Products, Inc.
590 St. Charles Avenue
Atlanta, Georgia

Magio Research & Manufacturing Co.
350 Cantor Avenue
Linden, New Jersey

Marcy Laboratories, Inc.
2161 North California Avenue
Chicago, Illinois

McRay's Company
1733 Grand Avenue
Des Moines, Iowa

The Mennen Company
Hanover Avenue
Morristown, New Jersey

Midwest Consultants, Inc.
3500 DeKalb Street
St. Louis 18, Missouri

Milwaukee Paint Products, Inc.
2600 West North Avenue
Milwaukee, Wisconsin

Minnesota Paints, Inc.
1101 South Third Street
Minneapolis, Minnesota

Mitann, Inc.
7240 Hinds Avenue
North Hollywood, California

Mohawk Finishing Products, Inc.
Route 30
Amsterdam, New York

Morton Pharmaceuticals, Inc.
1625-39 North Highland Street
Memphis 8, Tennessee

National Chemical Laboratories
Subsidiary of Osnaw Products
   Co., Inc.
44 Arnot Street
Lodi, New Jersey

National Pressure Filling & Cosmetic
   Laboratories, Inc.
1731 East Olympic Boulevard
Los Angeles 21, California

National Spray Can Filling Corp.
North & Dowd Avenues
Elizabeth, New Jersey

NEPCO Aerosol & Filling
  Company
700 Chestnut Street
Wyandotte, Michigan

Normand Co.
7190 Northwest 7th Avenue
Miami, Florida

North American Aerosol Company
34136 Myrtle Street
Wayne, Michigan

Norton, Inc.
2370 South 8th West
Salt Lake City, Utah

Nu-Vita Products, Inc.
910-912 Bowen Street
Pittsburgh, Pennsylvania

O'Neil Brands, Inc.
9130 State Road
Philadelphia, Pennsylvania

Orb Industries
Main & Race Streets
Upland, Pennsylvania

Ozon Products Division
The Bordon Company
50 Wallabout Street
Brooklyn, New York

Pacific Aerosol, Inc.
2424 Merced Street
San Leandro, California

Pactra Chemical Company, Inc.
6725 Sunset Boulevard
Hollywood, California

Pactra Paint Manufacturing Co., Inc.
420 South 11th Avenue
Upland, California

Par Industries, Inc.
2193 East 14th Street
Los Angeles 21, California

Pennsylvania Engineering Company
1119 North Howard Street
Philadelphia 23, Pennsylvania

PenRay Co.
1801 Estes Avenue
Elk Grove Village, Illinois

Peterson Filling & Packaging Corp.
Hegeler Lane
Danville, Illinois

Philips Investment Company
P. O. Box 1475
Studio City, California

Plasti-Kote
1000 Lake Road
Medina, Ohio

Plaze, Inc.
9401 Watson Industrial Park
St. Louis, Missouri

Power Pak Division
Paulisboro Chemical Industries, Inc.
145 Howard Avenue
Bridgeport, Connecticut

Pressure Pak, Inc.
Dyer Boulevard
West Palm Beach, Florida

Pro-Pak Company
23600 Corbin Drive
Cleveland, Ohio

Puritan Aerosol Corporation
Martin Street
Berkeley, Rhode Island

Puritan Aerosol Corporation
9101 South Sorenson
Santa Fe Springs, California

Quelcor, Inc.
Papermill Road & Baltimore Pike
Media, Pennsylvania

Raabe Paint Company
6430 West Fond de Lac Avenue
Milwaukee, Wisconsin

Ralm Laboratories
Division of Standard Pharmacal
  Company
1300 Abbott Drive
Elgin, Illinois

W. T. Rawleigh Company
223 East Main Street
Freeport, Illinois

Rayette-Faberge, Inc.
17 Chapel Avenue
Jersey City, New Jersey

Rayette-Faberge, Inc.
261 East Fifth Street
St. Paul, Minnesota

Red Devil Chemical Company
30 Northwest Street
Mount Vernon, New York

Reeves-Rockwin Company
117 Urban Avenue
Westbury, L. I., New York

Regal Chemical Corporation
115 Dobbin Street
Brooklyn, New York

Rexall Drug Company
8480 Beverly Boulevard
Los Angeles, California

Rexall Drug Company
3915 North Kingshighway Boulevard
St. Louis, Missouri

Richefleur Laboratories
1069 Rogers Avenue
Brooklyn, New York

Rochester Aerosols, Inc.
615 East Greendale
Detroit, Michigan

Sherwin-Williams Co.
Pigments, Color & Chemical Div.
11541 S. Champlain Avenue
Chicago 28, Illinois

Tube Manifold Corporation
415 Bryant Street
North Tonawanda, New York

Wheaton Plasti-Cote Corp.
Wheaton Avenue & G Street
Millville, New Jersey

Aerosol Loading Equipment Manufacturers

Aerosol Machinery Corp.
117 Urban Avenue
Westbury, New York

Andora Automation, Inc.
P. O. Box 10—Rugby Station
Brooklyn, New York

Cherry-Burrell Corp.
2400 6th Street, S. W.
Cedar Rapids, Iowa

Elgin Manufacturing Company
200 Brook Street
Elgin, Illinois

General Kinetics, Inc.
P.O. Box 4394
Atlanta, Georgia

J. G. Machine Works
14-28 Popular Avenue
Little Ferry, New Jersey

The Kartridg Pak Company
807 W. Kimberly Rd.
Davenport, Iowa

M.R.M. Company, Inc.
191 Berry Street
Brooklyn, New York

John R. Nalbach Engineering Co.,
Inc.
6139 W. Ogden Avenue
Chicago, Illinois

Oil Equipment Laboratories, Inc.
600 Pearl Street
Elizabeth, New Jersey

The Wallace Company
41 California Street
Bridgeport, Connecticut

Pressure Loading Equipment

Builders Sheet Metal Works, Inc.
108 Wooster Street
New York 12, New York

The Kartridg Pak Company
800 W. Central Road
Mt. Prospect, Illinois

John R. Nalbach Engineering Co., Inc.
6139 W. Ogden Avenue
Chicago, Illinois

Robert A. Foresman, Jr.
1690 Margaret Street
Philadelphia, Pennsylvania

Aerosol Container Closure Equipment

Builders Sheet Metal Works, Inc.
108 Wooster Street
New York 12, New York

Robert A. Foresman, Jr.
1690 Margaret Street
Philadelphia, Pennsylvania

The Kartridg Pak Company
800 W. Central Road
Mt. Prospect, Illinois

John R. Nalbach Engineering Co.,Inc.
6139 W. Ogden Avenue
Chicago, Illinois
(crimping   chinery, can coders)

C. A. Spalding Co.
2242 E. Venango Street
Philadelphia, Pennsylvania
(hand lever type)

Wheaton Plasti-Cote Corp.
Wheaton Avenue & G Street
Millville, New Jersey
(bottles)

Glass Compatibility Tubes

Fischer & Porter Co.
Warminster, Pennsylvania

Lab Glass, Inc.
Vineland, New Jersey

Sentinel Glass Co.
P. O. Box 58
Hatboro, Pennsylvania

Flexible Lines

Resistoflex Corporation
Roseland, New Jersey
(Neoprene tubing line with polyvinyl
alcohol—supplied in custom lengths
with end fittings attached)

Flexonics Corporation
Maywood, Illinois

Titeflex Metal Hose Co.
Springfield, Massachusetts

Line Couplers

Aeroquip Corporation
Jackson, Michigan

Wabash Speed Couplers
Wabash Manufacturing Co.
Chicago 8 Illinois
(gas tight knurled fittings)

Ever-Tite Coupling Co., Inc.
254 W. 54th Street
New York 19, New York

Aerosol Consultants

Battelle Memorial Institute
505 King Street
Columbus, Ohio

Robert A. Foresman, Jr.
1690 Margaret Street
Philadelphia, Pennsylvania

Arthur D. Little, Inc.
35 Acorn Park
Cambridge 42, Massachusetts

Midwest Div.—Miner Laboratories
130 N. Franklin Street
Chicago 6, Illinois

Lodes Aerosol Consultants, Inc.
730 Fifth Avenue
New York, New York

Foster D. Snell, Inc.
29 W. 15th Street
New York, New York

Aerosol Food Custom Packers

Aerosol Foods, Inc.
Corner of Anthony and Woodglen Rd.
Glen Gardner, New Jersey

Avoset Company
5131 Shattuck Avenue
Oakland, California

G. Barr Company
Div. of Pittsburgh Railways Co.
6100 West Howard Street
Niles, Illinois

Bodine's, Inc.
5757 West 57th Street
Chicago, Illinois

Dairy Products Southeast, Inc.
2198 West Beaver St.
Jacksonville, Florida

Lockport Canning Co.
Lake Avenue
Lockport, New York

Pet, Incorporated
Milk Products Division
P. O. Box 392
St. Louis, Missouri

INDEX